Proceedings of the 41st International Conference on Advanced Ceramics and Composites

Proceedings of the 41st International Conference on Advanced Ceramics and Composites

Ceramic Engineering and Science Proceedings, Volume 38, Issue 3, 2017

A Collection of Papers Presented at the 41st International Conference on Advanced Ceramics and Composites January 22–27, 2017, Daytona Beach, Florida

Edited by

Waltraud M. Kriven
Narottam P. Bansal
Mihails Kusnezoff
Tatsuki Ohji
Yanchun Zhou
Kyoung Il Moon
Josef Matyáš
Kiyoshi Shimamura
Soshu Kirihara

Volume Editors
Surojit Gupta
Jingyang Wang

The American Ceramic Society

WILEY

The rights of Waltraud M. Kriven, Narottam P. Bansal, Mihails Kusnezoff, Tatsuki Ohji, Yanchun Zhou, Kyoung Il Moon, Josef Matyáš, Kiyoshi Shimamura, Soshu Kirihara, Surojit Gupta, and Jingyang Wang to be identified as the authors of the editorial material in this work have been asserted in accordance with law.

Registered Office
John Wiley & Sons, Inc., 111 River Street, Hoboken, NJ 07030, USA

Editorial Office
111 River Street, Hoboken, NJ 07030, USA
For details of our global editorial offices, customer services, and more information about Wiley products visit us at www.wiley.com.

Wiley also publishes its books in a variety of electronic formats and by print-on-demand. Some content that appears in standard print versions of this book may not be available in other formats.

Library of Congress Cataloging-in-Publication Data is available
ISBN: 9781119474692
ISSN: 0196-6219

Cover design by Wiley

10 9 8 7 6 5 4 3 2 1

Contents

Introduction   ix

SOLID OXIDE FUEL CELLS

Mitigation of Compressor Stall/Surge in a Hybrid Solid Oxide Fuel   3
Cell-Gas Turbine System
M. Azizi and J. Brouwer

Characteristics of Protective Spinel Manganese Cobaltite   7
Coatings Produced by APS for Cr-Contained SOFC Interconnects
Chun-Liang Chang, Chang-sing Hwang, Chun-Huang Tsai, Sheng-Fu
Yang, Te-Jung Huang, Ming-Hsiu Wu, and Cheng-Yun Fu

Influence of Temperature and Steam Content on Degradation of   17
Metallic Interconnects in Reducing Atmosphere
Christoph Folgner, Viktar Sauchuk, Mihails Kusnezoff, and Alexander
Michaelis

Fabrication of the Anode-Supported Solid Oxide Fuel Cell with   35
Direct Pore Channel in the Cermet Structure to Improve the
Electrochemical Performance
Ming-Wei Liao, Tai-Nan Lin, Hong-Yi Kuo, Chun-Yen Yeh, Yu-Ming Chen,
Wei-Xin Kao, Jing-Kai Lin, and Ruey-Yi Lee

Critical Evaluation of Dynamic Reversible Chemical Energy   47
Storage with High Temperature Electrolysis
D. McVay, J. Brouwer, and F. Ghigliazza

Estimation of Polarization Loss Due to Chromium Poisoning of   59
LSM-Based Cathodes in Solid Oxide Fuel Cells
R. Wang, B. Mo, M. Würth, U. B. Pal, S. Gopalan, and S. N. Basu

ADVANCED PROCESSING AND MANUFACTURING TECHNOLOGIES

Synthesis and Tribological Behavior of Bi-Cr_2AlC Composites   69
F. AlAnazi, S. Ghosh, R. Dunnigan, and S. Gupta

Finite Element Analysis of Self-Propagating High-Temperature   75
Synthesis (SHS) of Silicon Nitride
Venkata V. K. Doddapaneni, Julia Lin, Ayako Hiranaka, Tomohiro
Akiyama, and Sidney Lin

TEM Analysis of Diffusion-Bonded Silicon Carbide Ceramics   87
Joined using Metallic Interlayers
T. Ozaki, Y. Hasegawa, H. Tsuda, S. Mori, M. C. Halbig, R. Asthana, and
M. Singh

Numerical Analysis of Inhomogeneous Behavior in Friction Stir   95
Processing by Using a New Coupled Method of MPS and FEM
Hisashi Serizawa and Fumikazu Miyasaka

ADVANCED MATERIALS AND INNOVATIVE PROCESSING IDEAS FOR THE INDUSTRIAL ROOT TECHNOLOGY

Study of Shielding Method to Reduce Leakage Magnetic Fields   105
of an Opening in a Magnetically Shielded Room
H. Sugiyama and K. Kamata

MATERIALS FOR EXTREME ENVIRONMENTS

Low-Temperature Synthesis of Hafnium Diboride Powder Via   119
Magnesiothermic Reduction in Molten Salt
Ke Bao, Joseph Massey, Juntong Huang, and Shaowei Zhang

Tribology Study of Novel Ti_3SiC_2 Matrix Composites Reinforced   131
with Ceramics (Al_2O_3, BN, B_4C) Particulates
J. Nelson, M. Olson, and S. Gupta

ADVANCED MATERIALS FOR SUSTAINABLE NUCLEAR FISSION AND FUSION ENERGY

Interfacial Reaction and Mechanical Properties of SiC Bonded   143
with Zircaloy-4 using Ni, Zr /Al Double Interlayers
Xin Geng, Guangwu Wen, and Xiaoxiao Huang

SINGLE CRYSTALLINE MATERIALS FOR ELECTRICAL, OPTICAL, AND MEDICAL APPLICATIONS

Characterization Approaches of Femtosecond Direct Laser Writing (DLW) Modifications inside Cubic YAG Crystals 157
W. Gebremichael, I. Manek-Hönninger, S. Rouzet, M. Chamoun, A. Fargues, V. Jubera, T. Cardinal, Y. Petit, and L. Canioni

ADDITIVE MANUFACTURING AND 3D PRINTING TECHNOLOGIES

Comparison of Dynamic Mask- and Vector-Based Ceramic Stereolithography 165
S. Baumgartner, M. Pfaffinger, B. Busetti, and J. Stampfl

Additive Manufacturing (3D Printing) of Ceramics: Microstructure, Properties, and Product Examples 175
P. Karandikar, M. Watkins, A. McCormick, B. Givens, and M. Aghajanian

GEOPOLYMERS, CHEMICALLY BONDED CERAMICS, ECO-FRIENDLY, AND SUSTAINABLE MATERIALS

Impact of Various Aluminosilicate Compounds in Geopolymer Foam Formation to a Si/M=0.7 of Silicate Solution 191
M. Arnoult, M. Perronnet, A. Autef, G. Gasgnier, and S. Rossignol

Geopolymers Based on Natural and Synthetic Metakaolin: A Critical Review 201
Joseph Davidovits

On the Durability Behavior of Natural Fiber Reinforced Geopolymers 215
A. C. C. Trindade, F. A. Silva, H. A. Alcamand, and P. H. R. Borges

Performance and Durability of Fe-Rich Inorganic Polymer Composites with Basalt Fibers 229
A. Peys, M. Peeters, A. Katsiki, L. Kriskova, H. Rahier, and Y. Pontikes

Wetting Angle: New Parameter Indicating the Reactivity of Alkaline Solutions and Geopolymer Binders 239
Ameni Gharzouni, Laeticia Vidal, Robin Stocky, Julie Peyne, and Sylvie Rossignol

Effect of TiO$_2$ and ZnO Nanopowders on Metakaolin-Sodium   251
Hydroxide Geopolymers
     D. Sarbapalli and P. Mondal

Eco-Friendly Geopolymer Composite for Winter Season   263
Pavement Pothole Patching
     M. Sarkkinen, K. Kujala, and S. Gehör

Introduction

This Ceramic Engineering and Science Proceedings (CESP) issue consists of 24 papers that were submitted and approved from select symposia held during the 41st International Conference on Advanced Ceramics and Composites (ICACC), held January 22–27, 2017 in Daytona Beach, Florida. ICACC is the most prominent international meeting in the area of advanced structural, functional, and nanoscopic ceramics, composites, and other emerging ceramic materials and technologies. This prestigious conference has been organized by the Engineering Ceramics Division (ECD) of The American Ceramic Society (ACerS) since 1977.

The 41st ICACC hosted more than 1,000 attendees from 41 countries that gave over 850 presentations. The topics ranged from ceramic nanomaterials to structural reliability of ceramic components, which demonstrated the linkage between materials science developments at the atomic level and macro level structural applications. Papers addressed material, model, and component development and investigated the interrelations between the processing, properties, and microstructure of ceramic materials.

The 2017 conference was organized into the following 15 symposia and 3 Focused Sessions and two Special Sessions:

Symposium 1	Mechanical Behavior and Performance of Ceramics and Composites
Symposium 2	Advanced Ceramic Coatings for Structural, Environmental, and Functional Applications
Symposium 3	14th International Symposium on Solid Oxide Fuel Cells (SOFC): Materials, Science, and Technology
Symposium 4	Armor Ceramics: Challenges and New Developments

Symposium 5	Next Generation Bioceramics and Biocomposites
Symposium 6	Advanced Materials and Technologies for Direct Thermal Energy Conversion and Rechargeable
Energy Storage	
Symposium 7	11th International Symposium on Functional Nanosmaterials and Thin Films for Sustainable Energy Harvesting, Environmental and Health Applications
Symposium 8	11th International Symposium on Advanced Processing & Manufacturing Technologies for Structural & Multifunctional Materials and Systems
Symposium 9	Porous Ceramics: Novel Developments and Applications
Symposium 10	Virtual Materials (Computational) Design and Ceramic Genome
Symposium 11	Advanced Materials and Innovative Processing ideas for the Production Root Technology
Symposium 12	Materials for Extreme Environments: Ultrahigh Temperature Ceramics (UHTCs) and Nano-laminated Ternary Carbides and Nitrides (MAX Phases)
Symposium 13	Advanced Materials for Sustainable Nuclear Fission and Fusion Energy
Symposium 14	Crystalline Materials for Electrical, Optical and Medical Applications
Symposium 15	Additive Manufacturing and 3D Printing Technologies
Focused Session 1	Geopolymers, Chemically Bonded Ceramics, Eco-friendly and Sustainable Materials
Focused Session 2	Advanced Ceramic Materials and Processing for Photonics and Energy
Focused Session 3	Carbon Nanostructures and 2D Materials and Composites
Special Symposium	3rd Pacific Rim Engineering Ceramics Summit
Special Symposium	6th Global Young Investigators Forum (GYIF)

The proceedings papers from this meeting are published in the below two issues of the 2017 Ceramic Engineering and Science Proceedings (CESP):

- CESP Volume 38, Issue 2 (includes 23 papers from Symposia 1, 2, 4, 5, and GYIF)
- **CESP Volume 38, Issue 3** (includes 24 papers from Symposia 3, 8, 11, 12, 13, 14, 15 and FS1)

The organization of the Daytona Beach meeting and the publication of these proceedings were possible thanks to the professional staff of ACerS and the tireless dedication of many ECD members. We would especially like to express our sincere thanks to the symposia organizers, session chairs, presenters and conference attendees, for their efforts and enthusiastic participation in the vibrant and cutting-edge conference.

ACerS and the ECD invite you to attend the 42nd International Conference on Advanced Ceramics and Composites (http://www.ceramics.org/icacc2018) January 21-26, 2018 in Daytona Beach, Florida.

To purchase additional CESP issues as well as other ceramic publications, visit the ACerS-Wiley Publications home page at www.wiley.com/go/ceramics.

SUROJIT GUPTA, University of North Dakota, USA
JINGYANG WANG, Institute of Metal Research, Chinese Academy of Sciences, China

Volume Editors
August 2017 ICACC

Solid Oxide Fuel Cells

MITIGATION OF COMPRESSOR STALL/SURGE IN A HYBRID SOLID OXIDE FUEL CELL-GAS TURBINE SYSTEM

M. Azizi[a] and J. Brouwer[b]

National Fuel Cell Research Center, University of California, Irvine, CA 92697, USA

ABSTRACT

One of the main purposes of an SOFC-GT hybrid system is for distributed power generation applications. This study investigates the possible use of an SOFC-GT hybrid system to power multi-MW dynamic loads. Based upon the integration of commercially available gas turbine technology, control strategies for the SOFC-GT hybrid system are investigated for different stationary power applications. Risk analysis of compressor stall/surge in the hybrid SOFC-GT power system as it is dynamically dispatched to meet demand is assessed in transient pre-load and post-load modes. Optimal control algorithm is proposed and applied to mitigate stall/surge in compressor as a response to sudden power demand change. This study aims to study compressor stall/surge mitigation assuming a connecting pipe that reduces the back pressure on the compressor in order to maintain the compressor mass flow rate at a specific setpoint.

INTRODUCTION

A better understanding of turbulent unsteady flows in compressor and gas turbine systems is a necessary step toward a breakthrough in compressor applications for hybrid fuel cell-gas turbine (FC-GT) systems transient operation. Hybrid fuel cell-gas turbines are among the many low emission power generation systems. In the previous studies at National Fuel Cell Research Center (NFCRC) at University of California, Irvine, compressor stall/surge analysis for a 4 MW locomotive hybrid solid oxide fuel cell-gas turbine (SOFC-GT) engine has been performed based on the 1.7 MW multi-stage air compressor similar to available commercial compressors[1]. Controls methods have been previously developed for these types of systems in order to avoid stall/surge in the compressor[2]. Computational fluid dynamics (CFD) tools can provide a better understanding of flow distribution and instabilities near the stall/surge line. In this study a mechanism is presented in order to mitigate stall/surge in the compressor assuming a connecting pipe between the compressor inlet and outlet that maintains constant air mass flow rate at the design condition of the compressor. This mechanism will avoid secondary stall/surge occurance in the compressor while the hybrid system is exposed to a sudden increase in power demand change from 3 MW to 3.5 MW.

TURBOMACHINERY MODELING

Shear stress transport (SST k-ω) fluid model is used for faster convergence in the turbomachinery problem. The pressure dynamics of the compressor outlet has been solved in the MATLAB/Simulink platform that was previously developed at NFCRC. Computational fluid dynamics analysis of the compressor is accomplished using ANSYS CFX software[3]. Power demand variation in the hybrid SOFC-GT system causes pressure change at the compressor outlet. The pressure variation is set as a boundary condition for the turbomachinery analysis. The results show that by using a pipe guiding the exit air flow rate to the compressor inlet, the compressor mass flow rate could be maintained at the design condition of 7 kg/s. 1.7 MW compressor is an appropriate choice among the industrial compressors to be used in a 4 MW hybrid locomotive SOFC-GT system with topping cycle design due to the enhanced ability to maintain air flow rate through the compressor during the sudden transient step-load change.

RESULTS AND DISCUSSION

Figure 1 shows the pressure variation contour on the compressor front and rear impellers post stall/surge. The pressure on the compressor outlet is reduced while the compressor mass flow converges to the design condition at a constant value.

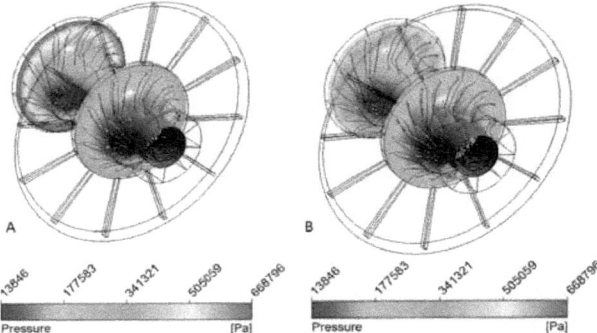

Figure 1. Pressure reduction and stall/surge mitigation of 1.7 MW multi-stage compressor based on controlled stall/surge assuming a connecting pipe between the compressor outlet and inlet that keeps the compressor mass flow rate at a constant rate.

Figure 2. shows the air mass flow rate increase post stall/surge on the rear impeller blades due to the controlled air flow rate to meet the mass flow rate at set point.

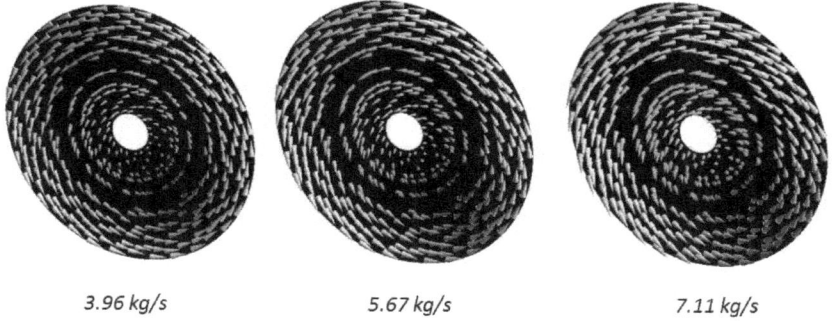

| 3.96 kg/s | 5.67 kg/s | 7.11 kg/s |

Figure 2. Air mass flow rate increase at the front impeller inlet due to the reduced back pressure on the compressor.

Figure 3. shows that it takes 10 rotor revolutions post stall/surge so that the air mass flow rate can be reached to the steady normal operating condition of the hybrid SOFC-GT system. Control algorithms are topics of future research in order to reduce the delay time between the stall/surge and the normal operating condition.

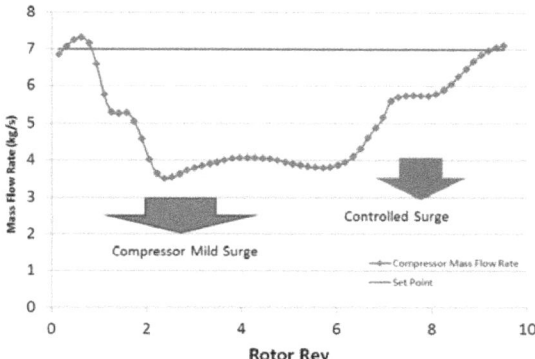

Figure 3. Compressor air mass flow rate increase post stall/surge to the normal design value of 7 kg/s.

CONCLUSION

In this study, analysis of post stall/surge of a 1.7 MW multi-stage compressor is investigated assuming a connecting pipe between the inlet and outlet of the compressor maintaining the compressor air mass flow rate at the 7 kg/s design condition. The reduced back pressure significantly reduces the risk of flow reversal in the 4 MW hybrid SOFC-GT system. As a result, a second stall/surge is less likely to occur after a sudden increase in the power demand is applied to the hybrid system. The response time of the air mass flow rate variation could help in future design of control systems for faster mitigation of compressor stall/surge.

ACKNOWLEDGEMENTS

The authors would like to thank Federal Railroad Administration (FRA) for its support during this research.

REFERENCES

[1] Azizi, Mohammad Ali, and Jacob Brouwer. "Transient Analysis of 220 kW Solid Oxide Fuel Cell-Gas Turbine Hybrid System Using Computational Fluid Dynamics Results." *ECS Transactions* 71.1 (2016): 289-301.

[2] McLarty, Dustin Fogle. *Thermodynamic Modeling and Dispatch of Distributed Energy Technologies including Fuel Cell--Gas Turbine Hybrids*. 2013.

[3] ANSYS® Academic Research, Release 16.0

CHARACTERISTICS OF PROTECTIVE SPINEL MANGANESE COBALTITE COATINGS PRODUCED BY APS FOR Cr-CONTAINED SOFC INTERCONNECTS

Chun-Liang Chang*, Chang-sing Hwang, Chun-Huang Tsai, Sheng-Fu Yang, Te-Jung Huang, Ming-Hsiu Wu, Cheng-Yun Fu

Physics Division, Institute of Nuclear Energy Research, Taiwan, ROC

ABSTRACT

The chromium-contained ferritic stainless steels are widely employed as metallic interconnects in intermediated temperature solid oxide fuel cells. However, the chromium content of these steels would cause obvious degradation phenomena such as the growth of Cr2O3, chromium poisoning and delamination of oxide scales. Due to the high electrical conductivity of about 60 S/cm at 800oC in air, the manganese cobaltite with spinel structures are employed as protective coatings on Cr-contained steel interconnects. In this study, Ce-doped manganese cobaltite (CMC) protective coatings are produced by the promising atmospheric plasma spraying technique on the pre-oxidized Crofer22H, Crofer22APU and SS441 substrates. The obtained CMC layers reveal relatively dense microstructure due to the employed process parameters. The ASR values of the coated interconnects were measured by a four-probe dc technique at 800°C in air. After about 4,343 hours ageing at 800oC in air, the initial and final ASR values of the coated Crofer22APU sample with pre-oxidation treatment are 1.50 and 1.86 $m\Omega cm^2$, respectively. The smallest increasing rate of ASR in this study is only about 0.083 $\Omega cm^2/hr$. X-ray diffraction analysis was adopted to identify the crystal structures of the obtained spinel coatings. The morphology and cross-sectional observations of the coatings on interconnects were characterized by scanning electron microscopy equipped with energy dispersive spectrometer.

INTRODUCTION

A fuel cell is an electrochemical device which can directly convert energy of applied fuel to electricity. The solid oxide fuel cells (SOFCs) have some unique advantages over other types of fuel cell or traditional power generation technologies, including inherently high efficiency, low gas pollution and high fuel flexibility.[1,2] SOFCs with reduced operation temperatures (500-700°C) or intermediate temperature SOFCs (ITSOFCs) provide numerous advantages, such as the application of low-cost component materials, the improvement of sealing capability, the reduction of the interfacial reaction during cell operation. Due to low material cost, good mechanical properties, high electrical conductivity, high thermal conductivity and easy manufacturing process to large area of metallic materials, traditional ceramic interconnectors such as LaCrO3 are replaced by high temperature alloys in SOFC stack. Many high temperature alloys such as Cr-based, Ni-based and Fe-based superalloys have been studied as metallic interconnector. Although Cr-based superalloy has good matching in thermal expansion with other SOFC components, the large amount of chromium trioxide formation and evaporation and further the degradation resulted from Cr-poison still need to be overcome. The Ni-based superalloys generally have a thermal expansion coefficient mismatch and an expansive material cost. The Cr-contained stainless steels with the chromium content about 20 wt.% (Fe-based superalloy) are commonly considered as interconnectors among these promising candidate materials due to their good electrical conducting oxide scales, low material cost, high thermal conductivity and compatible thermal expansion coefficients (TECs) with other cell components, etc.[3,4] However, the Cr-poison effect and growing of oxide scale still needing to be reduced for the Cr-contained interconnector. During the operating period of a SOFC stack, the oxide scale, Cr_2O_3, easily reacts with O_2 and/or H_2O and transforms to chromium trioxide (CrO_3) or chromium hydroxide (CrO_2OH) vapors. These gaseous species may transform back to solid chromium oxides at the triple phase boundaries (TPBs) of cathodes and cause dramatic performance degradations of SOFCs.[5,6] Furthermore, the oxide scales of Cr-contained steels are getting thicker with exposure time during operation, eventually result in large interfacial resistances. In general, the acceptable interfacial contact resistance of interconnector should be lower than 100 mΩcm^2 and the life-time of interconnector should be over 40,000 hours. Therefore, it is necessary to improve the performances of metallic interconnectors by applying effective protective coatings on surfaces of Cr-contained stainless steels to suppress the chromium poisoning and oxide scale growth. A protective coating acts as a mass barrier to chromium cation, oxygen anion and Cr-contained molecule to transport through it.[7,8] In addition, a protective coating should reveal a higher electrical conductivity than that of Cr_2O_3 to minimize interfacial contact resistances at the interfaces between electrodes and interconnectors. An excellent protective coating must be an excellent electron conductor with negligible oxygen ion conductivity and dense microstructure. In recent years, several studies suggest that the spinel coating is a promising protective coating due to the high electrical conductivity, excellent chromium retention and good thermal expansion coefficient (TEC) match with Cr-contained ferritic stainless steel.[9] According to the results published by of Petric and Ling et al.,[10] the electrical conductivity of Mn-Co spinel is 60 S/cm at 800°C which is almost highest value among the vast variety of binary spinels composed of Al, Mg, Cr, Fe, Co, Ni Cu and Zn. The TEC value of Mn-Co spinel is around 9.7×10^{-6} °C^{-1} which is closed to 11×10^{-6} °C^{-1} of Cr-contained steel. Yang's study shows that the Ce doped Mn-Co spinel coatings retained the advantages of the undoped Mn-Co spinel coatings, acting effectively as a Cr-outward and O-inward diffusion barrier and improving the electrical performance of the ferritic stainless steel. In addition, the RE addition to the spinel appeared to alter the scale growth behavior beneath the coating, leading to a more adherent scale/metal interface.[11] Atmospheric plasma spraying (APS) is a promising technique to produce ceramic coatings with thick films on various substrates materials. Therefore, APS process was employed to produce protective ceramic coatings of Cr-contained steel due to its flexibility, high deposition rate and low cost. In this study, $Ce_{0.05}Mn_{1.475}Co_{1.475}O_{4-\delta}$ (CMC) powders were synthesized by sol gel method and employed as feedstock powders for APS process to produce protective coatings on the substrates of Crofer22H, Crofer22APU and SS441 Cr-contained steels with pre-oxidation treatment. The properties of the coated samples were examined and discussed by means of X-ray diffractometer (XRD), scanning electron microscope (SEM) and DC four-point measurement.

EXPERIMENTAL

The CMC powders were synthesized using following starting materials: $Ce(NO_3)_3 \cdot 6H_2O$, $Mn(NO_3)_2 \cdot 4H_2O$, $Fe(NO_3)_3 \cdot 9H_2O$ and citric acid monohydrate ($C_6H_8O_7 \cdot H_2O$). Stoichiometric amounts of Ce-nitrate, Mn-nitrate and Co-nitrate were firstly dissolved into deionized water to obtain a cation solution. The citric acid monohydrate was further added into the cation solution to form complex solution and then the solution was heated to 80°C and dwelled for 1 hour to improve the chelating reaction. The molar ratio of citric acid to total metal ions and the concentration of the total metal ions in the deionized water were kept at 2 and 1.0 M, respectively. The solution was then evaporated until a gel was formed. This gel was further heated to 110 and 300°C in sequence until a brown spongy-like solid foam was formed. The solid foam was ground and heated to 600°C for 2 hours to remove the residual organic reagents.

Finally, the powders were heated at 600 and 800°C for 4hrs to complete the powder synthesis process.

As shown in Figure 1(a), the APS system consisted of a plasma torch (TriplexPro™-200, Sulzer Metco), a powder feeder system, a cooling system, a furnace, an IR (Infrared) detector and a Fanuc Robot ARC Mate 120iC system to scan plasma torch. The plasma torch was operated at medium current around 420 A and high voltage around 118 V. The mixed gas composed of argon, helium and nitrogen was used as plasma forming gas. Argon, helium and nitrogen flow rates were controlled by using mass flow controllers. Details of experimental apparatus and typical plasma spraying parameters were given in another published paper.[12,13] A re-granulation process of the CMC powders was conducted to form sphere-like agglomerates via the spray drying equipment. The morphology of re-granulated powders is shown in Figure 1(b). The agglomerated CMC powders sieved with the particle sizes between 20 to 45 μm were employed as feedstock. After re-granulation, feedstock powders keep its spinel crystalline structure and $Ce_{0.05}Mn_{1.475}Co_{1.475}O_4$ chemical composition respectively.

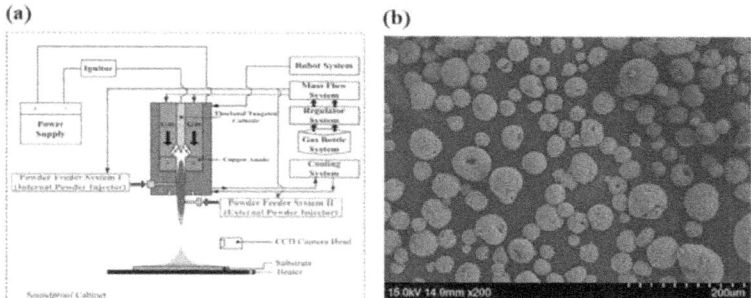

Figure 1. (a) Schematic diagram of the APS system. (b) Re-granulated CMC powders.

The Cr-contained steels, Crofer 22 H, Crofer 22 APU and SS441, were firstly cut into 1×1 cm² substrates and then their surfaces were blasted by abrasive Al_2O_3 powders via a sandblasting machine to increase the surface roughness. These substrates were heated to 800°C and held for 12 hrs to complete the pre-oxidization treatment.

The phase purity of coatings was determined by a XRD (Bruker D8) at room temperature. The scanning rate was 1 min/degree and the X-ray raddiation was Cu k_α. Surface morphology and cross-section observation were conducted by SEM (Hitachi S4800 and Jeol JSM-5310) equipped with an energy dispersive X-ray spectroscope (EDS). A DC four-point method was employed to measure the area specific resistance (ASR) values of the coated samples at 800°C.

RESULTS AND DISCUSSION

The ex-situ XRD patterns of the annealed CMC powders and CMC coatings are shown in Figure 2. Form Figure 2(a), it can be seen that cubic and tetragonal spinel crystalline structures are both exist in the annealed CMC powders at room temperature. These cubic and tetragonal spinel phases refer to cubic $MnCo_2O_4$ (JCPDF card No.23-1237) and Mn_2CoO_4 (JCPDF card No.77-0471), respectively. This observation is consistent with earlier work by Brylewski et al. and Naka et al showing the co-existence of the cubic- and the tetragonal-spinel for $Mn_{1+\delta}Co_{2-\delta}O_4$

with $0.3 < \delta < 0.9$.[14,15] According to the research results reported by Brylewski et al., the dual phase spinel has structural transformation with temperature. The tetragonal phase disappears at 400oC and only the cubic phase is then observed. As shown in Figure 2(b), it can be seen that the as-sprayed CMC coating still reveals dual phase as those of CMC powders. Even though a heat treatment of 800°C for 4343 hours was applied to the CMC coating, there is still no undesired phase shown in Figure 2(b). No obvious phase change is observed in the CMC coated samples. In addition, no CrO_2 phase signals are detected in these samples.

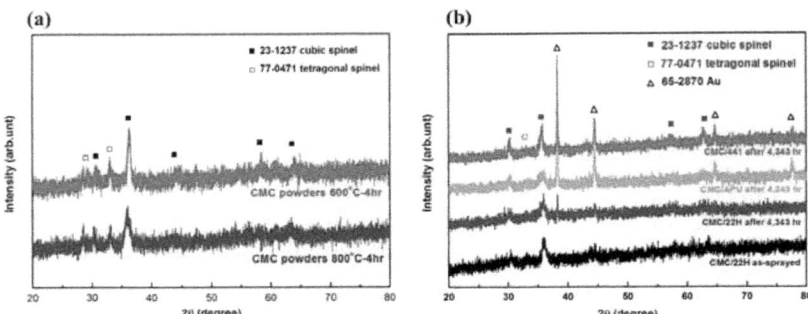

Figure 2. The XRD results of (a) annealed CMC powders with different heating temperature; (b) CMC coated samples after the heat treatment of 800°C for 4,343 hrs.

The microstructure of the CMC coating on a Crofer 22 H substrate is shown in Figure 3. Because the same APS spraying parameters are applied for preparing all the specimens, the same microstructures of these specimens can be assumed and a typical microstructure is shown here only. The observations show that the as-sprayed CMC coating reveals a relatively dense microstructure with few closed pores and without cracks. From cross-sectional observation, it is shown that the thickness of APS-LSM coating applied in this study is around 17 μm. The EDS results listed in Figure 3(b) show that the stoichimetric ratio of the obtained CMC coating is $Ce_{0.049}Mn_{1.488}Co_{1.463}MnO_{4-\delta}$, which is very close to $Ce_{0.05}Mn_{1.475}Co_{1.475}O_{4-\delta}$ of synthesised CMC powders. These EDS results imply that the APS process does not cause a serious evaporation for a particular element. From Figure 4, it can be found that the CMC coating still remains a dense microstructure without induced cracks and pores after a heat treatment of 800°C for 12 hours.

Figure 5 shows the surface morphologies of CMC coated Crofer22H specimen. The CMC coatings still remain crack-free microstructure after a long-term ASR measurement applied at 800°C for 4,343 hours. Comparing to Figure 3 and 4, it is found that the crystalline grain of annealed CMC coatings reveal unique pyramidal shape. According to the EDS results, it is indicated that the pyramidal CMC coatings were composed of Mn, Co and Cr. It implies that the Cr element diffuses from substrate to CMC coating and forms $(Mn, Co, Cr)_3O_4$ crystalline structure. Figure 6 shows the corresponding cross-sectional microstructures of CMC coatings on the aforementioned specimens. It is clearly found that the oxide scales was composed of CrO_2 layers between CMC spinel coating and metal substrates. Among these specimens the oxide scale of CMC-coated Crofer22APU specimen has the smallest thickness only about 2~3 μm after a heat treatment for 4,343 hours. Except CMC-coated SS441 specimen, CMC coatings were adhered well on the surfaces of Crofer22H and Crofer22APU substrates as shown in Figure 6(a) and 6(b). Figure 6(c) shows that many cracks paralleled to the interface between CMC coating and SS441 substrate are found in the SS441 specimen. These cracks are not induced by the APS process because the as-sprayed CMC coating on SS441 substrate reveals a crack-free morphology. The occurence of cracks might be due to the lack of rare earth element in SS441 steel so as to result in a poor oxide scale adhesion on SS441 substrate after 4,343 hours ASR measurement.

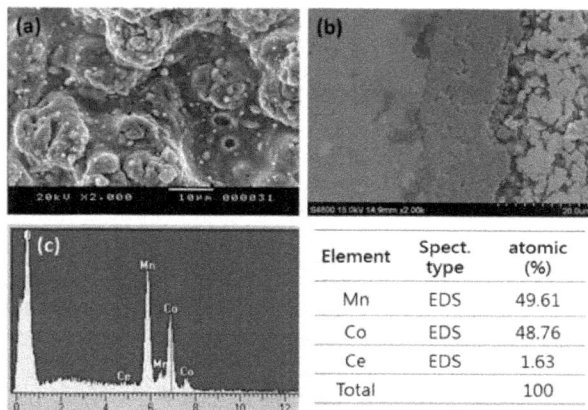

Element	Spect. type	atomic (%)
Mn	EDS	49.61
Co	EDS	48.76
Ce	EDS	1.63
Total		100

Figure 3. (a) Surface morphology of as sprayed CMC coatings on Crofer 22H and (b) EDS results obtained from the red square area in (a).

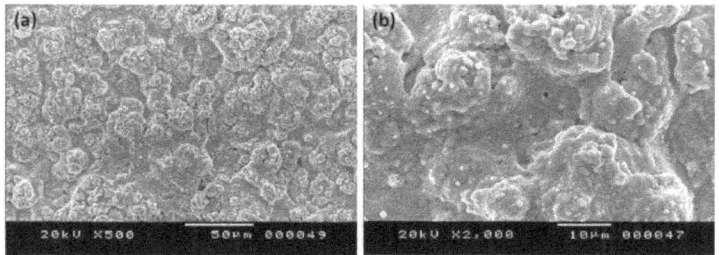

Figure 4. Micrographs in (a) low and (b) high magnifications of the sprayed CMC coating on a Crofer 22 H substrate and after the heat treatment at 800°C for 12 hours.

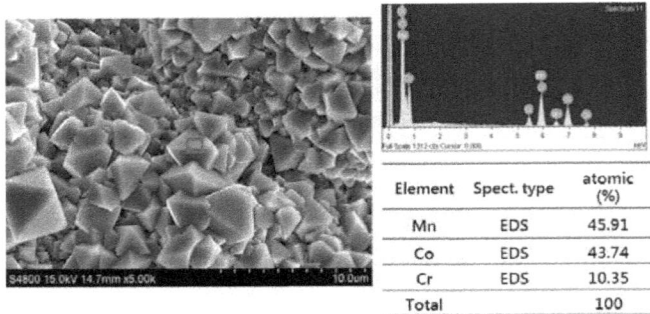

Element	Spect. type	atomic (%)
Mn	EDS	45.91
Co	EDS	43.74
Cr	EDS	10.35
Total		100

Figure 5. Surface morphology and its EDS analysis results of CMC/22H specimen after long-term durability test at 800oC for 4,343 hours in air.

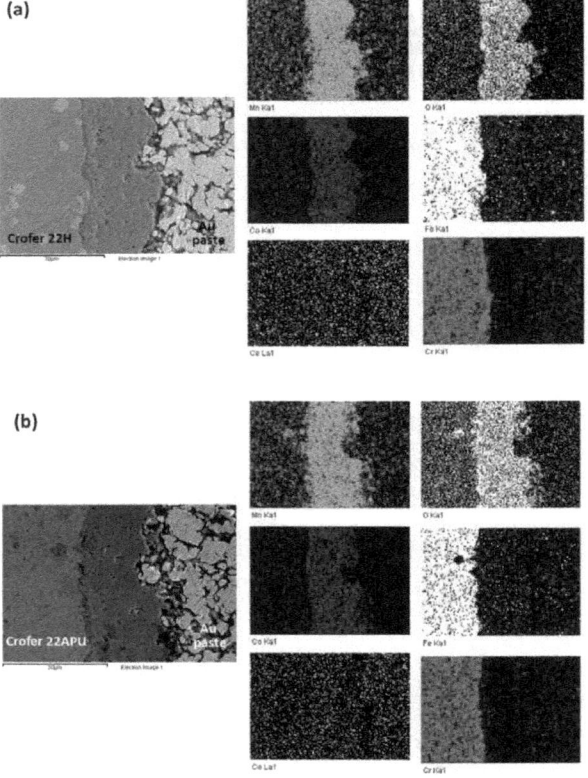

(c)

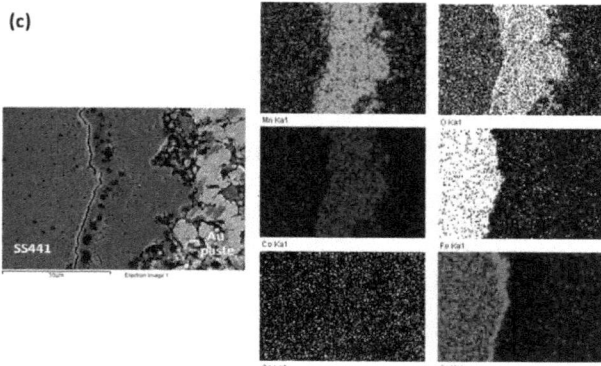

Figure 6. Cross-sectional micrographs and its element distribution images obtained from EDS analysis of (a) CMC/22H; (b) CMC/APU and (c) CMC/SS441 specimens after long-term durability test at 800°C for 4,343 hours in air.

From the ASR measurement, Arrhenius plots of the CMC coated specimens are given in Figure 7. The range of measurement temperature is from 600 to 800°C. The slopes of the CMC-coated Crofer22H and CMC-coated Crofer22APU specimens are similar. It implies that these specimens have the same conduction mechanism. It is attributed to the fact that similar oxide scales were formed at these specimens. The slope of CMC-coated SS441 specimen is significantly from the slopes of other specimens, this implies that the oxide scale or interface of this specimen has a different structure and conduction mechanism. The long-term ASR measurement results of the CMC-coated and LSM-coated specimens tested at 800°C in air for 4,343 hours are shown in Figure 8 and Table 1. These LSM coatings were prepared by APS process on the same substrates as that of CMC-coated specimens. It is clear to see that almost the CMC-coated specimens have lower initial ASR values than those of LSM-caoted specimens. It is attributed to the higher conductivity of CMC than those of LSM. The ASR increasing rates of CMC-coated specimens are all much lower than those of LSM-coated specimens. Among these specimens, the CMC-coated Crofer22APU specimen reveals the lowest ASR increasing rate of 0.083 Ωcm^2/hr, its ASR values varies from 1.50 to 1.86 mΩcm^2 after a long-term ASR measurement at 800°C for 4,343 hours in air. The final ASR values of CMC-coated Crofer22H, Crofer 22APU and SS441 specimens are only 3.75, 1.86 and 7.11 mΩcm^2, respectively, which are all much lower than 100 mΩcm^2. Their ASR increasing rates are all lower than the threshold value of 2.5 Ωcm^2/hr as well.The ASR results indicate that dense and crack-free CMC coatings with acurrate spinel phase and chmical composition can be produced by APS process and reveal outstanding electrical performances.

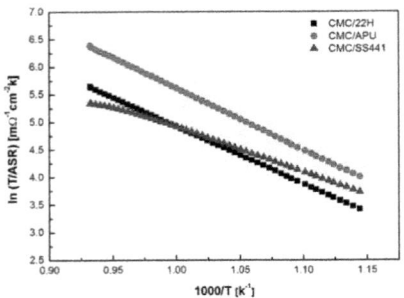

Figure 7. Arrhenius plots of CMC-coated Crofer 22 H, Crofer 22 APU and SS441 specimens.

Table I. Long-term ASR results of all specimens with elapsed time of 4,932 hours.

	Initial ASR [mΩcm²]	Increment, ΔR [mΩcm²]	Elapsed time, t [hr]	Increasing rate, ΔR/t [μΩcm²/hr]
CMC/Crofer 22H	2.21	1.54	4343	0.355
LSM/Crofer 22H	2.12	5.71	4343	1.315
CMC/Crofer 22APU	1.5	0.36	4343	0.083
LSM/Crofer 22APU	1.91	2.85	4343	0.656
CMC/SS441	2.7	4.41	4343	1.015
LSM/SS441	2.53	5.98	4343	1.377

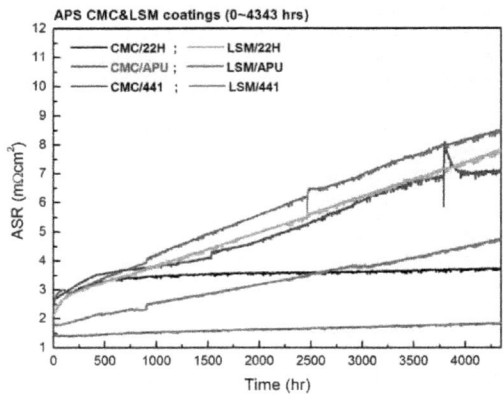

Figure 8. Long-term ASR measurement results of CMC- and LSM-coated Crofer22H, Crofer 22APU and SS441 specimens.

CONCLUSION

The CMC coatings are successfully produced on the Crofer22H, Crofer22APU and SS441 substrates by APS process. The obtained CMC coatings reveal dense and crack-free microstructure before and after a heat treatment of 800°C for 4,343 hours. The APS-CMC coatings with desired spinel structure and without X-ray detectable secondary phases are obtained in this study. The excellent thermal stability of the APS-CMC coatings is examined by applying a post-heat treatment at 800°C for 4,343 hours. After aging for 4,343 hours at 800°C in air, all specimens with APS-CMC coating reveal outstanding electrical performances in ASR increasing rate, their ASR increasing rates are all lower than the threshold value of 2.5 $\mu\Omega cm^2/hr$. Among these specimens, the coated Crofer22APU specimen with pre-oxidation treatment has the best performance, and its initial and final ASR values are 1.50 and 1.86 $m\Omega cm^2$, respectively and the measured ASR increasing rate is only about 0.083 $\mu\Omega cm^2/hr$.

REFERENCES

[1]T. H. Etsell and S. N. Flengas, Electrical Properties of Solid Oxide Electrolytes, *Chem. Rev.*, **70**, 339-376 (1970).

[2]H. Inaba and H. Tagawa, Ceria-based Solid Electrolytes, *Solid State Ion.*, **83**, 1-16 (1996).

[3]Z. Yang, K. S. Weil, D. M. Paxton and J. W. Stevenson, Selection and evaluation of heat-resistant alloys for SOFC interconnect applications, J. Electrochem. Soc., 150[9], A1188-A1201 (2003).

[4]G. Cabouro, G. Caboche, S. Chevalier and P. Piccardo, Opportunity of metallic interconnects for ITSOFC: Reactivity and electrical property, J. Power Sources, 156, 39-44 (2006).

[5]J. W. Fergus, Effect of cathode and electrolyte transport properties on chromium poisoning in solid oxide fuel cells, *Int. J. Hydrog. Energy*, **32**[16], 3664-3671 (2007).

[6]E. Konysheva, H. Penkalla, E. Wessel, J. Mertens, U. Seeling, L. Singheiser and K. Hilpert, Chromium poisoning of perovskite cathodes by the ODS alloy $Cr_5Fe_1Y_2O_3$ and the high chromium ferritic steel Crofer22APU, *J. Electrochem. Soc.*, **153**[4], A765-A773 (2006).

[7]C.-L. Chang, C.-s. Hwang, C.-H. Tsai, S.-F. Yang, W.-J. Shong, Z.-Y. J.-S. And T.-J. D. Huang, Characteristics Of Protective Lsm Coatings On Cr-Contained Steels Used As Metallic Interconnectors Of Intermediated Temperature Solid Oxide Fuel Cells, Advances in Solid Oxide Fuel Cells and Electronic Ceramics: A Collection of Papers Presented at 39th International Conference on Advanced Ceramics and Composites, 45-55 (2016).

[8]C.-L. Chang, C.-s. Hwang, C.-H. Tsai, S.-F. Yang, W.-J. Shong, T.-J. D. Huang and M.-H. Wu, Development of Plasma Sprayed Protective LSM Coating in INER, Advances in Solid Oxide Fuel Cells and Electronic Ceramics II: Ceramic Engineering and Science Proceedings, 37[3], (2017).

[9]F. Smeacetto, A. D. Miranda, S. C. Polo, S. Molin, D. Boccaccini, M. Salvo and A. R. Boccaccini, Electrophoretic deposition of $Mn_{1.5}Co_{1.5}O_4$ on metallic interconnect and interaction

with glass-ceramic sealant for solid oxide fuel cells application, J. Power Sources, 280, 379-386 (2015).

[10]A. Petric and H. Ling, Electrical Conductivity and Thermal Expansion of Spinel at Elecated Temperatures, J. Am. Ceram. Soc. 90, 1515-1520 (2007).

[11]Z. Yang, G. Xia, Z. Nie, J. Templeton and J. W. Stevenson, Ce-Modified $(Mn,Co)_3O_4$ Spinel Coatings on Ferritic Stainless Steels for SOFC Interconnet Application, Electrochem. Solid-State Lett., 11[8], B140-B143 (2008).

[12]C. S. Hwang, C. H. Tsai, C. H. Lo and C. H. Sun, Plasma Sprayed Metal Supported YSZ/Ni–LSGM–LSCF ITSOFC with Nanostructured Anode, J. Power Sources, 180, 132-142 (2008).

[13]C. H. Lo, C. H. Tsai and C. S. Hwang, Plasma-Sprayed YSZ/Ni-LSGM-LSCo Intermediate-Temperature Solid Oxide Fuel Cells, Int. J. of Appl. Ceram. Technol., 6[4], 513-524 (2009).

[14]T. Brylewski, A. Kruk, A. Adamczyk, W. Kucza, M. Stygar, K. Przybylski, Synthesis and characterization of the manganese cobaltite spinel prepared using two soft chemical methods, Mater. Chem. Phys., **137**, 310-316 (2012).

[15]S. Naka, M. Inagaki, T. Tanaka, On the formation of solid solution in Co3-xMnxO4 system, J. Mater. Sci., 7[4], 441-444 (1972).

INFLUENCE OF TEMPERATURE AND STEAM CONTENT ON DEGRADATION OF METALLIC INTERCONNECTS IN REDUCING ATMOSPHERE

Folgner, Christoph; Sauchuk, Viktar; Kusnezoff, Mihails; Michaelis, Alexander
Materials and Components, Fraunhofer IKTS, Dresden, Saxony, Germany.

ABSTRACT
        Three ferritic steels (Crofer 22 APU, Crofer 22 H, AISI441) were tested in a SOFC anode gas environment in a temperature range between 725 – 875°C. The experiments were performed by varying the amount of water vapor content in the gas mixture for different exposure times in order to create accelerated degradation testing conditions for metallic interconnectors (MICs) and to investigate the behavior of these materials caused by the formation of growing chromium oxide based scales. Gravimetric measurements and FESEM/EDX data were analyzed to characterize the oxidation kinetics and the microstructure of the oxide scales. A clear correlation between increasing temperatures and increasing oxide growth rate constants $k_{p,w}$ can be observed in all materials. This interrelation results in thicker surface oxide scales. The structures of the oxide layers are specific for each material and consist of Cr_2O_3 and $(Cr,Mn)_3O_4$ with different elemental distribution and thickness ratios. In addition, a zone of inner oxidation with Al-, Si- and Ti-rich oxide inclusions, can be seen in ferritic samples, whose microstructures differ depending on the analyzed material and temperature. Electrical measurements in dual gas atmosphere reveal also an increase in resistance within 1000 h material exposure. The results of oxide growth in tested samples are compared with data derived from operated SOFC stacks.

1. INTRODUCTION
        Despite continuous success in regards to the increase of operational durability of solid oxide fuel cells (SOFCs) [1], a demand for further improvements in the reliability and the robustness of SOFC stacks and their system components remains desirable. The metallic interconnect (MIC) is a crucial component, and plays an important role in terms of long term stability of SOFCs. Because of high operation temperatures (700 to 900 °C), there are number of well-established requirements pertaining to MIC materials for suitable application in SOFCs [2], [3], [4]. Besides good matching of the coefficient of thermal expansion (CTE) close to the adjacent cell components (MEA, glass bond and sealing) and a low ductility to provide a structural support within the stack, the interconnect material should be a good electrical conductor. Moreover, due to operation in the dual gas atmosphere, the interconnect material should be mechanically, thermally and chemically stable in both oxidizing and reducing environments. Chromium containing heat resistant ferritic steels are the preferred class of materials for metallic interconnects in relevant operation conditions because of their good mechanical properties, appropriate thermal expansion behavior, low cost and excellent manufacturability [5], [6]. Since the different metal suppliers try to match the parameters of interconnect materials to the SOFC requirements, there is variety of steel grades having a broad spectrum of compositions, manufacturing processes and material properties influencing MIC performance with regard to the corrosion behavior. One of the major problems arising from the use of Fe - Cr ferritic steels as SOFC interconnect is the formation of chromium containing oxide scales during their exposure to the SOFC operating conditions. On the one hand, these layers are desirable for passivation of the metal against chromium release. On the other hand, the growing oxide scale increases the contact resistance between the steel interconnect and the ceramic electrodes of the SOFC. The introduction of reactive additives like Al, Si, Ti, W, Yb, Hf, Zr, Y,

La, Ce, Mo can considerably reduce the rate of oxide scale formation and stabilize the MIC composition [4], [7], [8]. Moreover, the use of minor quantities of so called alloying elements - Manganese, Cobalt, Nickel - in steel compositions causes the formation of semiconducting oxide scales on the interconnect surface having much more lower resistance directly compared to pure chromium oxide and acting, in addition, as a protective layer to prevent the release of volatile chromium oxide species poisoning the other SOFC components [5], [9]. The interconnect corrosion and the formation of oxide scales after oxidation in SOFC cathode gas (ambient air) was intensively investigated over the entire period of SOFC development [4], [10], [11], [12]. Few prior investigations highlight the formation and development of oxide scales at metallic interconnects in reducing gas environments [13], [14], [15]. However, the oxidation behavior of MICs in anode gas conditions and its influence on the contact resistance between MIC and anode contacting remains scarcely represented in literature. An increase of operation times exceeding 40. 000 h is still necessary for a broad market introduction of SOFC stacks. This specification implies that both stacks and their components, including MIC materials, have to be stable over this duration. However, real time tests, generally used as a confirmation of reliability or failure of materials/layer modifications over such long time scales, are time consuming and cost intensive. A modification of standard material test procedures, with the aim to accelerate material degradation, is therefore necessary. For this purpose two ferritic steels, Crofer 22 APU and Crofer 22 H, optimized for application as MIC material in SOFC stacks, and the high temperature stainless steel AISI441, that has not been specially developed for application as an SOFC interconnector material, were analyzed with respect to oxidation kinetics and electrical properties in SOFC anode gas environment at varying temperatures and gas compositions.

2. METHODS & EXPERIMENTALS

To analyze the oxide scale growth behavior in the steam containing reducing SOFC anode gas atmosphere, three ferritic alloys Crofer 22 APU, Crofer 22 H and AISI 441 were tested between the temperatures 725°C and 875°C. The experiments were divided into two modes: (i) the currentless aging of the samples over a time period of 1000 h in which the mass gain and thickness changes were recorded, and (ii) the tests in which the evolution of resistivity was monitored and analyzed during 1000 h in reducing SOFC gas ambience (dual-gas atmosphere). The focus of the first (i) set of tests was to determine the kinetics and the microstructure changes of the inner and outer oxidation zones. Both blank and Ni[O] or Cu[O] coated (protected) samples were tested during the experiments. The contact/protection layer was applied via screen printing. The area of Ni[O]/Cu[O]- protection layers was 15×15 mm^2 for currentless experiments. The fuel gas composition (FG1) for the currentless tests (i) was N$_2$/H$_2$ = 80/20 vol. %+ 3 % H$_2$O (H$_2$/H$_2$O=87/13). The fuel gas was humidified by bubbling the dry fuel gas through a water reservoir. Crofer 22 APU and Crofer 22 H were purchased from VDM metals and AISI 441(1.4509) was obtained from JENSSEN GmbH&Co.KG. The compositions of the supplied materials were analyzed by X-ray fluorescence (XRF) analysis. The comparison of material compositions specified by the suppliers and the data from the XRF analysis is shown in Table 1. The samples for currentless tests were shaped into the square pieces (L×B×H = $20 \times 20 \times 2,5$ mm^3) by laser cutting. The sample size for resistivity measurements in dual gas atmosphere was chosen as $60 \times 30 \times 2,5$ mm^3. The contact area of Ni[O]/Cu[O] layers for these tests was 10×10 mm^2. The cold rolled and annealed material was de-flashed after laser cutting. The samples were then briefly (~10 s) polished with P-500 polishing fleece. The surface topography of the blank samples was monitored using a profilometer (FRT GmbH). The obtained values of the average surface roughness R_a were 2,4 - 2,5 µm for all three materials. Before coating, the samples were cleaned for 30 min with 2-propanol and 30 min with ethanol in an ultrasonic bath.

Table 1: Chemical composition of the tested MIC samples

MIC	Chemical composition [weight %]														
Crofer 22 H	Cr	Fe	C	N	S	Mn	Si	Al	W	Nb	Ti	La	P	Cu	Ni
min	20	Rest				0,3	0,1		1	0,2	0,02	0,04			
max	24		0,03	0,03	0,006	0,8	0,6	0,1	3	1	0,2	0,2	0,05	0,5	
XRF	22,35	70,87				0,39	0,35	0,17	1,81	0,57	0,09			1,07	2,3
Crofer 22 APU															
min	20	Rest				0,3					0,03	0,04			
max	24		0,03		0,02	0,8	0,5	0,5			0,2	0,2	0,05	0,5	
XRF	22,53	76,59				0,41	0,05	0,12			0,1				
AISI 441	17,5 - 18,5	Rest	≤0,0 3		0,015	≤1,0	≤1,0			Ti+Nb: 0,6			≤0,0 1		
XRF	16,85	80,93				0,26	0,71	0,13		Nb:0,36, Ti:0,15				0,09	0,31

Image analysis of microstructure of the inner and outer oxidation zones of the corroded MIC samples was done with FESEM tools from Zeiss AG (ULTRA 55/NVision 40). The qualitative analysis of the material composition (energy dispersive spectroscopy EDS scans /EDS mappings) were captured with a SDD-material characterization system from EDAX (EDAX-Trident XM4). For the determination of the layer thickness on the outer oxidation zone of blank samples, the arrays of 10 to 20 image frames with a magnification of 1500x – 3000x were analyzed by area analysis tool. A total image length of approximately 750 µm was characterized within this gray level analysis. The estimation of the depth of the inner oxidation zone was determined for each sample by linear analysis from 10 single images (measuring area: 76 µm × 57 µm), which were vertically segmented into 15 chords. Measurements began from the interface of the outer oxide scale and extended into the bulk of the samples until the end of the oxide inclusions. The weight gain of oxidation tests (i) were determined gravimetrically after annealing durations of 200, 400, 600, 800, 1000 h with a precision scale (Sartorius GmbH). After weighing, the samples were placed back into the oven and heated to peak temperature in a reducing atmosphere (N_2 atmosphere, with 2 K/min heating rate). Finally, steam was added to the corresponding atmosphere. Samples for the resistivity measurements in the dual gas atmosphere were assembled in a sandwich structure which is illustrated in Figure 1. The fuel gas mixture (FG2) of $H_2/H_2O =$ 50/50 vol. % was dispersed inside the glass sealed MIC samples, whereas the outer side of the assembly was surrounded by hot ambient air. The electrical resistivity was measured via 4-point probes method with a current of 0.5 A. The electrical contact between the upper and the lower MICs is connected via nickel net interconnection. Current flow contacts (Pt wires) are placed diagonally to provide uniform current distribution in the structure.

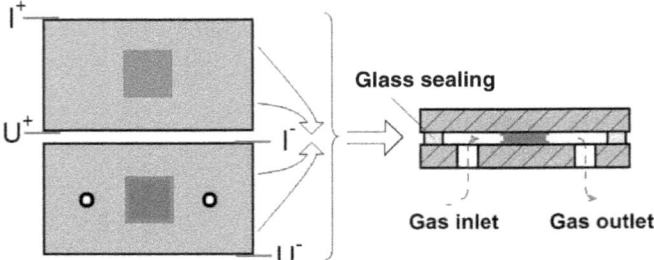

Figure 1. MIC sample layout for resistivity measurements in dual gas atmosphere: Green marked areas symbolize Ni-coated sections. Sealing from the outer environment was enforced with a special glass frame (yellow). Electrical contact between the upper and the lower steel samples was made with a nickel net interconnection.

3. RESULTS & DISCUSSION

3.1 Characterization of the oxide growth of the outer oxidation zone

Figure 2, 3, and Figure 4 illustrate the weight gain and thickness development during the oxidation tests (i). The tests are over a 1000 hr annealing period in a FG1 ambient environment (N_2/H_2 = 80/20 vol. %+ 3 % H_2O (H_2/H_2O=87/13)) with a temperature range of 725°C to 875°C. The growth rate of oxide scales during the high temperature oxidation can be described mathematically within the approach based on the Wagner's oxidations theory, in which the oxide growth is expressed with an exponent of 2 in ideal approximation.

$$\left(\frac{\Delta m}{A}\right)^2 = 2 \cdot k_{p,w} \cdot t \quad \text{Equation 1: Parabolic rate law}^{(16)}$$

Δm…mass gain; A…sample surface area; $k_{p,w}$…mass related parabolic oxidation constant, $k_{p,w}$ = f(T) [kg^2/m^4 s].

The origin of this approach is derived from Ficks First Law, because ion and electron diffusion is essential for oxide growth. In Wagner´s theory of parabolic oxide grow, the rate constant can be related to the self-diffusion coefficient of ions in the oxide layer [16]. The diffusion pathways through the growing oxide layer are getting longer over time with increasing oxide thickness so that the reaction at the phase boundary is decelerated. This process is accompanied by a deceleration of mass gain.

Development of the mass gains of the samples in Figure 2, 3, and Figure 4 illustrates nearly a parabolic oxidation behavior. Considering the influence of the Cu- and the Ni- coatings, one can see that the Ni-coated MIC samples show the smallest weight gain values for all three materials over the tested temperature range. The weight gain values of the Cu-coated samples are almost similar or only slightly smaller than that of the blank material (with exception of Crofer 22 APU at 825°C). Therefore, one can conclude that the oxide scale formation is better retarded with the screen printed Ni protection layer in comparison with the Cu-coating. A possible explanation of the weight gain reduction in the samples with the Ni coating could be done by presuming that Ni^{2+} ions can be incorporated in the Cr_2O_3 surface layer so that two Cr^{3+} cations and one cation vacancy are swapped against 3 Ni^{2+}, so that the total density of vacancies is reduced [17], and thus, the formation rate of Cr_2O_3 is reduced. With bivalent Cu^{2+} ions, the same mechanism should be observed. However, differences in interdiffusion and reaction capability between Cu^{2+} and Ni^{2+} in Cr_2O_3 could be responsible for the differences in mass gain.

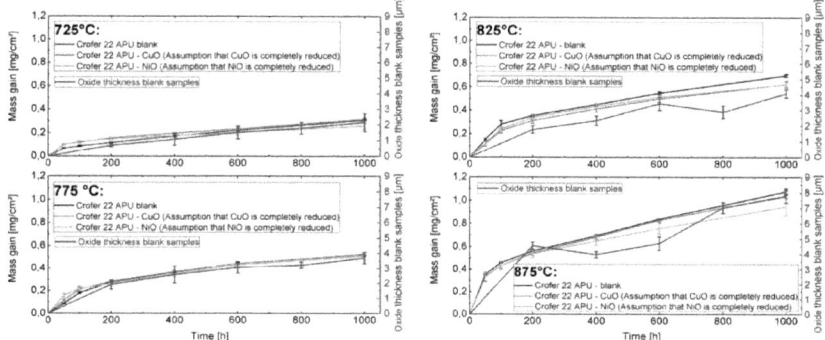

Figure 2. Mass- and thickness development over 1000 h oxidation in reducing atmosphere (FG1 = N_2/H_2 = 80/20 vol. % + 3 % H_2O (H_2/H_2O=87/13 vol. %)) for Crofer 22 APU in the temperature range 725°C (at the top left) to 875°C (bottom right). The results for the thickness development were gained from the analysis of microsections of blank samples.

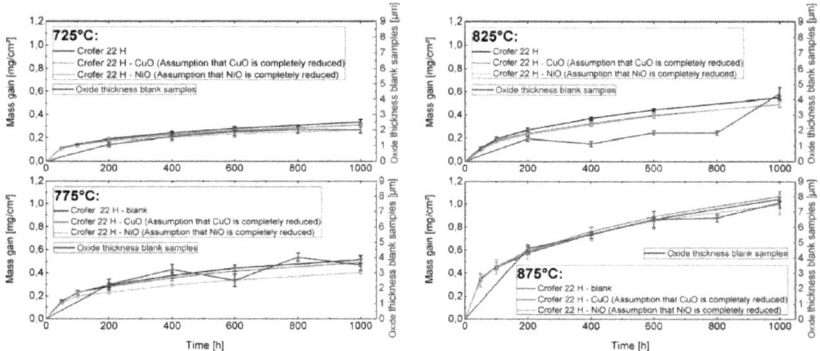

Figure 3. Mass- and thickness development over 1000 h oxidation in reducing atmosphere (FG1 = N_2/H_2 = 80/20 vol. % + 3 % H_2O (H_2/H_2O=87/13 vol. %)) for Crofer 22 H in the temperature range 725°C (at the top left) to 875°C (bottom right). The results for the thickness development were gained from the analysis of microsections of blank samples.

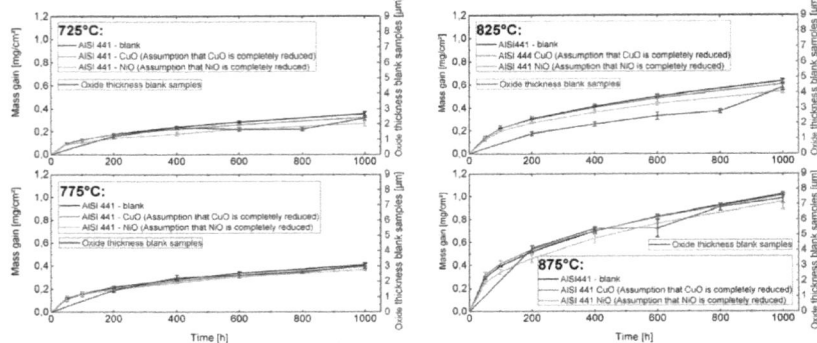

Figure 4. Mass- and thickness development over 1000 h oxidation in reducing atmosphere (FG1 = N_2/H_2 = 80/20 vol. % + 3 % H_2O (H_2/H_2O=87/13 vol. %)) for AISI 441 in the temperature range 725°C (at the top left) to 875°C (bottom right). The results for the thickness development were gained from the analysis of microsections of blank samples.

Analysis of the diagrams of thickness development of the blank samples in Figure 2 to Figure 4 (blue solid lines) indicate some graphs deviate from parabolic law behavior. These fluctuations originate due to samples being withdrawn at defined time marks (200 h, 400 h, 600 h, 800 h, 1000 h) for the microstructure analysis. Therefore, the same sample location for thickness monitoring was not available. The results of the mass gain and thickness development after the 1000 h annealing period in FG1 are summarized in Figure 5. A correlation between the temperature increase and increase in mass gain rate, and oxide scale thickness is evident for the investigated ferritic alloys. Furthermore, the comparison of the oxide thickness- and mass gain results between 725°C and 875°C after 1000 h oxidation indicates an acceleration in the range of Δm1000h: 2.9 ÷ 3.5 for the mass gain- and ΔTh1000h: 3.1 ÷ 3.8 for the thickness increase after 1000 h oxidation for the tested alloys.

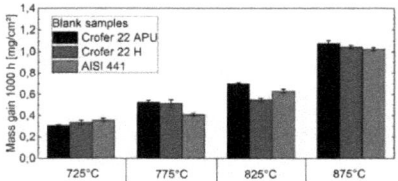 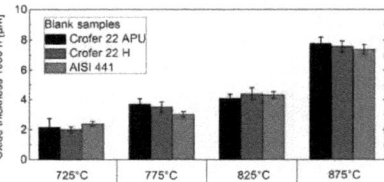

Figure 5. Comparison of mass- (left) and thickness (right) development of the blank samples after 1000 h oxidation period in reducing atmosphere (FG1 = N_2/H_2 = 80/20 vol. % + 3 % H_2O (H_2/H_2O=87/13 vol. %)) for tested ferritic alloys in the temperature range 725°C - 875°C.

The area specific mass gain values were also plotted as quadratic functions versus the oxidation time and the linearity of the plots was calculated to test whether the measured mass gains fulfill the parabolic rate law. As example, Figure 6 (left diagram) illustrates the plotted data for AISI 441 (1.4509) at 725°C fitted with linear regression. The value of the coefficient of determination

R^2 was 0.996 for this dependence (perfect linearity is given by value of 1.0). For the other tested alloys, we have calculated values of $R^2 > 0.96$ within the temperatures 725°C and 875°C, so that the mathematical description regarding the parabolic relationship should be sufficiently correct. For calculation of the $k_{p,w}$ values, characterizing the kinetics of the oxide scale growth, the slopes in the graphs were determined over the tested temperature range. Since the oxide growth is based on diffusion processes, the $k_{p,w}$ values can be expressed in form of an Arrhenius equation, which describes the temperature dependence of thermally activated growth processes:

$$k_{p,w} = k_{p,w}^0 \cdot exp\left(-\frac{E_a}{R \cdot T}\right) \quad \text{Equation 2: Arrhenius equation}$$

E_a...activation energy, R...universal gas constant, $k_{p,w}^0$...variable factor depending on grain boundary density

The $k_{p,w}$ values are specific for a certain gas composition, so that the oxygen partial pressure was held constant in FG1. For calculation of the activation energy E_a of the rate limiting step of surface oxide growth, the slope of the fitted linear regression curves of log10 $k_{p,w}$ values of the blank samples vs. 1/T values was analyzed. The results for the log10 $k_{p,w}$ values are shown in the right diagram in Figure 6. $k_{p,w}$ values of $\sim 1 \times 10^{-12}$ $g^2/cm^4 s$ at 900°C and $\sim 5 \times 10^{-13}$ $g^2/cm^4 s$ at 800°C in air for Crofer 22 APU, Crofer 22 H are known from literature [18], which are approximately one order of magnitude higher in comparison to values we found in reducing atmosphere. The results of the calculation of E_a are shown in Table 2.

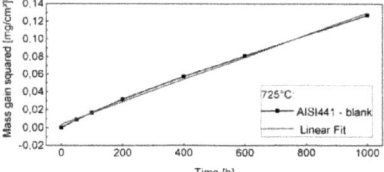 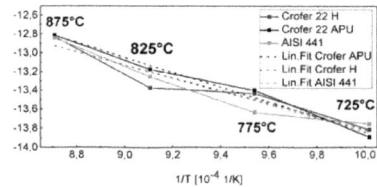

Figure 6. (Left) Squared mass gain for AISI 441 at 725°C in reducing atmosphere (FG1 = N_2/H_2 = 80/20 vol. % + 3 % H_2O (H_2/H_2O=87/13 vol. %)). Linear growth behavior with R^2=0,996 was found. (Right) Results for the logarithmized $k_{p,w}$ values which were calculated out of the slope of the squared mass development graphs of the blank samples(s. Figure 2 to Figure 4).

Table 2: Calculated activation energy of tested MIC samples oxidized over 1000 h in FG1
(N_2/H_2 = 80/20 vol. % + 3 % H_2O (H_2/H_2O=87/13 vol. %)).

Temperature range [°C]	Activation energy [kJ/mol]		
	Crofer 22 APU	Crofer 22 H	AISI 441
725 - 875	152	132	136
	Activation energy [eV]		
725 - 875	1,58	1,37	1,41

D.J. Young et al. [14] have reported a value of 200-280 kJ/mol in Crofer 22 APU for an oxidation period of 500 h in atmosphere composed of Ar-4%H_2-20%H_2O between temperatures 800 – 900 °C. Assuming, that the difference in the oxidation period in [14] has a minor importance, mainly the difference in the broader experimental temperature range and the differences in oxygen partial in the test gas could be responsible for the lower E_a values obtained in our experiments. At elevated oxygen pressures with linear phase controlled reactions, oxidation rates are often proportional to $\sqrt{p_{O_2}}$ [16]. At low values of pO_2, the reaction rate is also proportional to oxygen partial pressure. This could explain for the discrepancies between the literature data and the calculated E_a-value for Crofer 22 APU from our experiments, which was determined in an atmosphere with a lower pO_2. There are only minor differences between the E_a values at the tested alloys Crofer 22 APU, Crofer 22 H, and AISI 441. Under the assumption that there are no changes in the oxidation rate determining mechanisms over the test temperature range, the differences in the internal defect structure of the alloys may be responsible for this phenomenon. Materials possessing a complicated defect structure have contribution from alternative diffusion paths, e.g. via grain boundaries and/or lattice dislocations. This increases in addition to lattice diffusion the diffusion rate/or ion flux considerably. As a result, the activation energy decreases due to the contribution of alternative diffusion pathways. Based on the slightly higher E_a values in Crofer 22 APU, this can also indicate the defect concentration in Crofer 22 H and AISI 441 is higher.

3.2 Characterization of the oxide growth of the inner oxidation zone

The compositions of the grown oxides of the tested alloys (blank samples) were analyzed with EDS mappings after 600 h and 1000 h annealing for all tested temperatures (725-875°C). As examples, the results after 1000 h material oxidation at 875°C in FG1 are shown in Figure 7 to 11.

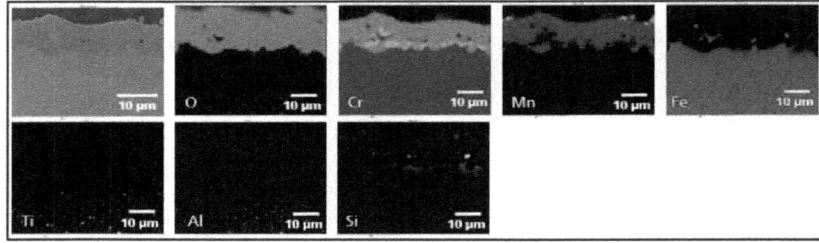

Figure 7. Element distribution detected by EDS-mapping in Crofer 22 APU samples after 1000 h oxidation period at 875°C in reducing atmosphere (FG1 = N_2/H_2 = 80/20 vol. % + 3 % H_2O (H_2/H_2O=87/13 vol. %)). An area of 57×45 μm was analyzed.

In general, the grown oxide scale was mainly composed of two oxides: Cr_2O_3 and spinel $(Mn,Cr)_3O_4$. In Crofer 22 APU (Figure 7), the upper layer is dominated by the Mn-rich spinel oxide. Analogous to the data from the 875°C test, the EDS mappings of the material tested at lower temperatures reveal also a higher content of the $(Mn,Cr)_3O_4$ versus Cr_2O_3 phases. Also some Al- and Ti- containing oxide domains or metal precipitates can be found in the bulk of the material. Furthermore, an increase of the inner oxidation zone with Al- and Ti- containing oxides can be observed with increasing temperatures. Some Si- /SiO_2 – containing inclusions can also be found in surface oxide layer. Similar to Crofer 22 APU, the spinel phase also dominates (Figure 8) in the upper oxide layer of the oxide scales grown on the surface of the Crofer 22 H samples. Moreover, there are some oxide scale regions in which Cr_2O_3 is completely embedded into the spinel layer. This fact indicates that the diffusivity and reactivity of Mn should be increased in comparison to Cr at the tested temperatures in FG1.

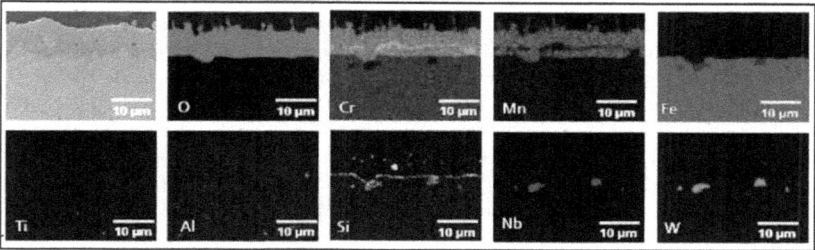

Figure 8. Element distribution detected by EDS-mapping in Crofer 22 H after 1000 h oxidation period at 875°C in reducing atmosphere (FG1 = N_2/H_2 = 80/20 vol. % + 3 % H_2O (H_2/H_2O=87/13 vol. %)). An area of 38 μm × 30 μm was analyzed.

Due to of the presence of Nb and W, Laves-phases with the composition AB_2 (A has a higher atomic radii and a lower electronegativity in comparison to B; examples for A,B: A = Mo, W, Ta, Nb, Ti; B = Cr, Fe, Mn, Si, Co) can be formed in this material. In general, such Laves-phases increase the hardness of the steel and provide a good electrical conductivity. Si also promotes Laves-phases generation [17]. However, we have also observed a Si-separation either in Nb(W)-rich regions or as the extended SiO_2 layer at the interface to the spinel/Cr_2O_3-scale (Figures 8 and 9, in more details Figure 9 B). Because of the insulating properties of SiO_2, its formation has a negative influence on the conductivity of the MIC.

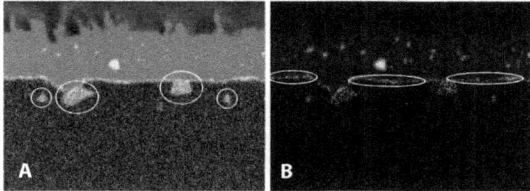

Figure 9. Overlay of added element distribution for O, Si and Nb (left, A) and superposition of Si with O (right, B). Laves phases are present at positions in which Si is dissolved in Nb (marked areas in A). SiO_2 is located at the interface between the outer oxide scale and the bulk of material (marked areas in B).

At the other temperatures tested, no clear phase separation was observed into the inner Cr_2O_3 and the outer spinel layer in the oxide scales grown on Crofer 22 H. Cr_2O_3 segments were detected between spinel rich layers in many sections of the oxide scales (Figure 8). The fraction of spinel phase increased with respect to Cr_2O_3 with increasing temperatures (+17 % at 875°C in comparison to 775°C). Figure 10 presents the element distribution in AISI 441 after a 1000 h oxidation period at 875°C in a reducing atmosphere. A clear separation of the oxide scale into a Cr_2O_3 bottom and a spinel top layer can be observed in the EDS mappings. An extended silicon separation area combined with oxygen (Figure 11 A) and a weak signal from wolfram (Figure 10) beneath the Cr_2O_3 interface was detected. In addition, some Si/Nb containing spots (Figure 11 B) can be also observed in this domain. Very likely that Si can be dissolved as Laves-phases mix with W/Nb or separately oxidized as continuous SiO_2 film. In the overlays of Figure 11, C and D, a significant amount of Ti oxide, as well as some Al containing oxide domains, can also be recognized. With reference to [19], titanium oxide particles can locally increase the mechanical strength of the materials and prevent the scale buckling by reduction of induced growth stress. The effect is strongly dependent on the overall Ti content. An enhanced internal oxidation at the alloy grain boundaries was observed in [20] at concentrations above 0.15 wt. % Ti. This can induce internal stress accompanied by buckling of the oxide scale grown. Furthermore, a positive effect on the conductivity due to Ti-doping of Cr_2O_3 is postulated [21]. On the other hand, the doping of Cr_2O_3 with Ti tends to increase the Cr_2O_3 growth rate. In comparison with the Crofer steels, the Ti concentration in the bulk of AISI 441 material was slightly higher (see Table 1).

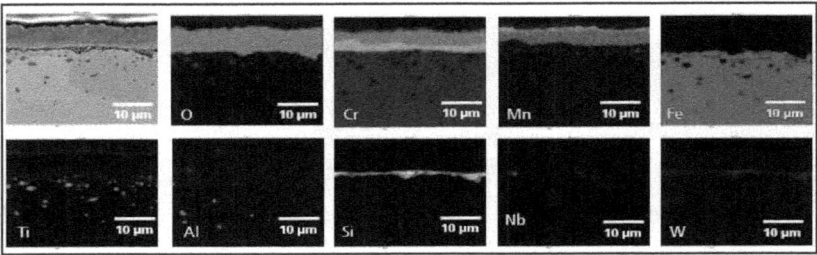

Figure 10. Element distribution detected by EDS-mapping in AISI 441 after 1000 h oxidation period at 875°C in reducing atmosphere (FG1 = N_2/H_2 = 80/20 vol. % + 3 % H_2O (H_2/H_2O=87/13 vol. %)). An area of 38 µm × 30 µm was analyzed. Cr_2O_3 and spinel are clearly separated from each other. Also TiO_2, SiO_2 and Al/Al_2O_3 can be found in the inner oxidation zone (for comparison, see Figure 11).

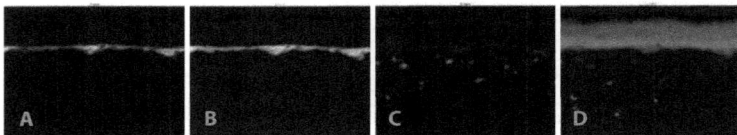

Figure 11. Overlay of implemented element distribution for Si with O (left, A), Si with Nb (B), Ti with O (C) and Al with O (right, D). SiO_2 is located at the interface to the outer oxide scale. Laves phases are present at positions in which Si is dissolved in Nb (violet spots in B). TiO_2 is located at reddish positions illustrated in C. Also some Al/Al_2O_3 can be found in D.

Analogous effects were observed in AISI 441 samples at lower annealing temperatures i.e. Si-separations in the form of Laves phases in combination with Nb as well as Si oxide domains, similarly to Si-distribution in Figure 11 B. Increased Al-oxide- and Ti-oxide- concentrations at the Cr_2O_3 interface are also visible at the 725°C samples. The methodology of linear analysis of the inner oxidation zone is illustrated in Figure 12 (left). The depth of the extent of the inner oxidation zone was analyzed. In all three of the ferritic alloys presented in Figure 12 (right), an increase in the precipitation of Al-, Ti- and Si- containing oxides with increasing temperatures can be seen (with the exception of Crofer 22 H 825°C).

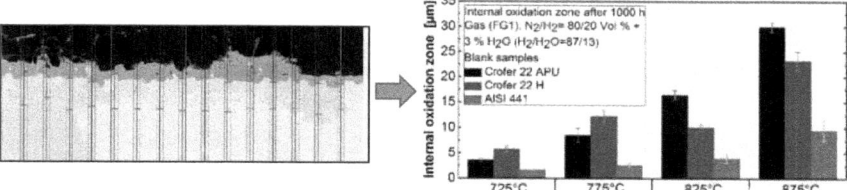

Figure 12. (Left) Schematic representation of the methodology of data analysis of the inner oxidation zone by linear analysis. Ten single images (measuring field: 76 µm × 57 µm) were segmented vertically into 15 chords. The length of the extension of the inner oxidation zone was measured in succession (distance from inner oxidation zone to the outer oxide layer). (Right) Results of linear analysis of the inner oxidation zone for the three tested ferritic steels after 1000 h oxidation in reducing atmosphere (FG1 = N_2/H_2 = 80/20 vol. % + 3 % H_2O (H_2/H_2O=87/13 vol. %) in the temperature range 725°C – 875°C.

The largest inner oxidation zone was measured in Crofer 22 APU samples, however the difference to Crofer 22 H was not significant. The measured Al- and Ti- content (see also Table 1) is almost identical in the both Crofer steels, but the Si content is higher in Crofer 22 H and AISI 441. It was shown in [22] that Nb additions to high Cr ferritic steels can bind residual Si in Laves-phases and reduce the formation of Nb oxide under ambient air, which increases the Cr_2O_3 growth rate due to a doping effect stimulating an enhanced diffusivity of Cr ions. Since Si containing oxides (probably SiO_2) were observed in Crofer 22 H and AISI 441 samples (see Figure 8 to 11), the differences in oxygen content could have an influence on the Si-binding behavior of Nb.

3.3 Investigation of the oxide growth at different water content in fuel gas

The mass gain data in FG 1 with H_2/H_2O = 87/13 vol. % humidification was compared with results gained with a higher level of humidification i.e. H_2/H_2O = 50/50 vol. %. The results of the mass gain and the oxide thickness increase are illustrated in Figure 13. The data for the greater water content have been marked as shaded bars. Except for AISI 441, no increase in mass gain and oxide thickness was observed with more steam content. Due to this fact, we can assume that the oxygen partial pressure is high enough in the tested humidification range thus, the oxygen adsorption cannot be the rate limiting parameter for oxide growth in the tested range of H_2O concentrations.

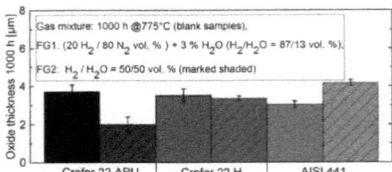
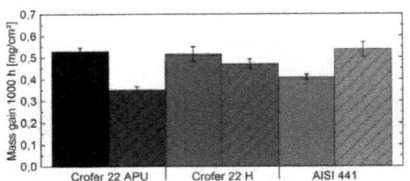

Figure 13. Comparison of the oxide scale thickness (left) and mass gain (right) after 1000 h oxidation for two different fuel gas compositions FG1= N_2/H_2 = 80/20 vol. % + 3 % H_2O (H_2/H_2O=87/13 vol. %), FG2 = H_2/H_2O = 50/50 vol. % (shaded bars).

3.4 Resistivity measurements in dual gas atmosphere

The results of the measurements from the area specific resistance (ASR) over a 1000 h oxidation period in H_2/H_2O = 50/50 vol. % in dual gas atmosphere are presented in Figure 14. This diagram illustrates the ASR development during the measurement at 878 °C with Ni[NiO] coated test samples. Since the samples are assembled symmetrically, the ASR values were calculated from the resistance:

$$R = \frac{1}{\sigma} \cdot \frac{L}{A} \; ; \; ASR = R \cdot A = \frac{L}{\sigma} = \rho \cdot L, \text{ and with symmetrical cell assembly: } ASR = \frac{R \cdot A}{2}$$

Equation 3: ASR calculation

R...resistance [Ω], σ...conductivity [1/(Ω*m)], L...length- equivalent to oxide thickness grown on the MIC, A...contact area (contact area of 1*1 cm² Ni-net), ρ...specific resistance [Ω*m]

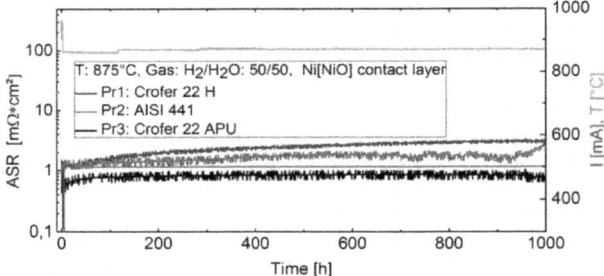

Figure 14. ASR development over 1000 h oxidation period in dual gas atmosphere. Inside the sealed test samples: a reducing atmosphere with FG2 = H_2/H_2O = 50/50 vol. % was induced.

An ASR increase of 33 % for Crofer 22 APU, 182 % for Crofer 22 H, and 123 % AISI 441 (compare with Table 3) was found between 10 h and 1000 h. A clear correlation between the ASR and the oxide scale thickness can be established from the data analysis in Table 3. Materials having a thicker oxide scale beneath the contact area show also comparatively higher absolute ASR-values, confirming the relationship from Equation 3 in which the ASR is proportional to oxide thickness (at constant values of conductivity σ and contact area A). Similar to the results of the analysis of the inner oxidation zone in FG1 gas composition, the EDS-mappings of the samples after ASR-measurements in FG2 have also shown higher silicon oxide concentrations in the contact areas of Crofer 22 H and AISI 441 in comparison to Crofer 22 APU. This could explain the ASR-differences between the three alloys in this measurement.

Table 3: ASR development in the samples during 1000 h oxidation at 875°C in FG2: H_2/H_2O = 50/50 vol. %

Material	ASR after 10 h oxidation [mΩ*cm^2]	ASR after 100 h oxidation [mΩ*cm^2]	ASR after 1000 h oxidation [mΩ*cm^2]	ASR increase 10 h vs. 1000 h in %	Oxide thickness after 1000 h oxidation in H_2/H_2O = 50/50 vol. % ; beneath Ni-mesh
Crofer 22 APU	0.6	0.8	0.8	33	5.7 ± 1.4
Crofer 22 H	1.1	1.6	3.1	182	10.0 ± 1.4
AISI 441	1.3	1.4	2.9	123	7.9 ± 1.4

In literature [23], an ASR value of ~ 4 mΩ*cm^2 was measured after 1000 h oxidation at 800°C in air for blank Crofer 22 APU. In [20], a range of about 6 - 20 mΩ*cm^2 is stated after 600 h in air for 800°C in blank Crofer 22 steel variants. For blank AISI 441, [24] stated an ASR of ~30 mΩ*cm^2 after 1000 h oxidation in air at 800°C and an oxide thickness on the MIC of about 5 µm after a 900 h duration. In comparison with the aforementioned literature data (800°C in air), the absolute ASR-values of our samples, with applied Ni-contact layer measured at 875°C in FG1, are considerably lower. In our AISI 441 samples, the grown oxide layer beneath the Ni-layer was only marginally thicker (7,9±1,4 µm) after testing under fuel gas (FG2) composition in comparison with 5 µm thickness in [24] (after 900 h@800°C, air), and a 10 times thinner ASR was measured. With the assumption, that the difference in oxide thickness is not mainly affecting the large difference between the measured and literature ASR values, other parameters e.g. temperature differences, the differences in the fuel gas composition (which has the influence on the microstructure of the oxidation zone) as well as the differences in contacting junction - could explain the discrepancies between the results. Chromium oxide has semiconducting properties, i.e. decreasing resistance with increasing temperatures. Nearly 40 % decrease of specific resistance was found in our measurements in bulk Cr_2O_3 samples between 800°C (26,1 Ω*cm) and 875°C (15,8 Ω*cm) in air (Data not shown).The growing oxide layer does not consist alone of pure Cr_2O_3 due to presence of alloying or stabilizing elements. Therefore, the resistivity drop due to temperature difference can even be larger. This finding could partly explain the lower ASR values which we measured in comparison to the results reported in the literature for annealing in air.

3.5 Comparison of the outer oxidation zone with data from SOFC stacks

The results of the oxide thickness measurements in the model samples from the oxidation in FG1 (Chapter 3.1) were also compared with data from Crofer 22 APU interconnects operated in different SOFC stacks. The stacks were operated at the following operation conditions: gas composition of $N_2/H_2/H_2O$ = 48/48/4 vol. %, fuel utilization = 70 %, current = 5 A, voltage = 27,5 V, power = 137,8 W. Furthermore, the stacks were either operated in a hot box (Figure 15 B) or in an oven at constant temperature (Figure 15 C, D). To make a comparison of the oxide thickness data from the model samples (Chapter 3.1) with stack sample based oxide thicknesses, the thickness related oxide growth diagrams of Crofer 22 APU (illustrated in Figure 15 A) were interpolated to the operation time of the stacks (i.e. 2000 h, 3300 h, 5000 h). Similar to Chapter 3.1, the thickness parameters of the stack samples were measured by area analysis of the image sections with linear extension of at least 750 µm. To assess the oxide growth rates at different steam contents in fuel gas, the samples for image analysis were sawn at different positions of a single cell interconnect, at the gas inlet (GI), and the gas outlet (GA) of a cell as it is illustrated in Figure 15 B,C. Because of the gradient of the fuel utilization, the gas composition (humidity) can vary between $p(H_2)/p(H_2O)$ = 95:5 vol. % for the gas inlet and $p(H_2)/p(H_2O)$= 20:80 vol. % for the gas outlet. In [18] the pO_2 ranges for the anode side of a SOFC were stated at 700°C: ~ 10^{-19} - 10^{-22} bar, at 900 °C: ~ 10^{-15} – 10^{-18} bar. Considering *Ellingham diagrams* [25] it is shown that at 800°C Cr_2O_3 is stable if $p(O_2)$ > ~10^{-27} bar (or $p(H_2)/p(H_2O)$ < ~$3*10^4$). This may be the reason why we can observe Cr_2O_3 based oxide growth at the anode side as well. The comparisons of results after a 2000 h operation time are shown in Figure 15 B. The measured oxide thickness values at the GI and GA for different stack cells versus the results of the artificially oxidized samples at different temperatures are presented in this diagram. Similar to experimental results with different steam content in fuel gas (Chapter 3.3), we cannot see a clear increase in oxide thickness at the gas outlet (side with higher H_2O content) in comparison to GI in the tested stack samples. Similar behavior is seen in the stack which was operated over 3300 h (Figure 15 C). This implies, that the reactions at the phase boundary are dominated by solid-state diffusion and the rate of supply of adsorbed oxygen should not be the limiting factor for oxide growth at the tested conditions. Furthermore, we can see that at comparable temperatures, our artificially oxidized samples show thicker oxide scales. This effect can be caused by an enhanced evaporation of constituents from the oxide scale under gas flow in the stack in comparison with our experimental conditions in oven (gas flow: 2.5 nml/min in 6 L oven volume).

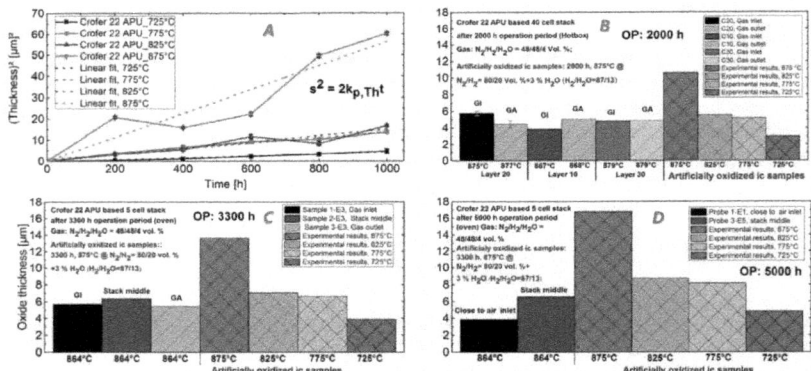

Figure 15. Comparison of the outer oxidation zone from artificially grown oxides on Crofer 22 APU with oxide thickness data of SOFC stacks measured at different stack layer positions. The stacks were operated over 2000 h (B), 3300 h (C) and 5000 h (D). The oxide thickness data was compared with oxide thickness data of artificially oxidized MIC samples interpolated to the operation time of the stacks, which is schematically illustrated in (A).

4. CONCLUSIONS

Despite implementing a reducing atmosphere, the formation of oxide scales was observed on the interconnector surface of three high temperature ferritic steels (i.e. Crofer 22 APU, Crofer 22 H, AISI 441). The process of the oxide scale formation can be described by the parabolic oxidation kinetics model with the gradual time dependent deceleration of the mass gain due to formation of a diffusion barrier at the interface between metal surface and ambient atmosphere.

An acceleration of the oxide scale formation rate was established with increasing temperatures in range between 725°C – 875°C. For the tested MIC materials Crofer 22 APU, Crofer 22 H, AISI 441, the acceleration factors for mass gain (Δm) and oxide thickness (ΔTh) increase were estimated under the SOFC anode gas conditions. It is shown that at temperatures between 725°C and 875°C they are in the range of: Δm1000h: 2.9 ÷ 3.5 and ΔTh1000h: 3.1 ÷ 3.8 after 1000 h oxidation. No clear evidence for an increase in oxide scale growth rate was detected in the tested fuel gas compositions with higher steam contents. Thus, it is concluded that reactions at phase boundaries are dominated by solid-state diffusion and adsorption rate of oxygen species should not be the rate limiting factor of oxide scale formation under testing conditions. The oxidation process at high temperatures can be separated into two pronounced formation zones; the zone of the outer oxide scale formation, consisting mainly of Cr_2O_3- and $(Mn,Cr)_3O_4$, and the inner oxidation zone, with randomly dispersed inclusions of Ti-, Al- and Si- containing oxide species and precipitates. The formation of an insulating SiO_2-layer as well as Si separation in intermetallic Laves phases was observed in AISI 441 and Crofer 22 H materials. Long term resistance measurements over an oxidation period of 1000 h at 875°C show the highest resistance increase values in Crofer 22 H and AISI 441. The formation of insulating oxide phases (e.g. SiO_2) beneath the Cr_2O_3- layer may be responsible for this behavior. Similar to experimental results with artificially oxidized samples at different steam contents in fuel gas, the analysis of samples taken from the real SOFC stacks show no clear difference between the oxide thickness at gas inlet (low H_2O content region) and gas outlet (high H_2O content region). This implies a degree of saturation with oxygen containing species at both GI and GA so that oxidation processes on the MIC surface, which should be dominated by solid state diffusion, can proceed.

5. ACKNOWLEDGEMENTS
Funding of the Federal Ministry of Education and Research (BMBF) within the 6. Energy Research Program of Federal Government – topic „Material research for energy development" – is kindly acknowledged (support code 03SF0494).

6. REFERENCES
1. L. Blum, L. G. J. de Haart, J. Malzbender, N. Margaitis, N. H. Menzler. Anode-Supported Solid Oxide Fuel Cell Achieves 70 000 Hours of Continuous Operation. 2016, *Energy Technology, Vol.4,* p. 939-942.

2. J. W.Fergus. Metallic interconnects for solid oxide fuel cells. *Materials Science and Engineering (A 397).* 17. February 2005, p. 271-283.

3. M. Lindner, Th. Hocker, L. Holzer, K. A. Friedrich, B. Iwanschitz, A. Mai, J. A. Schuler. Cr_2O_3 scale growth rates on metallic interconnectors derived from 40,000 h solid oxide fuel cell stack operation. *Journal of Power Sources 243.* 14. June 2013, p. 508-518.

4. G. Kuschert. Hochtemperatureigenschaften einer ODS-Fe26Cr Legierung zum Einsatz als Interkonnektor in einer SOFC-APU. Darmstadt : s.n., 2009.

5. W.J. Quadakkers, J. Piron-Abellan, V. Shemet, L. Singheiser. Metallic interconnectors for solid oxide fuel cells - a review. *Materials at High Temperatures, 20(2),* 2003, p. 115-127.

6. F. Tietz, H.-P. Buchkremer, D. Stöver. Components manufacturing for solide oxide fuel cells. *Solid State Ionics, Vol. 152-153,* 2002, p. 373-381

7. P. Huczkowski. Effects of geometry and composition of Cr steels on oxide scale properties relevant for interconnector applications in Solide oxide Fuel Cells (SOFCs). RWTH Achen: Hochschulbibliothek, 2005.

8. W.J.Quadakkers, L.Singheiser. Practical Aspects of the Reactive Element Effect. *Material Science Forum 369-372*, 2001, p. 77-92.

9. Zh. Yang, G.-G. Xia, J. W. Stevenson. Evaluation of Ni–Cr-base alloys for SOFC interconnect applications. *Journal of Power Sources , Vol. 160*, 2006, p. 1104–1110.

10. P. Huczkowski, N. Christiansen, V. Shemet, J. Piron-Abellan, L. Singheiser, W.J. Quadakkers. Growth Mechanisms and Electrical Conductivity of Oxide Scales on Ferritic Steels Proposed as Interconnect Materials for SOFC's. *Fuel Cells. 06/2006*, p. 93-99.

11. M. Stanislowski, Michael. Verdampfung von Werkstoffen beim Betrieb von Hochtemperaturbrennstoffzellen (SOFC). RWTH Achen : Hochschulbibliothek, 2006.

12. S. Megel. Kathodische Kontaktierung in planaren Hochtemperatur-Brennstoffzellen. Dresden : Hochschulbibliothek, 2009.

13. L. Niewolak, D.J. Young, H. Hattendorf, L. Singheiser, W. J. Quadakkers. Mechanisms of oxide scale formation on ferritic interconnect steel in simulated low and high pO_2 service environments of solid oxide fuel cells. *Oxid Met. 4. June 2014*, p. 82: 123-143.

14. D. J. Young, J. Zurek, L. Singheiser, W. J. Quadakkers. Temperature dependence of oxide scale formation on high-Cr ferritic steels in Ar-H_2-H_2O. *Corrosion Science 53.* 2011, p. 2131-2141.

15. M. R. Ardigo, I. Popa, S. Chevalier, S. Weber, O. Heintz, M.l Vilasi. Effect of Water Vapor on the Oxidation Mechanisms of a Commercial Stainless Steel for Interconnect Application in High Temperature Water Vapor Electrolysis. *Oxid Met 79.* 2013, p. 495-505.

16. P. Kofstad. High Temperature Corrosion. Department of Chemistry, University of Oslo, Norway : Elsevier Applied Science - London and New York, 1988.

17. R. Bürgel, H. J. Maier, Th. Niendorf. Handbuch Hochtemperatur-Werkstofftechnik. s.l.: Vieweg+Teubner, Springer Fachmedien Wiesbaden GmbH, 2011.

18. K. Kendall, M. Kendall. High-Temperature Solid Oxide Fuel Cells for the 21st Century - Fundamentals, Design and Applications. s.l. : Elsevier, 2016.

19. J. H. Froitzheim. Ferritic Steel Interconnectors and their Interactions with Ni Base Anodes in Solid Oxide Fuel Cells (SOFC). RWTH Achen: Hochschulbibliothek RWTH, 2008.

20. F.J. Pirón Abellán, W.J. Quadakkers. Development of Ferritic Steels for Application as Interconnect Materials for Intermediate Temperature Solid Oxide Fuel Cells (SOFCs). Jülich, Germany: *Berichte des Forschungszentrums Jülich, JUEL-4170, ISSN 0944-2952, 2005.*

21. A. Holt, P. Kofstad. Electrical conductivity of Cr_2O_3 doped with TiO_2. *Solid state ionics, Vol. 117*, 1999, p. 21-25.

22. J. Froitzheim, G.H. Meier, L. Niewolak, P.J. Ennis, H. Hattendorf, L. Singheiser, W.J. Quadakkers. Development of high strength ferritic steel for interconnect application in SOFCs. *Journal of Power Sources 178.* 2008, p. 163-173.

23. A. Venskutonis, W.Glatz, G.Kunschert. High volume fabrication of redy-to-stack components for planar SOFC concepts. 16TH, Plansee Seminar: Powder Metallurgical High Performance Materials: s.n., 2005.

24. Zh. Yang, G.-G. Xia, Ch.-M. Wang, Z. Nie, J. Templeton, J. W. Stevenson, P. Singh. Investigation of iron–chromium–niobium–titanium ferritic stainless steel for solid oxide fuel cell interconnect applications. *Journal of Power Sources 183,* 2008, p. 660-667.

25. H. J.T. Ellingham. Reducibility of Oxides and Sulfides in Metallurgical Processes. *Journal Chemical Society 63,* 125, 1944.

26. Menzler N. H. *et. al.,* Degradation phenomena in SOFCs with metallic interconnects. *Advances in solid oxide fuel cells IV.* Daytona Beach, Florida: *Ceramic Engineering and Science Proceedings, Vol. 29, Iss. 5,* 2008, p. 92-102.

27. G. Mallaiaha, P.R. Reddy, A. Kumar. Influence of titanium addition on mechanical properties, residual stresses and corrosion behaviour of AISI 430 grade ferritic stainless steel GTA welds. *Procedia Materials Science.* June 2014, p. 1740-1751.

FABRICATION OF THE ANODE-SUPPORTED SOLID OXIDE FUEL CELL WITH DIRECT PORE CHANNEL IN THE CERMET STRUCTURE TO IMPROVE THE ELECTROCHEMICAL PERFORMANCE

Ming-Wei Liao, Tai-Nan Lin*, Hong-Yi Kuo, Chun-Yen Yeh, Yu-Ming Chen, Wei-Xin Kao, Jing-Kai Lin, Ruey-Yi Lee

Nuclear Fuels and Materials Division
Institute of Nuclear Energy Research, Taiwan, R.O.C.

ABSTRACT

An anode-supported solid oxide fuel cell consisting of a NiO-YSZ anode, YSZ electrolyte, and YSZ-LSM ‖ LSM composite cathodes has been investigated. We use tape casting to produce thin strip of green tape anode substrate. In order to enhance the diffusivity of fuel gas to reach the triple phase boundary for electrochemical reaction, direct pore channel array perforation was created in the anode supporting substrate to improve performance behavior. The anode was subjected to a pore array perforation without causing irregular form or fracture at the perimeter of holes being perforated, so that the cell having low diffusion impedance in fuel gas electricity generation operation that can improve the ability of triple phase reaction at interfacial area among electrodes that can effectively enhance the output power density of the cell and provide power output with long term stability. By laser processing, the as-prepared $10 \times 10 \ cm^2$ cell was tested for comparison. The thicknesses of NiO-YSZ, YSZ, YSZ-LSM, and LSM layers are 450, 10, 10, and 25 μm, respectively. In an electrical performance test of the cell produced by perforation, it shows that the electricity generation efficiency increases from 369 to 406 mW cm^{-2} at 800 °C, and the gas diffusion impedance can be lowered by a percentage of 40 %.

INTRODUCTION

The solid oxide fuel cell (SOFC) is considered an environmental-friendly power generation device with fuel flexibility. As crude oil production is shrinking and environmental protection consciousness is rising in nowadays, seeking alternative energy source become an urgent task. Solid oxide fuel cell has high efficiency, low pollution, versatile modularized structure and sustainable power generation ability. It is the power generation device with the most potential. To obtain high performance SOFC with low degradation rate, both the materials selection and the cell structural morphology are critical issues. In the past decade, most studies focus on the development of advanced cathode materials which exhibit excellent properties for ionic and electronic conductions as well as the oxygen catalytic activity below 800 °C [1-2]. For anode-supported SOFC, the factors to evaluate cell performance in terms of electrode microstructure include electrode thickness, porosity, tortuosity, pore size distribution, electrical conductivity, triple phase boundary (TPB) length, and mechanical strength. It is difficult to obtain the real data directly via experimental approaches for some factors mentioned above. Simulation techniques are therefore used to interpret some consequential behaviors related to the electrode microstructure. Stefan-Maxwell equation and Knudsen diffusion mechanism can be used to describe the behaviors of the gas transportation, while Bulter-Volmer equation describes the electrochemical reaction and electro-neutrality in the reaction zone [3-4]. Recently, the issues of the green energy and carbon dioxide have attracted much attention due to the exhaustion of fossil fuels and the resulting green-house effect. Development of the new technology to produce energy is essentially important.

Most studies concentrate on an enhancement of the cell performance with a high-porosity anode microstructure. The main object of the present study is to provide a solid oxide fuel cell (SOFC) membrane electrode assembly (MEA) by using tape casting process to produce thin strip of green tape anode substrate. The thin strip of green tape anode is subject to a special treatment of pore array perforation, forming an anode supporting substrate through sintering, followed by convention cell fabrication process to produce the cell with low diffusion impedance in the fuel gas electricity generation operation [5]. However, the sintering shrinkage effect for the ceramic anode tapes with perforation treatment would be difficult to control the undesirable mechanical mismatch with the perforated holes in the cell structure, especially at the fabrication of the larger cell in size. Another approach to enhance the fuel gas diffusivity is using laser technique to produce pore-array holes in the anode substrate of a cell product. In this study, by means of above-mentioned perforation methods, the cell possesses pore array holes on the anode electrode side without causing irregular form or fracture at the perimeter of holes being perforated. These pore array structures in anode side provide greater fuel gas diffusion paths, that is, the low diffusion impedance so as to improve the ability of three phase reaction at interfacial area among electrodes of the unit cell that can effectively enhance the output power density of the cell and provide power output with long term stability.

EXPERIMENTAL DETAILS

The solid oxide fuel cell has been fabricated by series of patented ceramic processes in our labs [6-8]. For the preparation of tape casting slips, the solvents consisted of an azeotropic mixture of methyl ethyl ketone (MEK) and ethanol (EtOH). The dispersant was trietheanol amine (TEA, Merck) and the binary plasticizer system was a mixture of polyethylene glycol 2000 (PEG, Merck) and dibutyl phthalate (DBP, Merck) (1:1 weight ratio of PEG to DBP). The binder used was polyvinyl butyral 30000 (PVB, Merck). The slips were prepared following two-step mixing procedures. In the first mixing step, powders, solvents, and the dispersants were mixed and milled together for at least 5 days. In the second mixing step, the binder and plasticizers were added and mixed for 4 days. Prior to tape casting, the slurries were de-aired via vacuum operation. This removed entrapped/dissolved air in the slurries. The slip was casted on Mylar carrier film using the doctor-blade method. The green tape was dried at 80 °C before removing from tape caster. The casting was performed using a laboratory model tape casting system (ECS, Model CS-8). The anode green substrate was hot pressed in laminator for several times and co-fired at 1400 °C for 4 hours. The thin film electrolyte was manufactured by spin coating process. The electrolyte suspensions were coated onto anode substrate using a spin coater (Cee 200). For cathode layer, the LSM powders were mixed with the proper amount of terpineol-based solution, and further subjected to a milling process using the three-roll mill (EXAKT Model 80E) to obtain the cathode pastes. The LSM cathode pastes was screen-printed (EKRA XPRT1) onto the half-cell and fired at 1050 °C. After a sintering process, a membrane electrode assembly (MEA) of SOFC was obtained.

Two approaches were adopted to prepare the cell with anode pore array structures. One is to prepare the pore array structure in the tape-casted green tapes, which comprises (a) anode green tape fabrication, (b) pore array anode structure, (c) laminating layers of the anode green tape that include perforated and non-perforated anode green tape to form an anode supported substrate, (d) electrolyte layer built on an abrasive and polished surface of the anode electrode, and (e) fabrication of a cathode electrode (Fig. 1). Laminating multiple layers that include the perforated and non-perforated thin strips of the green tape anode through equalized water pressure lamination process to form green tapes for anode supported substrate and the thickness was set to range from 300 μm to 800 μm. The full size of the unit cell to be made is in the range from 5×5 to 10×10 cm^2, but in considering a sintering shrinkage effect during the fabrication process, the size of anode green tape is made in the range from 7×7 to 12×12 cm^2. Nevertheless, considering the shrinkage effect of the sintered tape-casted anode substrate, the other approach was to produce the perforated holes in an as-prepared cell on the anode substrate side by laser drilling processing, which will be easy to execute on a commercial available cell, avoiding the shrinkage related mechanical drawbacks. Microstructure of the MEA with YSZ film was examined by field emission scanning electron microscopy (FE-SEM, Hitachi S-4800). Optical microscope with image analyzing software (Leica DVM6 A Digital Microscope) was used to investigate the perforated hole. The gas permeability measurement was used to characterize the gas-tightness of the electrolyte layer, ensuring the fully dense electrolyte (PMI-CFP-1100A). The SOFCs were tested in a cell housing consisting of alumina base with alumina flanges for gas distribution, and platinum and nickel meshes for cathode and anode current collections, respectively [9]. Platinum wires were used as the current leads for cell voltage measurement. Air as an oxidant was available by the air supplier and pure hydrogen was used as a fuel. The flow rates of fuel and oxidant were 800 ml/min and 2000 ml/min, respectively. The electrochemical characteristics, such as theopen-circuit voltage

(OCV) and current-voltage (I-V) measurements were executed by using an Electronic load (Prodigit 3356F, Taiwan). While the electrochemical impedance spectroscopy (EIS) analyses were performed at OCV on an SI 1287 and SI 1260 (Solartron Instruments, Hampshire, UK) in the temperature range of 600 ~ 800 °C. The durability test was executed by a fixed current density treatment of 400 mA/cm^2 at 800 °C. The data acquisition system provided the information of V-t, P-t, I-t based on-line records.

RESULTS AND DISCUSSION

Fig. 2 shows the SEM microstructures of typical cell structure investigated in this study. The cell consists of a NiO-YSZ anode, YSZ electrolyte, and YSZ-LSM ∥ LSM composite cathodes and the thicknesses of NiO-YSZ, YSZ, YSZ-LSM, and LSM layers are 450, 10, 10, and 25 μm, respectively. According the cross-section pictures of cell, it shows that the YSZ electrolyte thin film with a thickness of 10 μm is fully dense. No significant cracks or pores are observed in the YSZ film, and the YSZ electrolyte thin film is uniformly continuous and adheres well to anode and cathode. The gas-tightness was measured to be less than 10^{-6} Darcy and it would be a critical issue in the cell's electrochemical performance because of an effective ionic transfer in the solid electrolyte. The average permeability should be achieved to approach an effective gas-tightness. For commercially available SOFCs, the measured gas permeability were ~ 1 × 10^{-6} darcy [10].

A preliminary investigation for green tape perforation approach, the as-prepared 5 × 5 cm^2 cell is subjected to power performance testing. The test result is compared with a unit cell fabricated without the pore array perforation on the anode green tape, as shown in Fig. 3 and Fig. 4. The open circuit voltage (OCV) has reached the theoretical standard value (> 1.1V), indicating that the YSZ electrolyte possesses dense structure without any cracks and pinholes as evidenced in the SEM result. The output power density improves from ~ 250 to 315 mW/cm^2, i.e., the electricity generation efficiency has been raised 25 %. From the EIS plots, it is coincident with the power behavior and the gas diffusion impedance has been lowered 40 % from Fig. 4. Fig. 5 also indicates a long duration electrical performance test result for a SOFC-MEA unit cell having pore array perforation on the anode green tape. The unit cell has increasing voltage rising rate under operation condition of constant electric current 400 mA/cm^2 for 60 hours and it indicates that the pore array perforation holes provide effective diffusion paths for fuel gas permeation, increasing the gas density throughout the anode substrate support to the triple phase boundaries (TPBs) at interfacial area between anodes and electrolyte, and thus improve the efficiency of electrochemical reaction to obtain higher power density. This increasing trend of voltage elevation also suggests that activation of the microstructure optimization. In our previous article, we reported a periodic oscillation phenomenon of the cell voltage under the constant current density operation. The low-porosity anode structure or structure without sufficient diffusion paths for fuel gas may result in the decrease of the effective diffusion coefficient and the accumulation of water vapor while electrochemical reaction occurs. The cell voltage oscillation was mainly caused by the concentration polarization as well as the boundary migration of the reaction zone, which will limit the electrochemical reaction efficiency [11].

As mentioned in the experimental details, green tape perforation would encounter issues of shrinkage related mechanical mismatch and it may not be precisely controlling the perforated holes in the designated array position by symmetry. There is another approach to increase the fuel gas diffusion paths for commercial available cells (10×10 cm²). Post treatment after membrane electrode assembly fabrication by laser perforation has been executed to investigate the performance enhancement expectations. The 15 W CO_2 Laser with a transverse mode of TEM_{00} is applied to drill a 6 x 6 pore array on the anode supporting layer. We investigate the laser perforation hole by OM and SEM. By the Leica DVM6 A Digital Microscope, the perforated depth was estimated to be 60 ~ 90 μm. It is noted that the structure of each pore is a reverse cone shape due to a Gaussian energy distribution of Lase TEM_{00} mode. For widening the bottom of a pore, each position is executed by second Laser pulse as shown in Fig. 6. It suggests that upon operation, the fuel gas would penetrate through this direct pore channel from outside the support surface all the way into one-fifth the substrate thickness, thus dominating the fuel supplying, which incorporating with the appropriate anode porosity to optimize the effective gas transport and electrochemical reactions. To evaluate the commercial available cells with dimensions of 10×10 cm², we compare both cells with and without post laser perforation treatment. Fig. 7 shows the performance data for cell with and without post laser perforation as a function of current density. If we simply compare the P_{max} for testing condition at alternating temperatures the power densities were 369, 286, 224 mW/cm^{-2} at 800, 750, and 700 °C for cell without laser perforation, while for cell with laser perforation, the power densities measured at 800, 750, and 700 °C were 406, 308, 202 mW/cm^{-2}, respectively. Notably, the power density enhancement from 369 to 406 mW/cm² at 600 mA/cm² / 800 °C can be observed. In terms of power generation per cell, it increases from 29.9 W to 32.9 W. However for alternative testing temperatures at 750 and 700 °C, it may not be precisely expression as suggested by reviewer. We compare the values of power density at fixed current density of 300 mA/cm², and the power density results are 243, 223, 196 mW/cm^{-2} at 800, 750, and 700 °C for cell without laser perforation, while for cell with laser perforation, the power densities measured at 800, 750, and 700 °C are 251, 223, 189 mW/cm^{-2}, respectively. From comparison of performances at 700°C it is clear that no increase of power density for cells with laser perforation was observed. It is presumably due to the fact that at low operating temperature other electrochemical steps (such as charge transfer) become rate determining and diffusion related polarization resistance has no big impact in the case. Notably, if we increase the pore array numbers from 36 to 144 by laser perforation, we observe a significant performance enhancement and the power density would enhanced to 44 W per cell at 800 °C which will be addressed in detailed elsewhere. By perforation in the anode support substrate, effective diffusion paths by the direct pore channel can be beneficial in providing sufficient fuel gas to permeating throughout anode and form effective electrochemical reactions in the triple phase boundaries.

CONCLUSIONS

The fabrication process for production of planar type solid oxide fuel cell with high electrical conductivity and low fuel gas impedance is executed. Approaches by green tape perforation and post laser perforation on the anode support substrate to form a direct pore channel structure essentially provide a good conduction effect for fuel gas and the solid oxide fuel cell with this treatment has features of high electrical conductivity and low fuel gas impedance to improve the performance of the cell. The pore array perforation holes provide effective diffusion paths for fuel gas permeation, increasing the gas density throughout the anode substrate support to the triple phase boundaries (TPBs) at interfacial area between anodes and electrolyte, and thus improve the efficiency of electrochemical reaction to obtain higher power density. For practical application, the fabrication and performance evaluation of 10×10 cm^2 cells were conducted. From the electrochemical investigation performance test by perforation, it shows that the power density increases from 369 to 406 mW cm^{-2} at 800 °C, and the gas diffusion impedance can be lowered by a percentage of 40 %.

REFERENCES

[1] J. Yan, H. Matsumoto, M. Enoki, T. Ishihara, "High-Power SOFC Using $La_{0.9}Sr_{0.1}Ga_{0.8}Mg_{0.2}O_{3-\delta}/Ce_{0.8}Sm_{0.2}O_{2-\delta}$ Composite Film", Solid-State Lett 8 A389-A391 (2005).

[2] W. Zhou, R. Ran, Z.P. Shao, "Progress in understanding and development of $Ba_{0.5}Sr_{0.5}Co_{0.8}Fe_{0.2}O_{3-\delta}$-based cathodes for intermediate-temperature solid-oxide fuel cells: a review", Journal of Power Sources 192 231-246 (2009).

[3] V. H. Schmidt, C.L. Tsai, "Anode-pore tortuosity in solid oxide fuel cells found from gas and current flow rates", Journal of Power Sources 180 253-264 (2008).

[4] C.L. Tsai, V. H. Schmidt, "Tortuosity in anode-supported proton conductive solid oxide fuel cell found from current flow rates and dusty-gas model", Journal of Power Sources 196 692-699 (2011).

[5] T.N. Lin, J.Y. Kuo, H.Y. Kuo, W.X. Kao, C.Y. Yeh, "A fabrication process for production of SOFC-MEA with a pore array anode structure for improving output power density", ROC Patent I509869 (2015.11.21).

[6] M.C. Lee, W.X. Kao, T.N. Lin, Y.C. Chang, C.H. Wang, "A Novel synergistic process and recipe for fabrication of a high integrity membrane electrode assembly of solid oxide fuel cell", European Patent EP2045858B1 (2010.04.07) / ROC Patent I326933 (2010.07.01) / US Patent

US7914636B2(2011.03.29) / Japan Patent 特許第 5099892 號 (2012.10.05).

[7]Y.C. Chang, M.C. Lee, W.X. Kao, C.H. Wang, T.N. Lin, J.C. Chang, and R.J. Yang, "Characterizations of the anode-supported solid-oxide fuel cells with an yttria stabilized zirconia thin film by the diagnosis of the electrochemical impedance spectroscopy", J. Electrochemical Soc., 158 (3) B259-B265 (2011).

[8]M.W. Liao, T.N. Lin, J.C Chang, M.C Lee, R.J. Yang, Y.C. Chang, W.X. Kao, L.S. Lee, R.Y.Lee, H.Y. Kuo, C.Y. Yeh, Y.M. Chen, "Investigation on the performance testing reliability by introducing current collection modification for the solid oxide fuel cell", Advances in Solid Oxide Fuel Cells and Electronic Ceramics II: Ceramic Engineering and Science Proceedings 37-3 137-149 (2016).

[9]W.X. Kao, T.N. Lin, M.C. Lee, "Fabrication and characterization of the anode-supported solid oxide fuel cell with Ni current collector layer", J. Ceramic Society of Japan 123 217-221 (2015).

[10]C. Zhang, H.L. Liao, W.Y. Li, G. Zhang, C. Coddet, C.J. Li, C.X. Li, X.J. Ning, "Characterization of YSZ solid oxide fuel cells electrolyte deposited by atmospheric plasma spraying and low pressure plasma spraying", Journal of Thermal Spray Technology 15-4 598-603 (2006).

[11]W.X. Kao, T.N. Lin, Y.C. Chang, M.C. Lee, "Oscillation phenomenon induced by a low-porosity anode structure in the solid oxide fuel cell", Key engineering Materials 656 124-128 (2015).

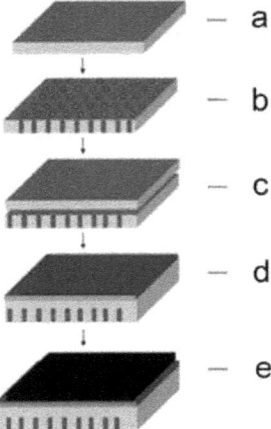

Fig. 1 The approach for anode pore array perforation of tape-casted green tapes: (a) anode green tape fabrication, (b) pore array anode structure, (c) laminating layers of the anode green tape that include perforated and non-perforated anode green tape to form an anode supported substrate, (d) electrolyte layer built on an abrasive and polished surface of the anode electrode, and (e) fabrication of a cathode electrode

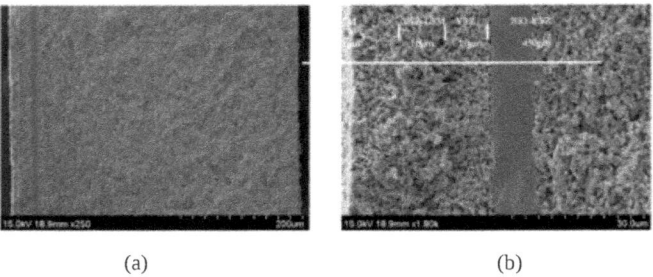

(a)   (b)

Fig. 2 SEM cross-section micrograph (a) lower magnification 250 ×, (b) higher magnification 1800 ×.

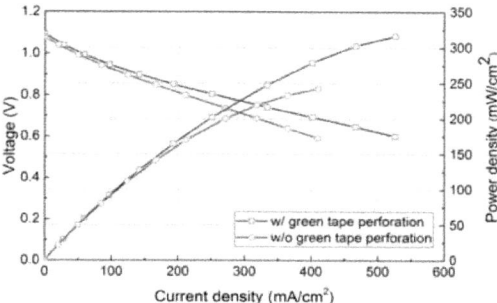

Fig. 3 Cell voltage and power density as a function of current density for the 5×5 cm^2 cell w/ and w/o green tape perforation.

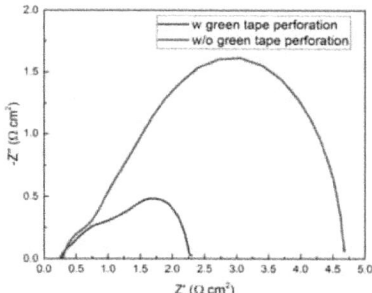

Fig. 4 The EIS impedance plots for the 5×5 cm^2 cell w/ and w/o green tape perforation at 800 °C and OCV condition (zero current loading).

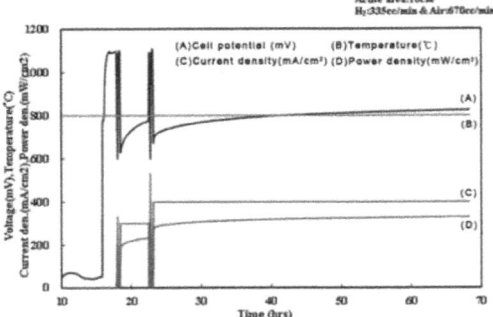

Fig. 5 The plots of cell potential, temperature, current density, and power density vs. time for the 5 × 5 cm² cell w/ green tape perforation.

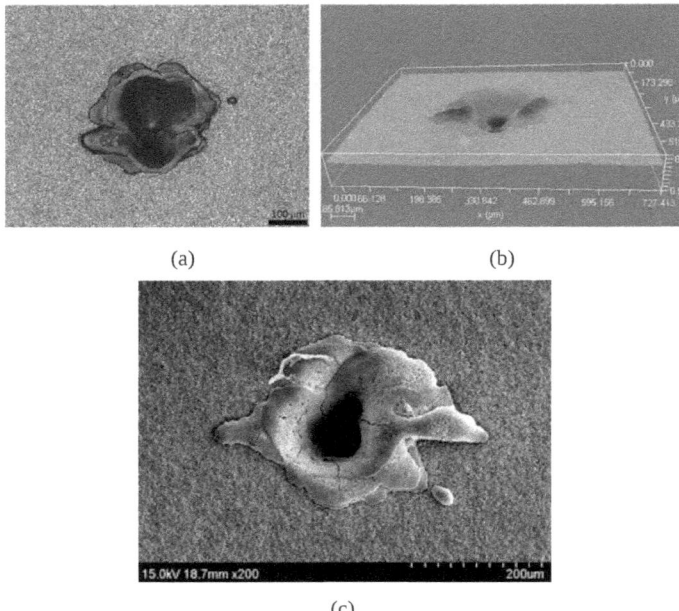

(a)   (b)

(c)

Fig. 6 The surface morphology for the post laser perforation hole observation by (a) top image and (b) Multifocus image by Leica DVM6 A Digital Microscope; (c) SEM image.

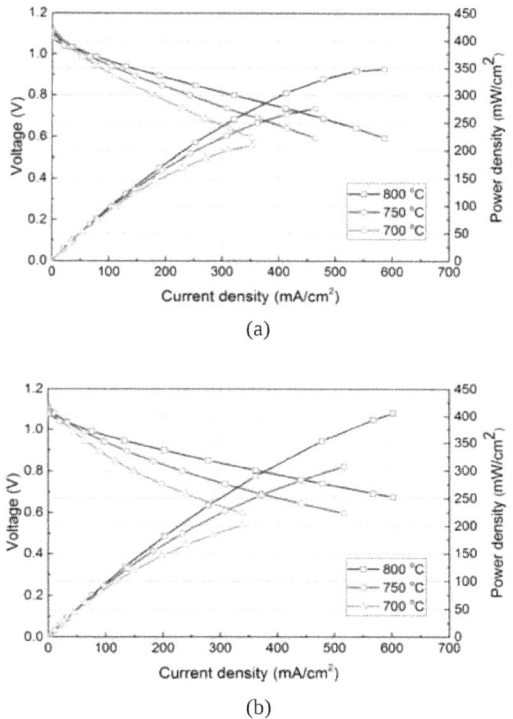

(a)

(b)

Fig. 7 Cell voltage and power density as a function of current density for the 10×10 cm^2 cell (a) w/o laser perforation, (b) w/ laser perforation.

CRITICAL EVALUATION OF DYNAMIC REVERSIBLE CHEMICAL ENERGY STORAGE WITH HIGH TEMPERATURE ELECTROLYSIS

D. McVay, J. Brouwer, and F. Ghigliazza
Department of Mechanical and Aerospace Engineering, UCI and SOLIDpower S.p.A.
Irvine, CA, USA and Mezzolombardo, Italy

ABSTRACT

A solid oxide electrochemical cell is capable of operating reversibly in both fuel cell and electrolysis modes with high efficiency compared to low temperature alternatives. However, thermodynamic integration and operation challenges exist for a system that can support a reversible solid oxide fuel cell stack (RSOFC). Key system components must be kept within a given temperature range while cycling between modes of operation and during dynamic operation. A six cell short stack with Ni/YSZ anode-supported cells, YSZ electrolyte, and LSCF oxygen electrode with a GDC interlayer was analyzed experimentally. Results from testing were used to verify a physical model of a 50 kW RSOFC system in order to study dynamic operating conditions in both fuel cell and electrolysis modes. Results suggest that electrolysis solar power input can be used directly whereas the highly dynamic fuel cell loads may need to be complemented with batteries or capacitors. Using measured renewable power dynamics, the dynamic RSOFC model was able to show load-following operation for realistic demands of RSOFC systems as applied to the electric grid that deploys high levels of renewable power.

BACKGROUND

For several reasons, sources of renewable power have become increasingly popular to utilities and utility customers alike. Increasing global temperatures as a result of injecting greenhouse gases into the atmosphere over the course of the industrial age is a major concern which has led to this increased adoption of renewable power sources. Furthermore, the availability of fossil fuels reduced as record amounts are consumed each year. These concerns, as well as other environmental concerns that exist as a direct result of burning fossil fuels, can be mitigated by introducing renewable power sources such as wind and solar farms. Renewable sources, which do not rely on fossil fuel input, can reduce overall greenhouse gas emissions, provide localized power generation, and improve environmental impacts if managed properly. Challenges exist in the deployment of these technologies which include: cost, availability of land, proximity to utilization, integration with the existing electrical grid, and variability of power output among other considerations[1–4]. A significant challenge to renewable power sources is that they cannot be appropriately dispatched in the case when there is less demand than available power. Renewable power plants must curtail their load when demand is not sufficient to accept the load. This results in a negative economic impact in terms of payback on capital investment. A significant contribution to renewable power sources can be made if sufficient technologies that enable utilization of curtailed power can be realized[5–7].

Fuel cell technology has been studied for over 150 years and various types of fuel cells have been developed commercially for different applications. Different types of fuel cells are usually distinguished by the materials from which they are made which often depend upon the operating temperature. Some of the more mature types of fuel cells include proton exchange membrane (80 °C), phosphoric acid (200 °C), alkaline (60 °C – 220 °C), molten-carbonate (650 °C), and solid oxide (600 °C – 1000 °C) [8]. Molten-carbonate (MCFC) and solid oxide fuel cells (SOFC) are interesting not only because of their high operating temperature, but also because they work with oxidizing ions that traverse the electrolyte. Fuel cells with these characteristics are generally more efficient due to lowered activation polarization and they are capable of being fuel flexible. The MCFC requires carbon dioxide in the cathode stream so that the carbonate ion ($CO_3^=$) can be generated to traverse the electrolyte. The SOFC has no such requirement since $O^=$ is the ion that goes across the electrolyte. Therefore, the balance of plant for an SOFC is somewhat simpler than for an MCFC. However, the thermal integration of an SOFC can be difficult due to the higher operating temperature and constraints on temperature gradients[8–10].

When current is applied to a fuel cell and enough water and air are supplied to the anode and cathode respectively, the fuel cell becomes an electrolyzer and will split water into hydrogen and oxygen gas. For an SOFC, the $O^=$ ions travel from the cathode to the anode. A solid oxide electrolysis cell (SOEC) has a current applied in the opposite direction, thus splitting water in the anode and sending $O^=$ ions from the anode to the cathode. Since the anode and cathode are designated depending on whether an oxidation or a reduction reaction is occurring, care must be taken when referring to a particular electrode. Therefore, the channel in which steam or fuel is supplied will be referred to as the fuel electrode. The channel where air is normally supplied will be referred to as the oxygen electrode. The resultant products are hydrogen and steam from the fuel electrode and oxygen enriched air from the oxygen electrode. Although commercial systems are typically designed to operate either as a fuel cell or an electrolyzer separately, research is ongoing to develop a reversible system with currently only one demonstration system having been built by Boeing for the U.S. Navy[11].

One possible application of curtailed power is to generate pure hydrogen and oxygen via electrolysis which can be then be separately stored and used at a different time when variable renewable power sources are not available. The pure hydrogen can be combusted in a heat engine or it can be used to generate electricity directly from fuel cells. The hydrogen could also be used in fuel cell vehicles, thus displacing petroleum related emissions and use. A problem with the latter paradigm is that no robust infrastructure currently exists that can transport large amounts of pure hydrogen over large distances. Efforts are being made, especially in California, to address this issue, but significant progress will take years to occur[12–18]. As an alternative, pure hydrogen can be pumped and stored for later use in the existing natural gas pipelines which already span the country. This existing infrastructure can provide a stop-gap solution until a robust hydrogen infrastructure is realized. Significant studies have been conducted and are ongoing that consider the impacts of hydrogen-enriched natural gas on existing pipelines as well as end-uses. For applications that require pure hydrogen (such as fuel cell vehicles), the hydrogen can be separated from the natural gas pipeline at centralized or distributed locations[19–24].

An important operating parameter for high temperature electrolysis cells is the thermoneutral voltage. Electrolysis of water is an endothermic reaction which requires energy input to move forward. This energy generally comes from electrical power input, but it can be supplanted by heat as well. High temperature electrolysis requires less electrical power input than low temperature electrolysis since more energy is available to the water molecules in the form of heat. As current is applied to the cell, joule heating occurs which is a result of ohmic resistance within the cell. The voltage at which the joule heating is supplying enough heat to the reaction to maintain a constant temperature is the thermoneutral voltage. Operation below the thermoneutral voltage requires additional heat input to maintain temperature whereas operation above the thermoneutral voltage requires cooling to maintain constant temperature. The actual value of the thermoneutral voltage depends upon pressure, temperature, species concentrations, and heat loss to the environment (which is specific to each system)[25–27].

The current work considers a reversible solid oxide fuel cell system (RSOFC) that is capable of producing renewable hydrogen and oxygen during periods of available solar power and electricity during periods of high electrical demand. A system that is capable of switching between consuming power and generating power may also need to be capable of idling. A challenge exists in maintaining the operating temperature of the stack and other high temperature components for each mode of operation. Figure 1 presents a possible configuration for a reversible solid oxide system. The combustor, typically needed for fuel cell operation, may also be necessary to heat streams for use in preheating inlet gases for electrolysis and idle modes. However, an alternative heat source may be used in place of a combustor to preheat fluids or provide heat to the stack itself in electrolysis operation that occurs below the thermoneutral voltage.

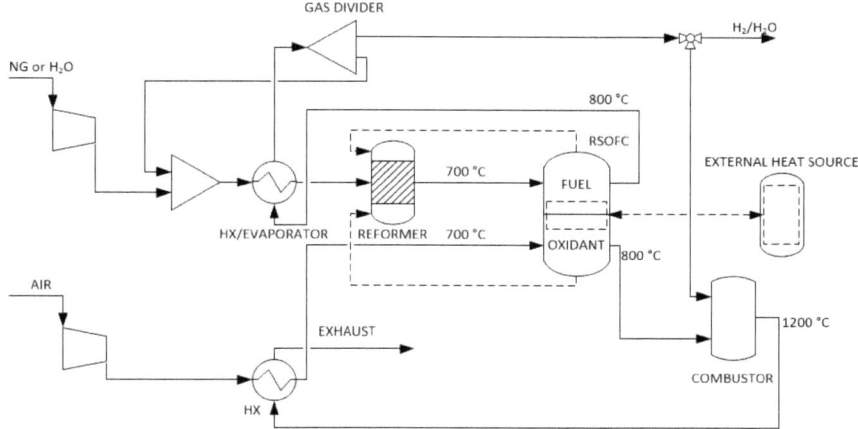

Figure 1: Configuration for reversible solid oxide system

EXPERIMENTAL

The National Fuel Cell Research Center (NFCRC) acquired an RSOFC short stack test stand from the Italian company SOLIDpower S.p.A. Since the anode and cathode change depending on the mode of operation, what is traditionally called the anode in fuel cell operation will be called the fuel electrode and the cathode will be called the oxygen electrode. The electrolyte is made up of 8 mol% yttria stabilized zirconia (YSZ) with ionic conductivity reported at $4.0 \times 10^2 S/cm$. The fuel electrode is made up of a nickel/YSZ cermet and the oxygen electrode is made up of gadolinium doped ceria and lanthanum strontium cobalt ferrite (GDC + LSCF). The stack is a fuel electrode supported stack with a fuel electrode thickness of 240 μm, an electrolyte thickness of 8 μm, and an oxygen electrode thickness of 40 μm. Experiments of RSOFC short stacks have been conducted, but there are very few published results that focus on any dynamic behavior of SOCs[28-37]. The short stack at the NFCRC was used to study the V-j characteristics of SOLIDpower's stack technology among several other parameters including operating temperatures, emissions, etc. A schematic detailing the flow configuration of the test stand is presented in Figure 2 below.

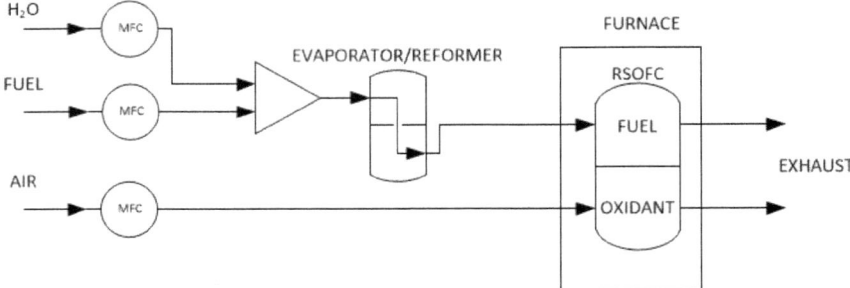

Figure 2: Test stand flow configuration

For purposes of dynamic system modeling, the V-j curve in fuel cell and electrolysis modes from the short stack is used for verification of the system stack model electrochemical performance. The operating temperature of the stack is also needed due to its direct effect on the cell overpotentials. The stack operating parameters used for obtaining the V-j curves are summarized Table 1 and Table 2 below.

Table 1: Fuel cell short stack operating parameters

Parameter	Value
Fuel	Natural gas
Fuel utilization	0.65
Furnace temperature	750 °C
Steam-to-carbon ratio	3.0

Table 2: Electrolysis short stack operating parameters

Parameter	Value
Fuel	Water and hydrogen
Inlet H_2 concentration	0.1
Furnace temperature	750 °C

According to SOLIDpower's recommended operation, the steam and hydrogen flow for electrolysis mode remained constant regardless of input current. As a result, the fuel utilization in electrolysis mode varied as a function of current according to Equation 1 below where U_f is fuel utilization, I is current, N is the number of cells, $\dot{n}_{H_2O}$ is the molar flow rate of water in mol/s, and F is Faraday's constant.

$$U_f = \frac{I \cdot N}{\dot{n}_{H_2O} \cdot 2 \cdot F} \tag{1}$$

At the electrolyzer operating conditions, the fuel utilization in electrolysis mode ranged linearly from 0 (at zero current) to 0.35 at maximum current. The variable fuel utilization affects the Nernst voltage of the cell due to the difference in reactant and product concentrations at the cell's outlet according to Equation 2 below where V_{Nernst} is the cell's Nernst voltage, E^0 is the standard electrode potential, R is the universal gas constant, T is the operating temperature, and p_i is the partial pressure of water or hydrogen dependent on the subscript.

$$V_{Nernst} = E^0 + \frac{RT}{2F}\ln\left(\frac{p_{H_2O}}{p_{H_2}}\right) \qquad (2)$$

Consideration of the variable fuel utilization for the electrolysis tests with respect to the theoretical Nernst voltage results in a voltage difference of approximately 82 millivolts between zero current and maximum current. Voltage versus current density (V-j) curves are presented for both fuel cell and electrolysis modes for the same stack below (Figure 3 and Figure 4).

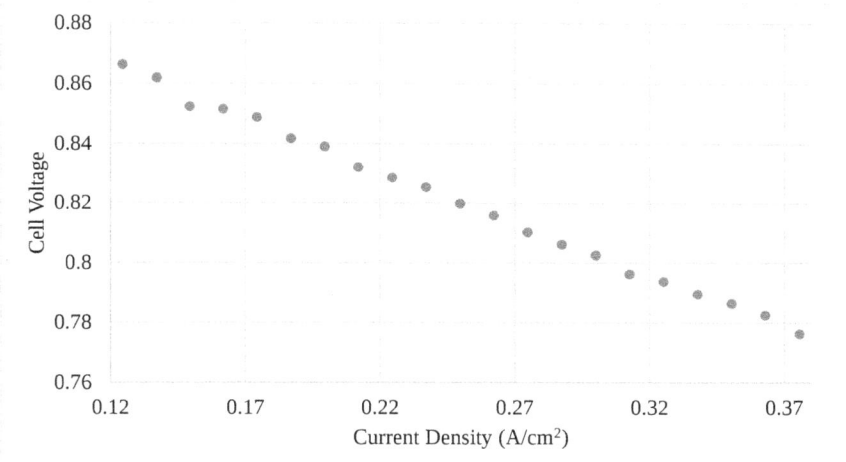

Figure 3: Fuel cell V-j performance at conditions listed in Table 1

The expected downward trend of voltage with respect to current is clearly observed. Furthermore, it should be noted that the minimum current density does not approach zero for these measurements since the water mass flow controller could not be operated below a specific threshold and the fixed fuel utilization and steam-to-carbon ratio would have been compromised with a potential for damage to the stack. As a result, the regime where activation polarization could be observed was not measured.

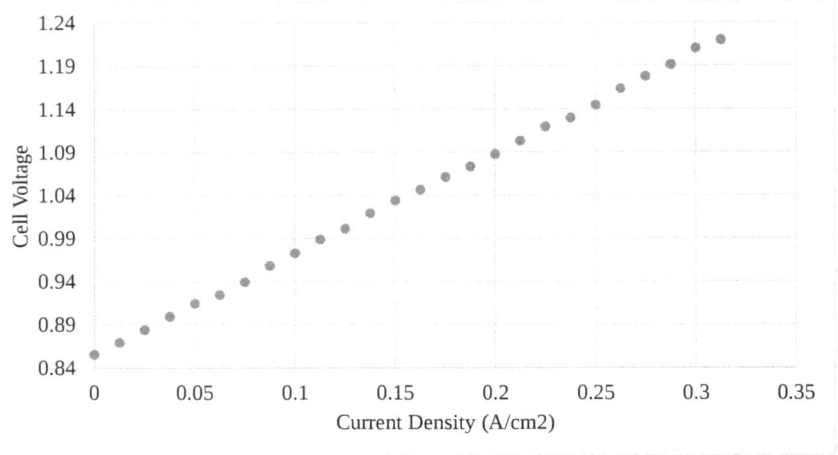

Figure 4: Electrolysis V-j performance at conditions listed in Table 2

In electrolysis mode, the fuel utilization varied, but the expected trend is observed. Namely a linear increase in voltage with respect to current as a result of largely ohmic losses. Furthermore, the voltage could be measured at each available set point between open circuit voltage and the maximum voltage. A unique characteristic evident in Figure 4 at the low current densities is that there does not appear to be any noticeable activation polarization. High temperature fuel cells generally have very low activation polarization due to the high temperature operation, the same holds true for electrolysis cells.

MODEL

A bulk fuel cell and electrolysis stack model was used in conjunction with a system balance of plant model designed to support both modes of operation in MATLAB Simulink. The model was previously developed at the NFCRC as a fuel cell system model and was converted to support reversible operation for this project. Details of the model are presented by McLarty et al.[38]. In order to support electrolysis operation, the polarization calculation was altered to account for the increase in voltage with increased current applied to the cell among other changes. In other words, the signs of the polarizations in the cell voltage equation (η) were made positive instead of negative and the ohmic polarization was tuned to match with the data.

$$V_{Cell} = V_{Nernst} + \eta_{act} + \eta_{ohm} + \eta_{conc} \qquad (3)$$

The electrochemical model was verified against data taken from the test stand primarily by tuning ohmic polarization parameters to meet the slope and magnitude of the V-j curves for both fuel cell and electrolysis modes. The given operating parameters (Table 1 and Table 2) were also used as inputs for the average cell temperature, fuel utilization, steam-to-carbon ratio, and fuel species composition. Bulk characterization of the system allows for a rough estimate of various operating parameters such as efficiency, outlet temperatures, outlet species concentrations, voltage, current, and flow rates. It is also capable of dynamic load following in fuel cell mode and dynamic power input in electrolysis mode. A microgrid that primarily relies on solar photovoltaics

(PV) for electrical power throughout the day and electrical energy storage throughout the night is an ideal case for RSOFC in that the system can absorb any otherwise curtailed power generated by the PV and can also dynamically provide power to the microgrid while the sun is down. Solar power generation and load data were collected from a microgrid located in Sonoma, California and were used as the dynamic inputs to the RSOFC system model.

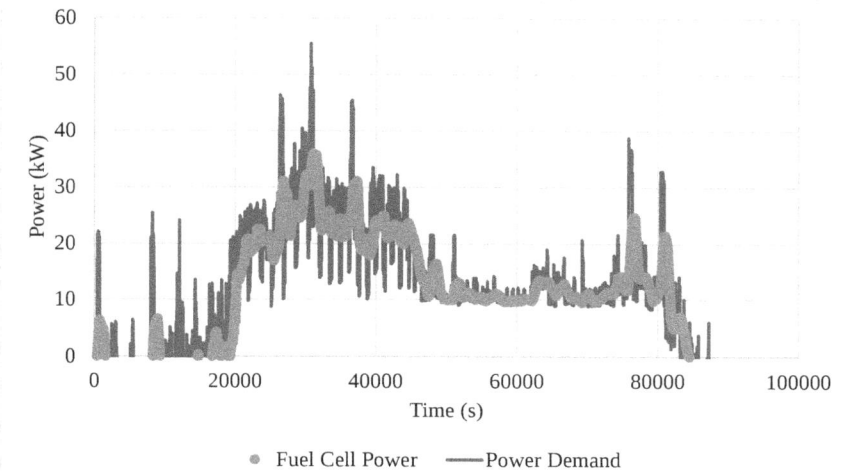

Figure 5: Microgrid power demand plotted with RSOFC power output

Figure 5 presents the power demand data of the microgrid for a 24 hour period in the blue series and the fuel cell's load following capabilities in the orange series. The primary loads in the microgrid consist of water pumps, water heaters, HVAC, lighting, and other typical residential loads. The peak demand for the day was 53 kW and the fuel cell power rating was 50 kW. The raw power demand data could not be supported by the fuel cell model, so a filtered version of the data was used with the assumption that a large capacitor bank or battery bank could make up the difference in the peaks. Since the model is based upon physical equations, it seems likely that a real system would have serious problems following the load dynamics as presented in Figure 5 without some large energy storage to make up the difference.

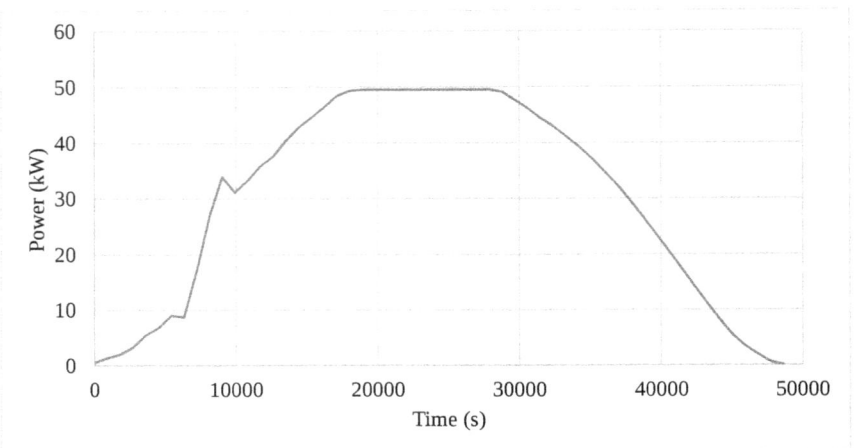

Figure 6: Solar PV power input to the RSOFC for hydrogen production

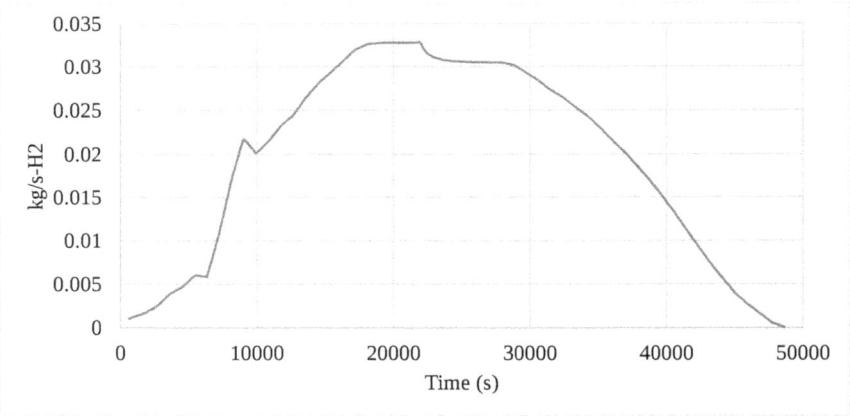

Figure 7: Hydrogen production by electrolysis of the same system with 50 kW maximum power input

Figure 6 and Figure 7 present the power input to the RSOFC for electrolysis and the resultant hydrogen production respectively. A system designed for absorbing curtailed PV power may not necessarily need to be operated the whole time the PV is producing power, but it should be expected to absorb the dynamics associated with the diurnal power input.

$$\dot{n}_{H_2} = \frac{I}{nF} \tag{4}$$

The hydrogen produced is directly proportional to the electrical current through the stack according to Faraday's law of electrolysis (Equation 4) where $\dot{n}_{H_2}$ is the molar flow rate of hydrogen in mol/s and F is Faraday's constant. Figure 7 should correlate exactly with the power input presented in Figure 6, but a slight dip in hydrogen generation at approximately 22,000 seconds occurs likely due to stack temperature control dynamics affecting the output hydrogen for the given power set point.

CONCLUSIONS
A dynamic RSOFC system model was developed which was verified with data from an experimental short stack capable of operating reversibly as a fuel cell or an electrolyzer. Load data from a microgrid located in Sonoma, CA was used for the power demand of the fuel cell mode while solar data from the same microgrid was used for power input into the electrolyzer. The fuel cell was not able to adequately follow the fast dynamics of the microgrid, without support of energy storage to smooth the dynamics, while the electrolyzer was able to follow the diurnal pattern of solar power production. Further work needs to be done to explore thermal management for stand-by modes.

REFERENCES
1. Georgilakis PS. Technical challenges associated with the integration of wind power into power systems. *Renew Sustain Energy Rev.* 2008;12(3):852-863. doi:10.1016/j.rser.2006.10.007.
2. Alsayegh O, Alhajraf S, Albusairi H. Grid-connected renewable energy source systems: Challenges and proposed management schemes. *Energy Convers Manag.* 2010;51(8):1690-1693. doi:10.1016/j.enconman.2009.11.042.
3. Chu S, Majumdar A. Opportunities and challenges for a sustainable energy future. *Nature.* 2012;488(7411):294-303. doi:10.1038/nature11475.
4. Armaroli N, Balzani V. The future of energy supply: Challenges and opportunities. *Angew Chemie - Int Ed.* 2007;46(1-2):52-66. doi:10.1002/anie.200602373.
5. Nyamdash B, Denny E, O'Malley M. The viability of balancing wind generation with large scale energy storage. *Energy Policy.* 2010;38(11):7200-7208. doi:10.1016/j.enpol.2010.07.050.
6. Schoenung S. *Economic Analysis of Large-Scale Hydrogen Storage for Renewable Utility Applications.*; 2011. http://prod.sandia.gov/techlib/access-control.cgi/2011/114845.pdf.
7. Brouwer J. On the role of fuel cells and hydrogen in a more sustainable and renewable energy future. *Curr Appl Phys.* 2010;10(2):S9-S17. doi:10.1016/j.cap.2009.11.002.
8. Lin C-K, Chen T-T, Chyou Y-P, Chiang L-K. Thermal stress analysis of a planar SOFC stack. *J Power Sources.* 2007;164(1):238-251. doi:10.1016/j.jpowsour.2006.10.089.
9. Selimovic A, Kemm M, Torisson T, Assadi M. Steady state and transient thermal stress analysis in planar solid oxide fuel cells. *J Power Sources.* 2005;145(2):463-469. doi:10.1016/j.jpowsour.2004.11.073.
10. Bossel U. Rapid startup SOFC modules. *Energy Procedia.* 2012;28:48-56. doi:10.1016/j.egypro.2012.08.039.
11. Carolina B, Mary McAdam. Boeing Delivers Reversible Fuel Cell-based Energy Storage System to U.S. Navy. Boeing Media Room. http://boeing.mediaroom.com/2016-02-08-Boeing-Delivers-Reversible-Fuel-Cell-based-Energy-Storage-System-to-U-S-Navy. Published 2016. Accessed August 22, 2016.
12. California Fuel Cell Partnership. A California Road Map. 2014;(July). http://cafcp.org/sites/files/Roadmap-Progress-Report2014-FINAL.pdf.
13. Nicholas M, Ogden J. Detailed Analysis of Urban Station Siting for California Hydrogen

Highway Network. *Transp Res Rec.* 2006;1983(1):121-128. doi:10.3141/1983-17.

14. Kang JE, Recker W. Strategic Hydrogen Refueling Station Locations with Scheduling and Routing Considerations of Individual Vehicles. *Transp Sci.* 2014;(June 2015):0-17.

15. Brown T, Stephens-Romero S, Scott Samuelsen G. Quantitative analysis of a successful public hydrogen station. *Int J Hydrogen Energy.* 2012;37(17):12731-12740. doi:10.1016/j.ijhydene.2012.06.008.

16. Stephens-Romero SD, Brown TM, Carreras-Sospedra M, et al. Projecting full build-out environmental impacts and roll-out strategies associated with viable hydrogen fueling infrastructure strategies. *Int J Hydrogen Energy.* 2011;36(22):14309-14323. doi:10.1016/j.ijhydene.2011.08.005.

17. Kang JE, Brown T, Recker WW, Samuelsen GS. Refueling hydrogen fuel cell vehicles with 68 proposed refueling stations in California: Measuring deviations from daily travel patterns. *Int J Hydrogen Energy.* 2014;39(7):3444-3449. doi:10.1016/j.ijhydene.2013.10.167.

18. Stephens-Romero SD, Brown TM, Kang JE, Recker WW, Samuelsen GS. Systematic planning to optimize investments in hydrogen infrastructure deployment. *Int J Hydrogen Energy.* 2010;35(10):4652-4667. doi:10.1016/j.ijhydene.2010.02.024.

19. Melaina MW, Antonia O, Penev M. *Blending Hydrogen into Natural Gas Pipeline Networks : A Review of Key Issues Blending Hydrogen into Natural Gas Pipeline Networks : A Review of Key Issues.*; 2013.

20. Schouten J, Janssenvanrosmalen R, Michels J. Modeling hydrogen production for injection into the natural gas grid: Balance between production, demand and storage. *Int J Hydrogen Energy.* 2006;31(12):1698-1706. doi:10.1016/j.ijhydene.2006.01.005.

21. Hamersma P, Janssenvanrosmalen R, MICHELS J, SCHOUTEN J. Effect of hydrogen addition on the route preference in natural gas flow in regular, horizontal T-junctions. *Int J Hydrogen Energy.* 2007;32(14):3059-3065. doi:10.1016/j.ijhydene.2006.12.016.

22. Schouten J, Janssenvanrosmalen R, Michels J. Condensation in gas transmission pipelines. *Int J Hydrogen Energy.* 2005;30(6):661-668. doi:10.1016/j.ijhydene.2004.10.010.

23. Schouten J. Effect of H2-injection on the thermodynamic and transportation properties of natural gas. *Int J Hydrogen Energy.* 2004;29(11):1173-1180. doi:10.1016/j.ijhydene.2003.11.003.

24. Tabkhi F, Azzaropantel C, Pibouleau L, Domenech S. A mathematical framework for modelling and evaluating natural gas pipeline networks under hydrogen injection. *Int J Hydrogen Energy.* 2008;33(21):6222-6231. doi:10.1016/j.ijhydene.2008.07.103.

25. Todd D, Schwager M, Mérida W. Corrigendum to "Thermodynamics of high-temperature, high-pressure water electrolysis" [J. Power Sources (2014) 424–429]. *J Power Sources.* 2015;289:184-186. doi:10.1016/j.jpowsour.2015.04.161.

26. LeRoy RL. The Thermodynamics of Aqueous Water Electrolysis. *J Electrochem Soc.* 1980;127(9):1954. doi:10.1149/1.2130044.

27. Sun X, Chen M, Jensen SH, Ebbesen SD, Graves C, Mogensen M. Thermodynamic analysis of synthetic hydrocarbon fuel production in pressurized solid oxide electrolysis cells. *Int J Hydrogen Energy.* 2012;37(22):17101-17110. doi:10.1016/j.ijhydene.2012.08.125.

28. Diethelm S, Herle J Van, Montinaro D, Bucheli O. Electrolysis and Co-electrolysis performance of SOE short stacks. *Fuel Cells.* 2013;13(4):631-637. doi:10.1002/fuce.201200178.

29. Ferrero D, Lanzini A, Leone P, Santarelli M. Reversible operation of solid oxide cells under electrolysis and fuel cell modes: Experimental study and model validation. *Chem Eng J.* 2015;274:143-155. doi:10.1016/j.cej.2015.03.096.

30. Kim S-D, Yu J-H, Seo D-W, Han I-S, Woo S-K. Hydrogen production performance of 3-

cell flat-tubular solid oxide electrolysis stack. *Int J Hydrogen Energy*. 2012;37(1):78-83. doi:10.1016/j.ijhydene.2011.09.079.

31. Li Q, Zheng Y, Guan W, Jin L, Xu C, Wang WG. Achieving high-efficiency hydrogen production using planar solid-oxide electrolysis stacks. *Int J Hydrogen Energy*. 2014;39(21):10833-10842. doi:10.1016/j.ijhydene.2014.05.070.

32. Mougin J, Chatroux A, Couturier K, et al. High Temperature Steam Electrolysis Stack with Enhanced Performance and Durability. *Energy Procedia*. 2012;29:445-454. doi:10.1016/j.egypro.2012.09.052.

33. Nguyen VN, Fang Q, Packbier U, Blum L. Long-term tests of a Jülich planar short stack with reversible solid oxide cells in both fuel cell and electrolysis modes. *Int J Hydrogen Energy*. 2013;38(11):4281-4290. doi:10.1016/j.ijhydene.2013.01.192.

34. Petitjean M, Reytier M, Chatroux A, et al. Performance and Durability of High Temperature Steam Electrolysis: From Single Cell to Short-Stack Scale M. *Electrochem Soc*. 2011;35(1):2905-2913.

35. Schefold J, Brisse A, Zahid M, Ouweltjes JP, Nielsen JU. Long Term Testing of Short Stacks with Solid Oxide Cells for Water Electrolysis. *Electrochem Soc*. 2011;35(1):2915-2927. doi:10.1149/1.3570291.

36. Stoots CM, Condie KG, Brien JEO, Herring JS, Hartvigsen JJ. Test results from the Idaho National Laboratory 15kw high temperature electrolysis test facility. In: *Proceedings of the 17th International Conference on Nuclear Engineering.* ; 2015:1-11.

37. Tietz F, Sebold D, Brisse A, Schefold J. Degradation phenomena in a solid oxide electrolysis cell after 9000 h of operation. *J Power Sources*. 2013;223:129-135. doi:10.1016/j.jpowsour.2012.09.061.

38. McLarty D, Brouwer J, Samuelsen S. A spatially resolved physical model for transient system analysis of high temperature fuel cells. *Int J Hydrogen Energy*. 2013;38(19):7935-7946. doi:10.1016/j.ijhydene.2013.04.087.

ESTIMATION OF POLARIZATION LOSS DUE TO CHROMIUM POISONING OF LSM-BASED CATHODES IN SOLID OXIDE FUEL CELLS

R. Wang[1], B. Mo[1], M. Würth[2], U. B. Pal[1,3], S. Gopalan[1,3], and S. N. Basu[1,3]
[1] Division of Materials Science and Engineering, Boston University, Brookline, MA 02446, USA
[2] Institute for Energy Systems, Technische Universität München, Boltzmannstraße 15, 85748 Garching, Germany
[3] Department of Mechanical Engineering, Boston University, Boston, MA 02215, USA

ABSTRACT

Chromium (Cr) vapor species from chromia-forming alloy interconnects are known to cause performance degradation of cathodes in solid oxide fuel cells (SOFCs). To understand the impact of Cr-poisoning on cathode performance, it is important to determine its effect on the changes in the cathodic polarization losses. In this study, anode-supported SOFCs, with a Sr-doped $LaMnO_3$ (LSM) + yttria-stabilized zirconia (YSZ) cathode active layer and a LSM cathode current collector layer, were fabricated. Electrochemical tests of the cells with their cathodes in direct contact with the chromia-forming alloy interconnect were performed at 800 °C. Under open circuit condition, the cell performance had no observable degradation. However, with a constant cathodic current (0.5 A/cm^2), significant performance degradation was observed. The cathode performances with and without chromium (Cr) poisoning were analyzed by estimating the polarization losses at the cathode. This is done with the help of curve-fitting the experimentally measured current-voltage (C-V) plots to a polarization model. The modeling results indicate that the degradation of cathode performance is mainly due to the increase of cathodic activation polarization loss. Microstructures of the cathode cross sections were characterized to help understand the analyzed results.

INTRODUCTION

Performance degradation during long-term operation is one of the greatest challenges to overcome for commercialization of solid oxide fuel cells (SOFCs). At SOFC operating temperature, Chromium (Cr) vapor species that evaporate over Fe-Cr alloy interconnect, can transport and deposit within the cathode and thereupon cause degradation of the cathode performance [1,2]. Although extensive studies have been conducted on the Cr-poisoning phenomena, the mechanism of cathode performance degradation caused by Cr-poisoning is still unclear [3]. In the present work, Sr-doped $LaMnO_3$ (LSM), a well-developed cathode material, has been chosen for studying the effect of Cr-poisoning on cathode performance.

Cathode performance can be characterized by identifying the contributions of individual polarization losses. These polarization losses are associated with different cathode processes [4]. For understanding the mechanism of cathode performance degradation caused by Cr-poisoning, it is a necessity to identify what is the most affected cathode process, namely, polarization loss in the cathode. To separate the contributions of polarization losses, deconvolution of the impedance spectra by complex nonlinear least-square (CNLS) fitting the data into equivalent circuits is usually performed [4,5]. However, it can be difficult to interpret the data from impedance spectra, since the correlation between the equivalent circuit elements and the physical or chemical parameters are not straightforward [6].

In this study, the current voltage (C-V) traces measured with and without Cr-poisoning effect were curve-fitted to an analytical polarization model. Polarization losses associated with different cathode processes were successfully evaluated and compared. Polarization loss that is closely correlated with the effect of Cr-poisoning is determined, and a physical interpretation of this effect is proposed to help understand the mechanism of cathode performance degradation.

EXPERIMENTAL
 Cell structures consisting of a Ni/8YSZ (8 mol% Y_2O_3–92 mol% ZrO_2) anode substrate, a Ni/8YSZ anode interlayer and an 8YSZ electrolyte, were commercially purchased from Materials and Systems Research Inc (Salt Lake City, UT). Two layers of cathodes were screen printed over the electrolyte: LSM/8YSZ composite cathode active layer and LSM cathode current collector layer. Slurries for cathode layers were prepared by mixing and milling LSM+8YSZ or LSM powders overnight in alpha-terpineol with the desired amount of pore former and binder. Details of cathode slurry preparation can be found in our previous work [7]. After deposition of each cathode layer, the structure was sintered at 1200 °C for 2 hours. The cathode area of a cell was 2 cm^2 after fabrication.

 The cell test setup was comprised of two alumina tubes, with the cell sandwiched between them. A gold gasket on the cathode side and a mica gasket on the anode side were used to seal and prevent direct contact between the cell and the ceramic tubes. Glass paste was applied outside the alumina tubes around the mating circumference in order to obtain a tight seal. Crofer 22 H was used as the interconnect materials in the present work. Commercially available Crofer 22 H mesh (Fiaxell Sàrl, Switzerland) was pre-attached on cathode with LSM paste for cathode current collection. Nickel mesh was pre-attached on the anode with nickel paste for anode current collection. Two silver wires on the cathode side and two nickel rods on the anode side were firmly pressed on the corresponding Crofer 22 H mesh and nickel mesh, to ensure good contacts. On each side, one wire/rod was used for current application, and the other for voltage measurement. 98% H_2 – 2% H_2O was circulated over the anode to simulate low fuel utilization. The fuel flow rate was 300 cm^3/min, which provided a flooded fuel condition and negligible fractional fuel utilization. Dry air was circulated over the cathode at a flow rate of 1000 cm^3/min also providing a flooded condition with negligible fractional oxidant utilization.

 Two identical cells were electrochemically tested at 800 °C. After the cells fully equilibrated, the initial performances of the cells were characterized by performing current voltage (C-V) measurements. These two cells were then operated under different current conditions: (a) under open circuit condition (Cell A); and (b) under a constant cathodic current density of 0.5 A/cm^2 (Cell B). These current conditions were interrupted by C-V measurements every 24 hours on both cells, for a total duration of 120 hours. Electrochemical measurements were performed with Princeton Applied Research PARSTAT® 2273 potentiostat, and KEPCO power amplifier.

 Cross-sectional microstructures of the tested cells were observed using scanning electron microscopy (SEM, Zeiss Supra 55) and energy-dispersive X-ray spectroscopy (EDS, EDAX).

RESULTS AND DISCUSSION

Polarization Modeling
 In order to separate the contribution of individual polarization losses, a polarization model developed in our group was employed for analyzing the C-V measurement characteristics [7]. This polarization model is based on an electric potential balance equation that relates the operating cell potential (V_{cell}) to the open-circuit potential (V_o) and various polarization (ohmic, activation, and concentration) losses:

$$V_{cell} = V_o - iR_i - \eta_{act} - \eta_{a,conc} - \eta_{c,conc} \qquad (1)$$

where i is the current density, R_i is the area specific ohmic resistance of the cell, η_{act} is the activation polarization, $\eta_{a,conc}$ is the anodic concentration polarization, and $\eta_{c,conc}$ is the cathodic concentration polarization.

Activation polarization, η_{act}, is caused by slow charge transfer reactions between the electronic and ionic conductors at the triple phase boundaries (TPB's), and it is related to the current density by the Butler-Volmer equation:

$$i = i_o \exp\left(\frac{\alpha n \eta_{act} F}{RT}\right) - i_o \exp\left(\frac{(1-\alpha)n\eta_{act}F}{RT}\right) \tag{2}$$

where i_o is exchange current density, α is the transfer coefficient, n is the number of electrons transferred per reaction. R is the gas constant, T is the temperature, and F is Faraday constant. n could be 1 or 2 depending on the reaction mechanism, and the value chosen in this study was 1 because of a better fit to the polarization model. α is set equal to 0.5 with the assumption of a symmetric activation energy barrier for the fuel cell application [8]. Thus, activation polarization, η_{act}, can be expressed as [7, 9]:

$$\eta_{act} = \frac{2RT}{F} \ln\left\{\frac{1}{2}\left[\left(\frac{i}{i_o}\right) + \sqrt{\left(\frac{i}{i_o}\right)^2 + 4}\right]\right\} \tag{3}$$

In η_{act}, the activation polarizations occurring at both the cathode ($\eta_{act,c}$) and the anode ($\eta_{act,a}$) are lumped together.

Concentration polarization, η_{conc}, is caused by slow mass transport of gaseous reactants and product species through the porous anode and cathode. The anodic concentration polarization, $\eta_{a,conc}$, with H_2-H_2O gas mixture as fuel in this study, can be expressed as [10]:

$$\eta_{a,conc} = -\frac{RT}{2F} \ln\left(\frac{p_{H2}^{(i)}p_{H2O}^o}{p_{H2}^o p_{H2O}^{(i)}}\right) = -\frac{RT}{2F} \ln\left(1-\frac{i}{i_{as}}\right) + \frac{RT}{2F}\left(1+\frac{p_{H2}^o i}{p_{H2O}^o i_{as}}\right) \tag{4}$$

where R is the gas constant, F is Faraday constant, $p_{H2}^{(i)}$ and $p_{H2O}^{(i)}$ are the partial pressure of hydrogen and water vapor at the anode/electrolyte interface, respectively, p_{H2}^o and p_{H2O}^o are the partial pressure of hydrogen and water in the bulk anode, respectively, and i_{as} is the anodic saturation current density. The anodic saturation current density is defined as the current density at which the $p_{H2}^{(i)}$ becomes zero.

The cathodic concentration polarization, $\eta_{c,conc}$, with air (O_2 and N_2 mixture), can be expressed as [10]:

$$\eta_{c,conc} = -\frac{RT}{4F} \ln\left(\frac{p_{O2}^{(i)}}{p_{O2}^o}\right) = -\frac{RT}{4F} \ln\left(1-\frac{i}{i_{cs}}\right) \tag{5}$$

where $p_{O2}^{(i)}$ is the partial pressure of oxygen at the cathode/electrolyte interface, p_{O2}^o is the partial pressure of oxygen in the bulk cathode, and i_{cs} is cathodic saturation current density. The cathodic saturation current density is defined as the current density at which the $p_{O2}^{(i)}$ becomes zero.

Finally, the polarization model which relates the operating cell potential to the current density can be obtained by substituting equation (3), (4), and (5) into equation (1) [7]:

$$V_{cell}(i) = V_o - iR_i - \frac{2RT}{F} \ln\left\{\frac{1}{2}\left[\left(\frac{i}{i_o}\right) + \sqrt{\left(\frac{i}{i_o}\right)^2 + 4}\right]\right\} + \frac{RT}{2F} \ln\left(1-\frac{i}{i_{as}}\right) - \frac{RT}{2F}\left(1+\frac{p_{H2}^o i}{p_{H2O}^o i_{as}}\right) + \frac{RT}{4F} \ln\left(1-\frac{i}{i_{cs}}\right) \tag{6}$$

Cell Performance Analysis

Figure 1 shows the C-V curves and the corresponding power density data of the two cells operated under the two different current conditions. Cell A, which was operated under open circuit condition, had no observable degradation. The initial maximum power density of this cell was 0.43 W/cm², and the maximum power density after 120 hours was 0.44 W/cm². In contrast, a significant degradation of performance was observed on Cell B, which was operated under a constant cathodic current density of 0.5 A/cm². The maximum power density of this cell decreased from 0.44 W/cm² to 0.25 W/cm² after 120 hours, indicating serious Cr-poisoning within the cathode.

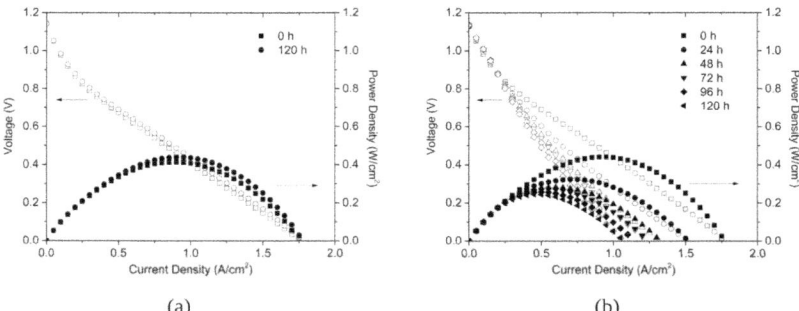

(a)   (b)

Figure 1. Cell test results of the cell (a) under open circuit condition, and (b) under 0.5 A/cm² constant cathodic current condition.

The experimentally measured C-V curves were curve fitted to the polarization model discussed previously. The fitting parameters, including the area specific ohmic resistance (R_i), exchange current density (i_o), anodic limiting current density (i_{as}), and cathodic limiting current density (i_{cs}) were obtained for both cells and listed in Table I. No significant difference was found in the polarization modeling results of Cell A, due to the stable cell performance. Based on polarization modeling result of Cell B, however, a significant change on exchange current density (i_o) and a noticeable change on cathodic limiting current density (i_{cs}) were found.

Table I. Curve fitting results of the cells before and after 120-hour exposure to the Cr environment, under open circuit condition (Cell A) and under 0.5 A/cm² cathodic current (Cell B).

Fitting Parameters	Cell A		Cell B	
	0 hour	120 hours	0 hour	120 hours
R_i (Ohm·cm²)	0.155	0.164	0.169	0.130
i_o (A/cm²)	0.115	0.137	0.146	0.043
i_{as} (A/cm²)	2.146	2.155	2.210	1.785
i_{cs} (A/cm²)	1.760	1.765	1.760	1.065

Evaluation of Cr-Poisoning Effect

Exchange current density is associated with activation polarization through equation (3), and cathodic limiting current density is associated with the cathodic concentration polarization through equation (5). To evaluate the effects of Cr-poisoning on these two polarizations, it is important to determine which of them is playing the dominant role on the performance degradation.

As mentioned before, the contributions of cathode and anode are lumped together in the activation polarization obtained from the polarization modeling. The activation polarization (η_{act}) is a sum of the cathodic and anodic contributions. In this work, since the anodic fuel composition simulated negligible fuel utilization (98 H_2 - 2% H_2O) under flooded condition, the activation polarization was dominated by cathodic contribution and the anodic contribution was negligible [11]. Therefore, the total activation polarization can be approximated as due to the cathodic activation polarization ($\eta_{act,c}$) [12]:

$$\eta_{act} = \frac{2RT}{F} \ln\left\{\frac{1}{2}\left[\left(\frac{i}{i_o}\right) + \sqrt{\left(\frac{i}{i_o}\right)^2 + 4}\right]\right\} \approx \eta_{act,c} = \frac{2RT}{F} \ln\left\{\frac{1}{2}\left[\left(\frac{i}{i_{o,c}}\right) + \sqrt{\left(\frac{i}{i_{o,c}}\right)^2 + 4}\right]\right\} \qquad (7)$$

where $i_{o,c}$ is the cathodic exchange current density.

Using equation (5) and (7), and the fitting parameters from polarization modeling, the cathodic concentration polarizations and the cathodic activation polarizations of Cell B before and after Cr-poisoning were calculated and plotted in Figure 2. It shows that the effect of Cr-poisoning on the cathodic concentration polarization was negligible and the increase of cathodic activation polarization was indeed the dominant loss. Thus the result indicates that the performance degradation caused by Cr-poisoning is primarily due to the significant increase of the cathodic activation polarization, and it is considered to be due to the decreasing electrochemically active sites for O_2 reduction at the cathode/electrolyte interface.

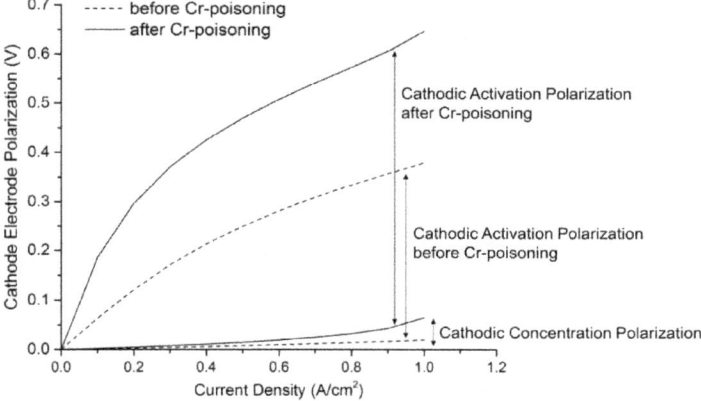

Figure 2. Cathodic electrode polarization of Cell B before and after Cr-poisoning as a function of current density.

Microstructures

SEM and EDS were used for the microstructure characterization of the cathode cross sections of both the tested cells. Figure 3 shows the SEM micrographs of the cathode cross sections of both Cell A and Cell B after electrochemical tests. The cathode cross section of Cell A appears relatively clean with no apparent Cr-containing deposits. In contrast, at the cathode/electrolyte interface of Cell B, it was observed that the triple phase boundaries were almost completely covered by a layer of contamination.

The cross-sectional microstructures were also examined by EDS analysis. In the absence of Cr-containing deposit, $La_{L\alpha}/La_{L\beta}$ ratio of pure LSM is ~5.10 [13]. However, when La and Cr containing sample is examined by EDS, overlap between $La_{L\beta}$ peak and $Cr_{K\alpha}$ peak in the spectrum is commonly observed. In this study, the relative intensity ratio of the $La_{L\alpha}/(La_{L\beta} + Cr_{K\alpha})$ was taken as an effective criterion for Cr deposition on La-containing cathodes. A smaller $La_{L\alpha}/(La_{L\beta} + Cr_{K\alpha})$ ratio indicates a higher quantity of Cr-containing deposits. In Cell A, $La_{L\alpha}/(La_{L\beta} + Cr_{K\alpha})$ ratio measured at cathode/electrolyte interface is 4.95, close to the $La_{L\alpha}/La_{L\beta}$ ratio of pure LSM, indicating trace amount of Cr deposition. In Cell B, however, $La_{L\alpha}/(La_{L\beta} + Cr_{K\alpha})$ ratio measured at cathode/electrolyte interface is 0.19, which indicates major Cr deposition. The $LSM/YSZ/O_2$ triple phase boundaries covered by the Cr-containing deposits are considered to be less electrochemically active for oxygen reduction, and therefore result in the observed degradation of the cathode performance.

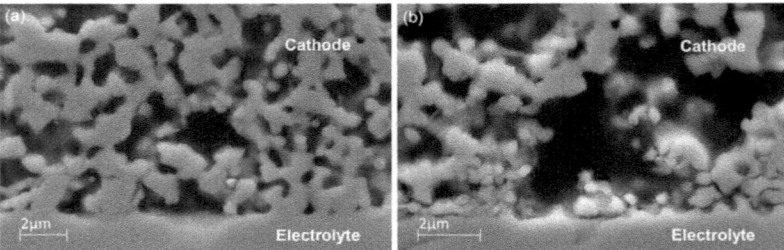

Figure 3. SEM micrographs of cathode cross sections in cell (a) tested under open circuit condition, and (b) tested under 0.5 A/cm^2 cathodic current condition.

CONCLUSIONS

In this work, anode-supported solid oxide fuel cells with LSM-based cathode were tested at 800 °C with and without cathodic current. When there was no cathodic current, the cell performance had no observable degradation and only trace Cr-containing deposits at TPB's were observed. However, in the presence of a constant 0.5 A/cm^2 cathodic current, the cell performance significantly degraded and the cathode/electrolyte interface was found to be covered by Cr-containing deposits. Employing a polarization model, it was determined for the first time that the Cr-poisoning primarily impacts the cathodic activation polarization and degrades the cell performance. Microstructural investigation showed that the Cr-containing deposits covered the electrochemically active sites at the TPB's and thus impeded the oxygen reduction reaction and degraded the cathode performance.

ACKNOWLEDGMENTS

Financial support from U.S. Department of Energy, Office of Fossil Energy, through Award # FE0023325 is gratefully acknowledged.

REFERENCES

[1] Badwal, S. P. S., Deller, R., Foger, K., Ramprakash, Y., & Zhang, J. P. (1997). Interaction between chromia forming alloy interconnects and air electrode of solid oxide fuel cells. *Solid State Ionics*, *99*, 297-310.

[2] Hilpert, K., Das, D., Miller, M., Peck, D. H., & Weiss, R. (1996). Chromium vapor species over solid oxide fuel cell interconnect materials and their potential for degradation processes. *J. Electrochem. Soc.*, *143*, 3642-3647.

[3] Jiang, S. P., & Chen, X. (2014). Chromium deposition and poisoning of cathodes of solid oxide fuel cells–a review. *Int. J. Hydrogen Energy*, *39*, 505-531.

[4] Jørgensen, M. J., & Mogensen, M. (2001). Impedance of solid oxide fuel cell LSM/YSZ composite cathodes. *J. Electrochem. Soc.*, *148*, A433-A442.

[5] Barfod, R., Mogensen, M., Klemensø, T., Hagen, A., Liu, Y. L., & Hendriksen, P. V. (2007). Detailed characterization of anode-supported SOFCs by impedance spectroscopy. *J. Electrochem. Soc.*, *154*, B371-B378.

[6] Bieberle, A., & Gauckler, L. J. (2002). State-space modeling of the anodic SOFC system Ni, H$_2$–H$_2$O|YSZ. *Solid State Ionics*, *146*, 23-41.

[7] Yoon, K. J., Zink, P., Gopalan, S., & Pal, U. B. (2007). Polarization measurements on single-step co-fired solid oxide fuel cells (SOFCs). *J. Power Sources*, *172*, 39-49.

[8] Chan, S. H., Khor, K. A., & Xia, Z. T. (2001). A complete polarization model of a solid oxide fuel cell and its sensitivity to the change of cell component thickness. *J. Power Sources*, *93*, 130-140.

[9] Li, P. W., & Chyu, M. K. (2005). Electrochemical and transport phenomena in solid oxide fuel cells. *J. Heat Transfer*, *127*, 1344-1362.

[10] Kim, J. W., Virkar, A. V., Fung, K. Z., Mehta, K., & Singhal, S. C. (1999). Polarization effects in intermediate temperature, anode-supported solid oxide fuel cells. *J. Electrochem. Soc.*, *146*, 69-78.

[11] Yoon, K. J., Gopalan, S., & Pal, U. B. (2009). Analysis of electrochemical performance of SOFCs using polarization modeling and impedance measurements. *J. Electrochem. Soc.*, *156*, B311-B317.

[12] Yoon, K. J., Gopalan, S., & Pal, U. B. (2007). Effect of fuel composition on performance of single-step cofired SOFCs. *J. Electrochem. Soc.*, *154*, B1080-B1087.

[13] Chen, X., Zhang, L., Liu, E., & Jiang, S. P. (2011). A fundamental study of chromium deposition and poisoning at (La$_{0.8}$Sr$_{0.2}$)$_{0.95}$(Mn$_{1-x}$Co$_x$)O$_{3\pm\delta}$ ($0.0 \leq x \leq 1.0$) cathodes of solid oxide fuel cells. *Int. J. Hydrogen Energy*, *36*, 805-821.

Advanced Processing and Manufacturing Technologies

SYNTHESIS AND TRIBOLOGICAL BEHAVIOR OF Bi-Cr$_2$AlC COMPOSITES

F. AlAnazi, S. Ghosh, R. Dunnigan, and S. Gupta*

Department of Mechanical Engineering
University of North Dakota, ND 58201
*Corresponding Author – surojit.gupta@engr.und.edu

ABSTRACT

In this paper, we report the synthesis of Bi-Cr$_2$AlC composites for the first time by hot pressing of Bi and Cr$_2$AlC particulates. Detailed inspection of the samples by SEM analysis showed that the Cr$_2$AlC particulates are well dispersed in the Bi-matrix. The addition of Cr$_2$AlC particulates had a beneficial effect on the mechanical and tribological behavior. For example, the yield strength increased by 1.8 times from ~40 MPa in Bi to ~72 MPa in Bi-30%Cr$_2$AlC. The addition of 10 vol% Cr$_2$AlC was able to decrease the WR of Bi-composites by ~100 times as compared to the Bi. Detailed SEM investigations showed that the antiwear properties of these composites is due to the formation of smooth and lubricious tribofilms.

INTRODUCTION

Bi is an important metal for solid lubrication and hot forming as it can be considered a green alternative to Pb, and has been studied for solid lubrication and hot forming [1-2]. Recently, Al-Anazi et al. [3] reported that Ti$_3$SiC$_2$ can be added as reinforcing additive in Bi which can further improve the mechanical and tribological behavior of Bi. As a background, Ti$_3$SiC$_2$ belongs to a family of solids known as M$_{n+1}$AX$_n$ (MAX) phases (over 70+ phases), where n =1,2,3; M is an Early Transitional Metal, A is a Group A element (mostly groups 13 and 14); and X is C and/or N. These solids have attracted a lot of attention due their excellent properties like solid lubrication, damage tolerance, thermal shock resistance and machinability [4-8]. Recently, Gupta et al. [9-11] also demonstrated that the addition of Ti$_3$SiC$_2$ particulates can enhance the mechanical and tribological performance of different technologically important metals like Al, Sn, and Zn. The authors had referred to these new generation of composites as MRM (Metal Reinforced with MAX) as 5-30 vol% Ti$_3$SiC$_2$ was used to reinforce the metal matrix [3]. In this paper, the authors will study the effect of Cr$_2$AlC particulates on the mechanical and tribological behavior of Bi.

EXPERIMENTAL

Cr (-325 mesh, Sigma-Aldrich, St. Louis, MO), Al (-325 mesh, Alfa Aesar, Haverhill, MA), and C (-325 mesh, Alfa Aesar, Haverhill, MA) powders were mixed in the molar ratio of 2:1.1:1 in a ball mill (8000 M mixer Mill, SPEX SamplePrep, Metuchen, NJ) for 5 minutes. The powders were cold pressed then heated at 10 °C/min to the desired temperature, and then sintered at 1350 °C for 4 h in a tube furnace with Ar gas flowing though the furnace. The phase pure Cr$_2$AlC powder was then mixed with calculated concentrations of Bi powders (-100 mesh, Sigma Aldrich, St. Louis, MO)) by dry ball milling for 5 minutes. All the powders were then poured in a die. Two different types of Bi-based compositions were fabricated by using ~12.7 mm (EQ-Die-12D-B, MTI Corporation, Richmond, CA) and ~6.35 mm (EQ-Die-06D, MTI Corporation, Richmond, CA) dies for hot pressing. The Bi-based compositions were sintered in atmospheric air by hot pressing (HP) with a uniaxial compressive stress of ~201 MPa (~12.7 mm die) or ~251 MPa (~6.35

mm) at 290 °C for 5 minutes [3]. Samples from the former set was used for tribology studies where samples from the latter set was used hardness and mechanical performance. Composites were allowed to cool in the HP to room temperature (RT) before characterization. Samples of pure Bi was also fabricated by following the above mentioned method. Bi-based MRM composites were fabricated by adding 10 vol% (Bi-10%Cr$_2$AlC), 20 vol% (Bi-20%Cr$_2$AlC), and 30 vol% (Bi-30% Cr$_2$AlC) Cr$_2$AlC in the Bi- matrix.

The methodology for determining relative density and porosity of the compacts is reported in Ref. 3. Briefly, the relative density was determined by normalizing the experimental density with theoretical density. A mechanical testing unit (Shimadzu AD-IS UTM, Shimadzu Scientific Instruments Inc., Columbia, MD) was at a deflection rate of 1 mm/min was used for evaluating the strength of samples during compression by testing a set of 3 samples for composition. In the paper, stress versus displacement plots are reported due to the experimental limitations of the authors which did not allow for accurate measurement of the actual strain during mechanical testing. During this study, the yield strength is defined as the stress at which the stress verse displacement plot transitions from the linear to non-linear regime where the linear region of the composites had a regression fitting of $R^2 > 0.95$. In addition, an average of 3 yield strength results was used to calculate average strength in the text [3]. Vicker's micro-hardness indentor (Mitutoyo HM-112, Mitutoyo Corporation, Aurora, IL) was used to measure the hardness of the polished samples ($R_a < 1$ µm). Bi-based MRMs were tested by loading the samples at 0.49 N for 12 s and an average of five readings for each composite system is reported in the paper.

Block (tab)-on-disc tribometer (CSM Instruments SA, Peseux, Switzerland) was used to study the tribological behavior of the samples at 5 N (~0.3 MPa), 50 cm/s linear speed, 1000 m sliding distance, and ~10 mm track radius against alumina disks (AL-D-42-2, AdValue Technology, Tucson, AZ). Bi samples were tested for only ~80 m as they had higher wear as compared to Bi-based MRMs which were cycled for ~1000 m [3]. For tribological testing, the blocks (~4 mm x ~ 4 mm x ~3 mm) and alumina disks were polished to a ~1 µm finishing. A surface profilometer (Surfcom 480A, Tokyo Seimitsu Co. Ltd., Japan) was used to confirm that all the samples had a $R_a < 1$ µm. The average of the individual mean results reported from the three data set for each composition was calculated and reported in the text as μ_{mean}. The specific wear rate (WR) was then calculated from:

$$WR = (m_i - m_f)/(\rho Nd) \text{ -------------------------(I)}$$

where, m_i is the initial mass, m_f is the final mass, ρ is density of the composite, N is the applied load, and d is the total distance traversed by the tab during the tribology testing [23-25].

Secondary electron (SE) and Backscattered Electrons (BSE) images were obtained by using a JEOL JSM-6490LV Scanning Electron Microscope (JEOL USA, Inc., Peabody, Massachusetts.) and X-ray information was obtained via a Thermo Nanotrace Energy Dispersive X-ray detector with NSS-300e acquisition engine. For microscopy analysis, alumina samples were coated with Au/Pd by using a Balzers SCD 030 sputter coater (BAL-TEC RMC, Tucson AZ USA), and then mounted on aluminum mounts. The tribosurfaces, especially sub-stoichimetric oxides, could very well contain C which is very difficult to determine experimentally by X-ray analysis. If a region is determined to be chemically uniform at the micron level then it will identified with two asterisks as *microconstituent* to emphasize that these areas are not necessarily single phases. In addition, the presence of C in these tribofilms will be shown by adding {C$_x$} in the composition [3, 12].

RESULTS AND DISCUSSION

Figures 1 shows the microstructure of Bi-Cr$_2$AlC composites. In all the cases, Cr$_2$AlC particles are well dispersed in the Bi-matrix with minimal reaction. Figure 2a plots the variation of hardness as a function of Cr$_2$AlC and Ti$_3$SiC$_2$ content. In both cases, the addition of harder MAX phase constituent increased the hardness of the matrix. Figure 2b shows the porosity of the composites as a function of the MAX phase content. In both cases, the porosity increased with the addition of higher vol% of MAX phases which shows that it is difficult to densify the compacts as the concentration of both Cr$_2$AlC and Ti$_3$SiC$_2$ [3] is increased inside the matrix. The same behavior was observed during the processing of Al [9], Sn [10], and Zn [11] based MRM composites. Thus, it can be concluded that higher pressures and/or temperature are needed as compared to the pristine metal samples when the amount of MAX phases are increased in the metal matrix.

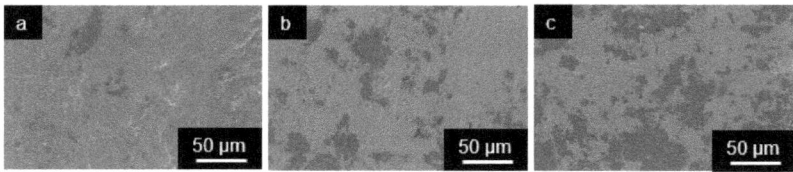

Figure 1: SE SEM images of polished, (a) Bi-10%Cr$_2$AlC, (b) Bi-20%Cr$_2$AlC, and (c) Bi-30%Cr$_2$AlC surfaces.

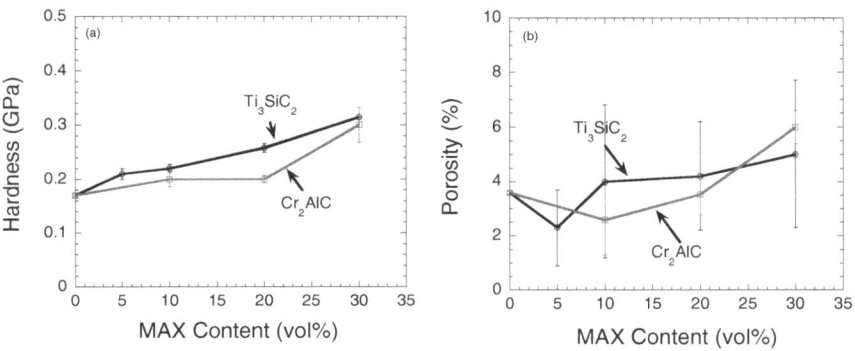

Figure 2: Plot of, (a) hardness, and (b) porosity versus MAX content.

Figure 3a plots the compressive strength versus displacement of Bi-Cr$_2$AlC composites. The yield strength gradually improved as the concentration of Cr$_2$AlC was increased in the Bi-matrix. Figure 3b compares the yield strength of Bi-Ti$_3$SiC$_2$ and Bi-Cr$_2$AlC composites. Both the composites showed similar trends, and the addition of MAX phases enhanced the yield strength of Bi-matrix, for example, the yield strength of Bi-Cr$_2$AlC composites increased from ~40 MPa in Bi to ~72 MPa in Bi-30%Cr2AlC (~1.8 times enhancement). Furthermore, this study shows that Cr$_2$AlC particulates can be as effective as Ti$_3$SiC$_2$ in enhancing the yield strength of metal matrix.

Figure 4a plots the μ$_{mean}$ of Bi-Cr$_2$AlC and Bi-Ti$_3$SiC$_2$ composites. In general, Bi-Ti$_3$SiC$_2$ composites showed lower friction coefficient that Bi-Cr$_2$AlC composites. Figure 4b plots the comparison of WR of Bi-Cr$_2$AlC and Bi-Ti$_3$SiC$_2$ composites. Both the composites showed similar trend in WR. For example, the WR of Bi decreased from ~0.02 mm^3/N.m to ~2 x 10^{-4} mm^3/N.m in Bi-10%Cr$_2$AlC, then increased to ~9 x 10^{-4} and ~7 x 10^{-4} mm^3/N.m in

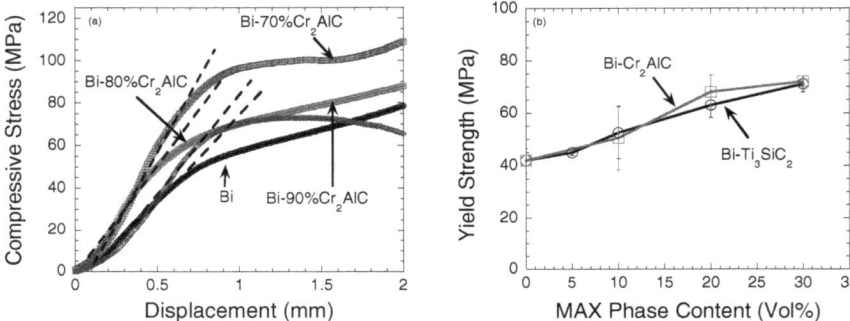

Figure 3: Plot of, (a) compressive stress versus displacement, and (b) yield strength versus MAX phase content of different Bi-MAX composites.

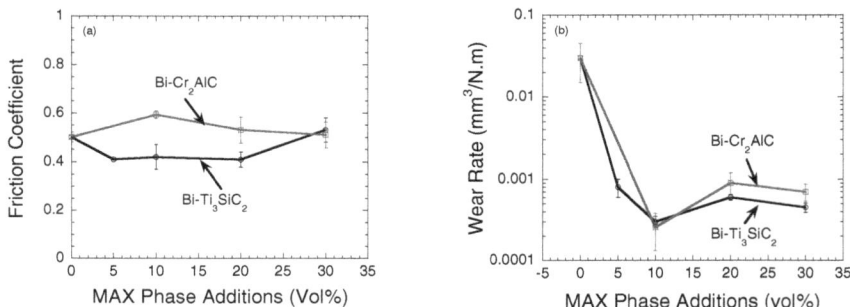

Figure 4: Plot of, (a) friction coefficient, and (b) wear rate versus MAX phase additions.

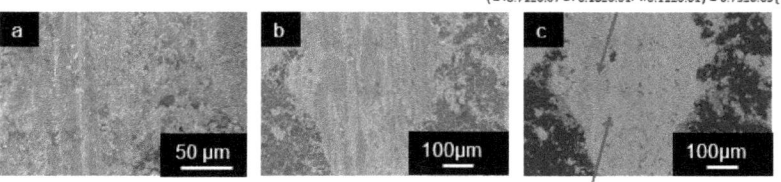

$(Bi_{0.71\pm0.07}Cr_{0.18\pm0.01}Al_{0.11\pm0.01})O_{0.75\pm0.09}\{C_x\}$

$(Bi_{0.54\pm0.01}\ Cr_{0.19\pm0.02}\ Al_{0.14\pm0.01})O_{0.93\pm0.03}\{C_x\}$

Figure 5: SEM SE micrographs of, (a) Bi-20%Cr$_2$AlC, (b) alumina surface, and (c) BSE image of the same region after tribological testing.

Bi-20%Cr$_2$AlC and Bi-30%Cr$_2$AlC, respectively. In other words, the addition of 10 vol% Cr$_2$AlC particulates can decrease the WR of the composites by ~100 times. This study demonstrates that Cr$_2$AlC can be an effective anti-wear additive like Ti$_3$SiC$_2$. At this juncture, it is not clear why the μ_{mean} of Bi-Cr$_2$AlC is higher as compared to Bi-Ti$_3$SiC$_2$ composites. Detailed investigations are needed to understand the exact mechanism. Figure 5a shows the surface of Bi-20%Cr$_2$AlC after tribological testing. Figures 5b-c show the the tribofilms formed due to triboxidation of Bi-20%Cr$_2$AlC on the alumina surface. The tribofilms were uniform, for example the chemistry of two regions are *(Bi$_{0.71\pm0.07}$Cr$_{0.18\pm0.01}$Al$_{0.11\pm0.01}$)O$_{0.75\pm0.09}$\{C$_x$\}* and *(Bi$_{0.54\pm0.01}$Cr$_{0.19\pm0.02}$Al$_{0.14\pm0.01}$)O$_{0.93\pm0.03}$\{C$_x$\}*, respectively. Like Bi-Ti$_3$SiC$_2$ composites [3], the Bi-Cr$_2$AlC composites are mainly contributing towards the formation of tribofilms. According to the classification proposed by Gupta and Barsoum [12], the tribofilms formed between Bi-based MRMs and alumina can be classified as Type IVa as the tribofilms are lubricious and chemically homogenous.

CONCLUSIONS

Bi-Cr$_2$AlC composites were fabricated for the first time by hot pressing. The addition of Cr$_2$AlC particulates increased the yield strength of Bi-Cr$_2$AlC composites increased from ~40 MPa in Bi to ~72 MPa in Bi-30%Cr$_2$AlC (~1.8 times increased in enhancement). The WR of Bi decreased from 0.02 mm^3/N.m to ~2 x 10^{-4} mm^3/N.m in Bi-10%Cr$_2$AlC, then increased to ~9 x 10^{-4} and ~7 x 10^{-4} mm^3/N.m in Bi-20%Cr$_2$AlC and Bi-30%Cr$_2$AlC, respectively. In other words, the addition of 10 vol% Cr$_2$AlC particulates can decreased the WR of the composites by ~100 times. Type IV a (smooth and lubricious) tribofilms was observed on the alumina surfaces.

ACKNOWLEDGEMENTS

One of the authors (SG) would like to acknowledge the University of North Dakota startup funding, NASA EPSCoR under the NASA grant number NNX13AB20A, and NSF EPSCoR for support. Authors would like to thank Kanthal Inc. for supplying the Ti$_3$SiC$_2$ powders. NDSU Electron Microscopy Center core facility is also acknowledged for the microscopy. This material is also based upon work supported by the National Science Foundation under Grant No. 0619098, and 1229417. Any opinions, findings, and conclusions or recommendations expressed in this material are those of the author(s) and do not necessarily reflect the views of the National Science Foundation. Kanthal Inc. is acknowledged for the supply of Ti$_3$SiC powders.

REFERENCES

1. "Bismuth – the new ecologically green metal for modern lubricating engineering", O. Rohr, Industrial Lubrication and Tribology **54**,153 – 164 (2002).
2. "Tribochemistry of Bismuth and Bismuth Salts for Solid Lubrication", P. Gonzalez-Rodriguez, K. J. H. van den Nieuwenhuijzen, W. Lette, D. J. Schipper, and J. E. ten Elshof, ACS Appl. Mater. Interfaces **8**, 7601–7606 (2016).
3. "Synthesis and Tribological Behavior of Novel Ag- and Bi-based Composites Reinforced with Ti_3SiC_2 ", F. AlAnazi, S. Ghosh, R. Dunnigan, and S. Gupta (accepted for publication in Wear).
4. "MAX Phases: Properties of Machinable Ternary Carbides and Nitrides", M.W. Barsoum, John Wiley & Sons (2013).
5. "Elastic and Mechanical Properties of the MAX Phases", M.W. Barsoum and M. Radovic Annu. Rev. Mater. Res. **41**, 195-227 (2011).
6. "Synthesis and characterization of a remarkable ceramic: Ti_3SiC_2", M.W. Barsoum, T. El-Raghy, J. Am. Ceram. Soc. **79**, 1953–1956 (1996).
7. M.W. Barsoum, The $M_{n+1}AX_n$ phases: a new class of solids; thermodynamically stable nanolaminates, Prog. Solid State Chem. **28**, 201–281 (2000).
8. S. Amini, M.W. Barsoum, T. El-Raghy, Synthesis and mechanical properties of fully dense Ti_2SC, J. Am. Ceram. Soc. **90** (12), 3953–3958 (2007).
9. "Synthesis and Characterization of Novel Al-Matrix Composites Reinforced with Ti_3SiC_2 Particulates", S. Gupta, T. Hammann, R. Johnson, and M.F. Riyad, Journal of Materials Engineering and Performance, 24, 1011-1017 (2014).
10. "Effect of Ti_3SiC_2 Particulates on The Mechanical and Tribological Behavior of Sn Matrix Composites", T. Hammann, R. Johnson, M. F. Riyad, and S. Gupta, Proceedings of 39th Int'l Conf & Expo on Advanced Ceramics & Composites (ICACC 2015).
11. "Synthesis and Characterization of Ti_3SiC_2 Particulate-Reinforced Novel Zn Matrix Composites", S. Gupta, Habib, M.A., Dunnigan, R. et al. J. of Materi Eng and Perform (2015) 24: 4071.
12. "On the tribology of the MAX phases and their composites during dry sliding: A review", S. Gupta and M.W. Barsoum, Wear **271**, 1878– 1894 (2011).

FINITE ELEMENT ANALYSIS OF SELF-PROPAGATING HIGH-TEMPERATURE SYNTHESIS (SHS) OF SILICON NITRIDE

Venkata V. K. Doddapaneni
Dan F. Smith Department of Chemical Engineering, Lamar University
Beaumont, TX 77710, U.S.A.

Julia Lin
Dan F. Smith Department of Chemical Engineering, Lamar University
Beaumont, TX 77710, U.S.A.

Ayako Hiranaka
Department of Chemical Engineering, Hokkaido University
Sapporo, Japan

Tomohiro Akiyama
Department of Chemical Engineering, Hokkaido University
Sapporo, Japan

Sidney Lin
Dan F. Smith Department of Chemical Engineering, Lamar University
Beaumont, TX 77710, U.S.A.

ABSTRACT

Self-propagating High-temperature Synthesis (SHS) is a cost effective process to synthesize advanced ceramic materials. In this work, the effect of reactant stoichiometry on the adiabatic temperature of the SHS of silicon nitride is studied and a 3-D time-dependent mathematical model based on the finite element analysis is developed to study the effects of the dimensions of the reaction system and ignition condition on the SHS process, including temperature and composition distributions as well as the production rate and conversion. Momentum transfer, heat transfer, reaction kinetics and temperature dependent thermodynamic properties are integrated in this model. The effects of pellet diameter, ignition time, ignition flux, green density of the reactant pellet, and composition of silicon nitride diluent on the initiation of the self-sustained reaction, temperature history and velocity of the reaction front movement are calculated using this model. The results of this work can be used to improve commercial production of nitride materials.

INTRODUCTION

Silicon nitride has outstanding thermal and mechanical properties. It has wide variety of applications in the engineering, biomedical fields include bearings, turbines, heat exchangers, hip-balls, and artificial knee joints[1-4]. Synthetic silicon nitride (Si_3N_4) was first synthesized by Devill and Wöhler in 1859, but the commercial production of silicon nitride based ceramics did not accelerate until the 1960s[1]. From then, it has been successfully synthesized by conventional methods such as carbothermal reduction, diimide decomposition and direct nitridation[5-7]. However, these methods compromise on the time of production, purity of the final product and require high-temperature furnaces[8]. An exothermic self-sustained reaction called Self-propagating High-temperature Synthesis (SHS) solved the concerns of the conventional production of silicon nitride.

Self-propagating High-temperature Synthesis was developed by Merzhanov et al. in the late 1960s[9, 10]. It is a self-sustained exothermic reaction and does not need external heat for the propagation of the reaction. Thus, external heat is only required to initiate the reaction. The reaction wave then moves through the reactants mixture converting reactants to products. Different materials such as oxides, nitrides, hydrides, and phosphides have been successfully synthesized using this method[11].

EXPERIMENTAL

In our previous work, the reactants (silicon and sodium azide) were mechanically mixed overnight, pressed into cylindrical pellets, and placed into a reactor pressurized under nitrogen. The pellets were ignited by a graphite ignitor. The heat output from the ignitor was controlled by a variable autotransformer. The heat source was turned off after one end of the pellet was ignited. Then the reaction propagated through the whole pellet like a wave from the ignition end[12].

The reaction mechanism at 1 atm of nitrogen was studied using thermal analyses (TG/DTA). It was concluded that the exothermal decomposition of NaN_3 occurred at 673 K to form nitrogen gas and molten sodium. The molten sodium vaporized at about 1,156 K. The TG curve of silicon in nitrogen showed a slow weight gain after the NaN_3 decomposition and a broad exothermic peak in the DTA curve starting at 673 K confirmed the nitridation of silicon initialized at 673 K[12]. In this work, these results are used to calculate the adiabatic temperature of the reaction. Furthermore, Rossetti et al. and Yin et al. studied the silicon nitridation to determine the first order reaction rate constant and the activation energy of silicon nitridation[13-14]. Their results showed the silicon nitridation reaction rate increased significantly when the temperature is above 1,623 K[13-14].

ADIABATIC TEMPERATURE CALCULATIONS

For an exothermic reaction to be self-sustaining after ignition, its adiabatic temperature should be greater than 2,073 K[15-16]. The adiabatic temperature can be calculated by Equation (1).

$$\int_{298}^{T_{ad}} \{\Sigma_i (n_i C_{pi})\} \ dT \ = \ -\Delta H_r^o \tag{1}$$

Reaction (2) and reaction network proposed in Figure 1 are used to calculate the adiabatic temperature of SHS of silicon nitride in this work. In this reaction scheme, silicon nitride acts as a diluent to regulate the reaction temperature and sodium azide acts as the solid nitrogen source to facilitate the synthesis process.

$$mSi_{(s)} + yNaN_{3(s)} + z \ Si_3N_{4(s)} + wN_{2(v)} \rightarrow$$
$$(\tfrac{m}{3} + z)Si_3N_{4(s)} + yNa_{(v)} + \left(\tfrac{3y}{2} - \tfrac{2m}{3} + w\right) N_{2(v)} \tag{2}$$

FINITE ELEMENT ANALYSIS MATHEMATICAL MODEL

A three dimensional finite element analysis model is developed to study the SHS of silicon nitride. Momentum transfer, heat transfer, reaction kinetics and temperature dependent thermodynamic properties are integrated in the model to study the effect of different parameters on the SHS reaction based on Reaction (3).

$$3Si_{(s)} + \tfrac{4}{3}NaN_{3(s)} + 0.2Si_3N_{4(s)} \rightarrow (1 + 0.2)Si_3N_{4(s)} + \tfrac{4}{3}Na_{(v)} \tag{3}$$

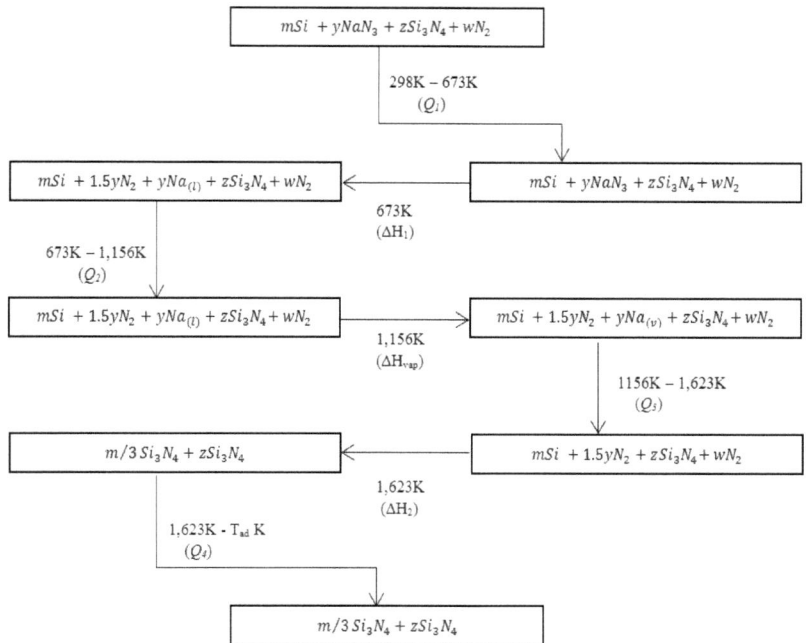

$$mSi + yNaN_3 + zSi_3N_4 + wN_2$$

298K – 673K
(Q_1)

$$mSi + 1.5yN_2 + yNa_{(l)} + zSi_3N_4 + wN_2 \quad \longleftarrow \quad mSi + yNaN_3 + zSi_3N_4 + wN_2$$

673K
(ΔH_1)

673K – 1,156K
(Q_2)

$$mSi + 1.5yN_2 + yNa_{(l)} + zSi_3N_4 + wN_2 \quad \longrightarrow \quad mSi + 1.5yN_2 + yNa_{(v)} + zSi_3N_4 + wN_2$$

1,156K
(ΔH_{vap})

1156K – 1,623K
(Q_3)

$$m/3\,Si_3N_4 + zSi_3N_4 \quad \longleftarrow \quad mSi + 1.5yN_2 + zSi_3N_4 + wN_2$$

1,623K
(ΔH_2)

1,623K - T_{ad} K
(Q_4)

$$m/3\,Si_3N_4 + zSi_3N_4$$

Figure 1. Reaction mechanism proposed for SHS of silicon nitride.

Model Description

The model geometry (Figure 2) consists of two cylinders. The large cylinder represents the nitrogen gas and the small cylinder represents the reactant pellet.

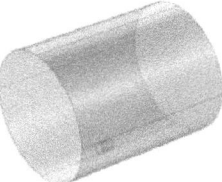

Figure 2. Model geometry sketches of the reaction system.

The steady state Navier-Stokes equation and the continuity equation are first used to solve the nitrogen flow profile inside the reactor before the SHS reaction. The steady state velocity profile then is used as the initial value for the time dependent study. A constant velocity (0.05 ms⁻¹) of nitrogen gas is used at the reactor inlet and constant pressure (1 atm) boundary condition applied to the outlet of the reactor. The reactor inside wall and the outside surface of the pellet are set to the no slip boundary condition.

$$\rho \frac{\partial u}{\partial t} + \rho u \cdot \nabla u = -\nabla p + \nabla \cdot (\mu(\nabla u) + (\nabla u)') - \frac{2}{3}\mu((\nabla \cdot u)I) + F \tag{4}$$

$$\frac{\partial \rho}{\partial t} + \nabla \cdot (\rho u) = 0 \tag{5}$$

Conduction and convection heat transfers are considered in nitrogen gas which are included in Equation (6). The initial temperature of nitrogen gas is assumed to be the room temperature and the reactor inside wall is set to heat insulation. At the reactor inlet, the room temperature is assumed and the convective boundary condition is used at the outlet of the reactor.

$$\rho Cp \left(\frac{\partial T_f}{\partial t} + u \cdot \nabla T_f\right) = \nabla \cdot (k\nabla T_f) \tag{6}$$

Equation (7) is used to solve time dependent temperature profile in the reactant pellet. The initial temperature of the pellet is assumed to be the room temperature.

$$\rho Cp \left(\frac{\partial T_p}{\partial t}\right) = \nabla \cdot (k\nabla T_p) + Q \tag{7}$$

An external heat flux of 5×10^7 W/m² is used at the front end of the reactant pellet for 5 seconds to initiate the reaction.

$$-n \cdot (-k\nabla T_p) = 5 \times 10^7 \; (W/m^2) \tag{8}$$

The heat conducted to the outer surface of the pellet is equal to the radiation and convective heat losses from the pellet outer surface to the surrounding nitrogen gas.

$$-n \cdot (-k\nabla T_p) = -\varepsilon \times \sigma \times (T_p{}^4 - T_\infty{}^4) - h \times (T_p - T_f) \tag{9}$$

The total heat generation rate is calculated by the reaction rates and molar reaction heats of silicon nitridation, sodium azide decomposition and sodium vaporization, respectively.

$$Q = \Delta H_{Si_3N_4} \times \frac{1}{3}\left(\frac{-dC_{Si}}{dt}\right) + (\Delta H_{NaN_3}) \times \left(\frac{-dC_{NaN_3}}{dt}\right) + \Delta H_{Na,v} \times (-k_{Na} \times C_{Na}) \tag{10}$$

Kinetic parameters of silicon nitridation are calculated from the data reported by Yin et al.[14] In this case, the estimated values of activation energy and pre-exponential factor are 94.58 kJ/mole and 2.215 sec⁻¹, respectively. According to our previous experimental observations, the pre-exponential factor is adjusted from 2.215 sec⁻¹ to 5 sec⁻¹ in this work.

$$3Si_{(s)} + 2N_{2(g)} \rightarrow Si_3N_{4(s)} \tag{11}$$

$$\frac{-dC_{Si}}{dt} = 3 \times C_{Si} \times k_{Si,o} \times exp\left(\frac{-E_{Si}}{RT_p}\right) \tag{12}$$

Jacobs et al.[17] studied the reaction kinetics of decomposition of sodium azide. In this study, the reported data are used to estimate the activation energy, and pre-exponential factors. The estimated values are 62.83 kJ/mole and 898.435 sec⁻¹, respectively.

$$NaN_{3(s)} \rightarrow 1.5N_{2(v)} + Na_{(l)} \tag{13}$$

$$\frac{-dC_{NaN_3}}{dt} = C_{NaN_3} \times k_{NaN_3,o} \times exp\left(\frac{E_{NaN_3}}{RT_p}\right) \tag{14}$$

Yu et al. studied the evaporation of sodium and potassium in silicate melts[18]. It is concluded that the rate of vaporization of sodium follows the first order kinetics.

$$Na_{(l)} \rightarrow Na_{(v)} \tag{15}$$

From our previous experiments, it was observed that 90% of sodium evaporates around 50sec when the temperature reaches 1,156 K. Using the above information, the rate constant is estimated to be 0.046 sec^{-1}. The rate of formation of sodium from the sodium azide decomposition and the rate of vaporization of sodium can be calculated using Equations (16) and (17).

$$\frac{dC_{Na}}{dt} = \frac{-dC_{NaN_3}}{dt} \qquad\qquad if\ Tp < 1{,}156\ K \tag{16}$$

$$\frac{dC_{Na}}{dt} = \frac{-dC_{NaN_3}}{dt} + (-k_{Na} \times C_{Na}) \qquad\qquad if\ Tp > 1{,}156\ K \tag{17}$$

RESULTS AND DISCUSSION

Reaction (3) $(3Si_{(s)} + \frac{4}{3}NaN_{3\ (s)} + 0.2Si_3N_{4\ (s)})$ is used as the baseline to study the impacts of silicon nitride and sodium azide composition on adiabatic temperature. Figure 3 (a) shows that the adiabatic temperature of SHS of silicon nitride decreases from 7,596 K to 5,767 K when the silicon nitride varies from 0 to 0.8 moles because heat density decreases with increase in silicon nitride. From Figure 3 (b), it can be seen that the adiabatic temperature decreases from 7,022 K to 5,106 K when the sodium azide varies from 1.33 to 3 moles. This is because a higher sodium azide concentration produces more sodium during the reaction thus, a larger amount of heat is used to vaporize the sodium. A reaction using excess sodium azide $(3Si_{(s)} + 2NaN_{3\ (s)} + 0.2Si_3N_{4\ (s)})$ is used as the baseline to study the impact of silicon on adiabatic temperature. Figure 3 (c) shows that the adiabatic temperature increases from 5,821 K to 7,201 K when silicon varies from 2.5 to 4.5 moles because the heat density increases with increase in silicon.

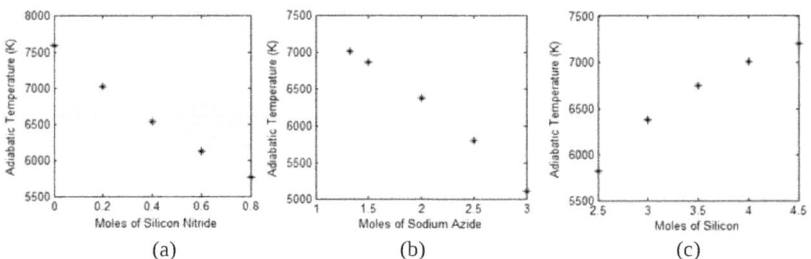

(a)   (b)   (c)

Figure 3. The change of the adiabatic temperature when different amounts of reactants are used in the reactant pellet.

A finite element analysis model is used to study the temperature and composition distribution and velocity of reaction front movement of a reaction pellet of a diameter of 25.4 and a length of 127 mm using an ignition heat flux of 5×10^7 W/m^2 for of 5 seconds.

Figure 4 shows the calculated properties in the pellet 25 seconds after the ignition. The highest temperature reached is 4,440 K and the decomposition of sodium azide started at around 650 K, which is close to the decomposition temperature of sodium azide reported by Lin[12]. It can also be seen that a fast increase of silicon nitride mole fraction and decrease in silicon mole fraction at around 1,630 K indicates the silicon nitridation process becomes more significant at temperatures higher than 1,623 K, which is close to the nitridation temperature of silicon reported by Yin et al. and Rosette et al. The mole fraction of sodium is higher at x = 0 mm than at the x = 7 mm due to severe convection heat losses at the ignition surface of the pellet.

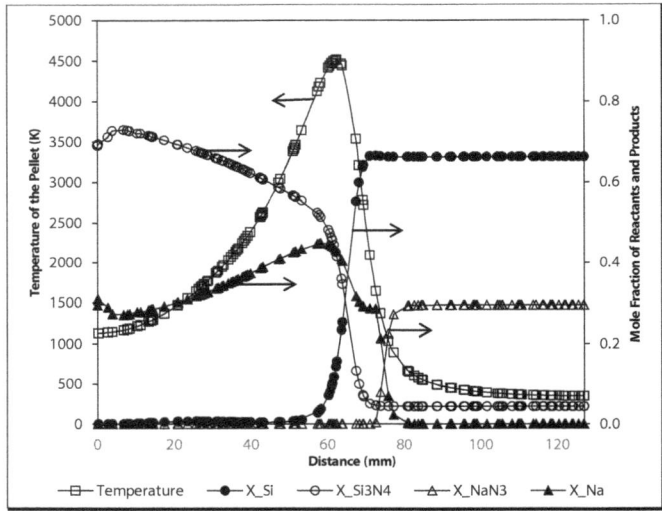

Figure 4. The temperature and composition distributions of species at the center line of the pellet at 25 seconds after the ignition. Reactant composition is 3 Si + 4/3 NaN$_3$ + 0.2 Si$_3$N$_4$ and the ignition heat flux = 5×10^7 W/m^2, ignition time = 5 s, ϕ = 0.7, pellet diameter = 25.4 mm, and pellet length = 127 mm.

The velocity of combustion front movement during the reaction along the pellet center line is computed from dividing the distance between two known points (2.5 mm ahead and after the point) by the time difference for these two points to reach the same temperature. Figure 5 shows the velocity of reaction front movement reaches maximum at the midpoint of the pellet (63.5 mm) because the lower axial heat losses at midpoint of the pellet compared to other positions.

The effects of standard Self-propagating High-temperature Synthesis variables, including pellet diameter, ignition time, and ignition flux, amount of silicon nitride diluent, and green density on the SHS of silicon nitride are critical for a successful large scale production in which a large reaction pellet with minimum external heat source (ignition flux and ignition time) is desired.

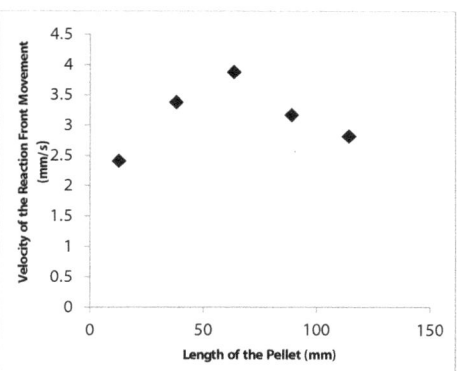

Figure 5. The velocity of reaction front movement along the length of the pellet. Reactant composition is 3 Si + 4/3 NaN$_3$ + 0.2 Si$_3$N$_4$ and the ignition heat flux = 5× 10^7 W/m^2, ignition time = 5 s, ϕ = 0.7, pellet diameter = 25.4 mm, and pellet length = 127 mm.

Figure 6 shows that the maximum temperature at the midpoint of the pellet centerline (x= 63.5 mm) increases from 477 K to 5,941 K when the pellet diameter varies from 12.7 mm to 50.8 mm due to the lower convection heat losses from the pellet outer surface.

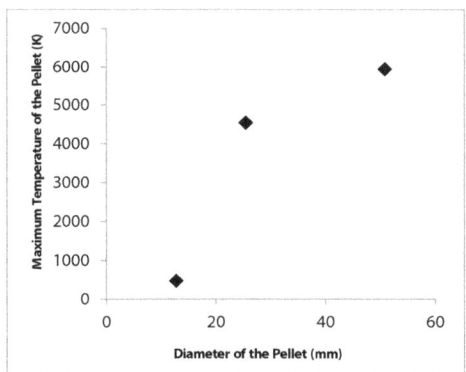

Figure 6. The change of maximum temperature at the midpoint of the pellet (x = 63.5 mm) when different pellet diameters are used. Reactant composition is 3 Si + 4/3 NaN$_3$ + 0.2 Si$_3$N$_4$ and the ignition heat flux = 5× 10^7 W/m^2, ignition time = 5 s, ϕ = 0.7, (12.7, 25.4 and 50.8 mm), and pellet length = 127 mm.

Form Figure 7, it can be seen that with an ignition time of 5 seconds the maximum temperatures exceed 4,500 K when the ignition flux is set to be 5 × 10^7 W/m^2 or 1 × 10^7 W/m^2. However, the maximum temperature reached is less than the sodium azide decomposition temperature (673 K) when a flux of 5 × 10^6 W/m^2 is used. Therefore, the reaction front does not propagate after ignition. A longer ignition time is needed for the reaction front to propagate at room temperature when a heat flux of 5 × 10^6 W/m^2 is used.

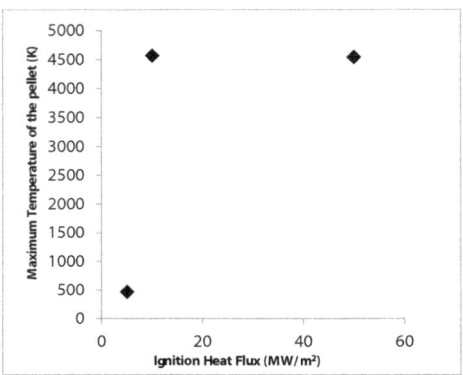

Figure 7. Change of maximum temperature at the midpoint of pellet (x = 63.5 mm) when different heat flux is used. Reactant composition is 3 Si + 4/3 NaN$_3$ + 0.2 Si$_3$N$_4$ and the ignition heat flux = (5× 10^6, 1× 10^7, 5× 10^7 W/m^2), ignition time = 5 s, φ = 0.7, pellet diameter = 25.4 mm, and pellet length = 127 mm.

Figure 8 shows that when a heat flux of 5× 10^7 W/m^2 is used, the maximum temperature can be reached is below 673 K when the ignition time is less than 3 seconds. Therefore, the decomposition of sodium azide, the first exothermic reaction in our proposed reaction network cannot complete and therefore the reaction front will not propagate.

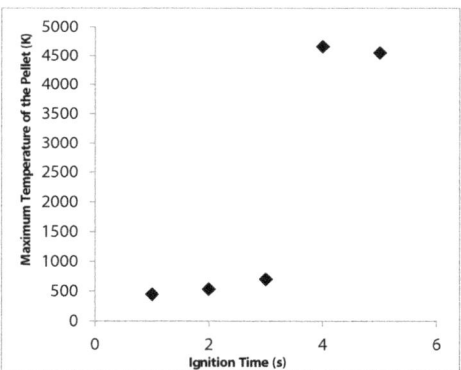

Figure 8. The change of maximum temperature at the midpoint of pellet (x = 63.5 mm) when different ignition times are used. Reactant composition is 3 Si + 4/3 NaN$_3$ + 0.2 Si$_3$N$_4$ and the ignition heat flux = 5× 10^7 W/m^2, ignition time = (1, 2, 3, 4 and 5 s), φ = 0.7, pellet diameter = 25.4 mm, and pellet length = 127 mm.

Figure 9 shows the impact of the amount of diluent on the maximum temperature reached. The maximum temperature remains at about 4,800 K when less than 0.3 moles of silicon nitride is used and decreases sharply below 1,000 K when there is no propagation when more than 0.3 moles of silicon nitride is used. This is because silicon nitride absorbs heat, which lowers the pellet temperatures below the critical temperature for the reaction front to propagate (2,073 K). Thus, the nitridation reaction cannot complete.

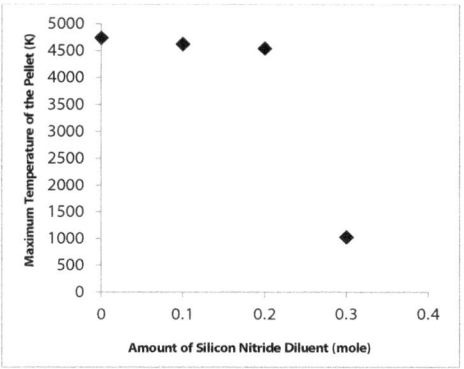

Figure 9. The change of maximum temperature at the midpoint of pellet (x = 63.5 mm) when different amounts of silicon nitride diluents are used. Reactant composition is 3 Si + 4/3 NaN$_3$ + z Si$_3$N$_4$ (z = 0, 0.1, 0.2 and 0.3) and the ignition heat flux = 5× 10^7 W/m^2, ignition time = 5 s, φ = 0.7, pellet diameter = 25.4 mm, and pellet length = 127 mm.

Figure 10 shows that reaction propagates when pellet green density varies from 0.7 to 0.5 of the theoretical density, while there is no propagation when the ratio is above 0.7. This can be explained by the increase in thermal conductivity with increase in green density, which results in a faster heat conduction loss that makes it impossible to reach the ignition temperature.

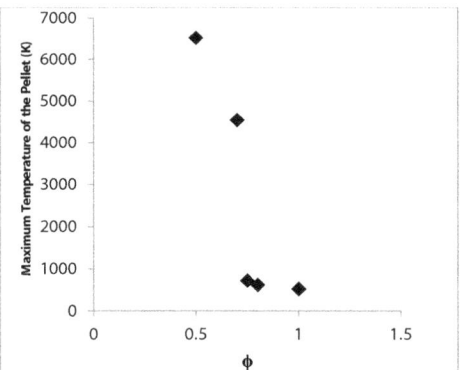

Figure 10. The change of maximum temperature at the midpoint of pellet (x = 63.5 mm) when different green densities are used. Reactant composition is 3 Si + 4/3 NaN$_3$ + 0.2 Si$_3$N$_4$ and the ignition heat flux = 5× 10^7 W/m^2, ignition time = 5 s, φ = (1, 0.8, 0.75, 0.7 and 0.5), pellet diameter = 25.4 mm, and pellet length = 127mm.

CONCLUSIONS

A finite element analysis model of Self-propagating High-temperature Synthesis (SHS) of silicon nitride is successfully developed to investigate the reaction at a normal pressure nitrogen environment (1 atm). It is shown that the velocity of reaction front movement is highest at the midpoint of the pellet and the maximum temperature of the pellet is strongly dependent on the green density and amount of silicon nitride added because of the variation of heat density and thermal conductivity with both green density and silicon nitride diluent.

ACKNOWLEDGMENT

This work is partially supported by the Technology Transfer Initiative (TTI) Program of Center for Advanced Research of Energy and Materials (CAREM) of Faculty of Hokkaido University, Japan. The authors also wish to thank T. Wei and S. Gohar for helpful assistance and discussions.

NOMENCLATURE

T_{ad}	Adiabatic temperature [K]
n_i	Stoichiometric coefficient of reactants and products
C_{pi}	Specific heat of species i [J/mole/ K]
ΔH_r^o	Standard heat of reaction (J)
ρ	Density of fluid, [kg/m^3]
u	Velocity of fluid, [m/s]
p	Fluid pressure, [Pa]
μ	Viscosity of the fluid, [Pa-s]
I	Identity matrix
F	External force applied on the fluid, [N]
m	Moles of silicon
y	Moles of sodium azide
z	Moles of silicon nitride
w	Moles of nitrogen
I	Identity matrix
F	External force applied on the fluid, [N]
$\frac{-dC_{Si}}{dt}$	Rate of consumption of silicon nitride, [mol/m^3/s]
C_{Si}	Concentration of silicon nitride at any time (t), [mol/m^3]
$k_{Si,o}$	Pre-exponential factor for formation of silicon nitride, [sec^{-1}]
$-E_{Si}$	Activation energy for the formation of silicon nitride, [J/mole]
R	Gas constant, [8.314 J/mole/K]
T_p	Temperature of the pellet at any time (t), [K]
$\frac{-dC_{NaN_3}}{dt}$	Rate of consumption of sodium azide, [mol/m^3/s]
C_{NaN_3}	Concentration of sodium azide at any time (t), [mol/m^3]
$k_{NaN_3,o}$	Pre-exponential factor for the decomposition of sodium azide, [sec^{-1}]
$-E_{NaN_3}$	Activation energy for the decomposition of sodium azide, [J/mole]
$\Delta H_{Na,v}$	Heat of vaporization of sodium, [J/mole]
C_{Na}	Concentration of sodium azide at any time (t), [mol/m^3]
$-k_{Na}$	Vaporization rate constant of sodium, [mol/m^3/s]
Φ	Ratio of green density to theoretical density
X	Distance from the ignition surface, [mm]

REFERENCES

[1] F.L. Riley, "Silicon Nitride and Related Materials", *J. Am. Ceram. Soc.* **83**, 245–265 (2000).

[2] Berroth, K. Silicon Nitride Ceramics for Product and Process Innovations, *Advances in Science and Technology*, **65**, 70-77 (2010).

[3] S. Hampshire, "Silicon Nitride Ceramics – Review of Structure, Processing and Properties", *Journal of Achievements in Materials and Manufacturing Engineering*, **1**, 43-50 (2007).

[4] H. Vladimir, J.A. Puszynski, "Chemical Engineering Aspects of Advanced Ceramic Materials", *Ind. Eng. Chem. Res.*, **35**, 349-377 (1996).

[5] M. Ekelund, B. Forslund, "Carbothermal Preparation of Silicon Nitride: Influence of Starting Material and Synthesis Parameters", *J. Am. Ceram. Soc.*, **75** (3), 332-339 (1992).

[6] A. Atkinson, A.J. Moulson, E.W. Roberts, "Nitridation of High-Purity Silicon", *J. Am. Ceram. Soc.*, **59**, 285 (1976).

[7] T. Yamada, "Preparation and Evaluation of Sinterable Silicon Nitride Powder by Imide Decomposition Method", *Am. Ceram. Soc. Bull.*, **72**, 99-106 (1993).

[8] W.P. Shen, F. Wang, W.U. Zhuohui, G.E. Changchun, "Effect of Auxiliary Gases on Combustion Synthesis of Si_3N_4", *J. Mater. Sci.* Technol., **21** (5) (2005).

[9] G. Xanthopoulou, G. Vekinis, "An Overview of Some Environmental Applications of Self-propagating High-temperature Synthesis", *Adv. Environ. Res (Oxford, U. K.)*, **5**, 117-128 (2001).

[10] S. Lin, J. Selig, H.T. Lin, H. Wang, "Self-propagating High-temperature Synthesis of Calcium Cobaltate Thermoelectric Powders", *J. Alloys Compd.*, **503**, 402-409 (2010).

[11] A. Varma, S.R. Alexander, S.X. Alexander, S. Hwang, "Combustion Synthesis of Advanced Materials: Principles and Applications", *Adv. Chem. Eng.*, **24**, 79-226 (1998).

[12] S. Lin, R. Wilkins, Z. Henry, "High-pressure Self-propagating High-temperature Synthesis (SHS) of Silicon Carbide-Silicon Nitride Composites", *Ceram. Trans.*, **103**, 51-61 (2000).

[13] G.A. Rossetti, R.P. Denkewicz, "Kinetics Interpretation of α- and β-Si_3N_4 Formation from Oxide Free High-purity Silicon Powder", *J. Mater. Sci.*, **24**, 3081-3086 (1989).

[14] S.W. Yin, L. Wang, L.G. Tong, F.M. Yang, Y.H. L, "Kinetic Study on the Direct Nitridation of Silicon Powders Diluted with α-Si_3N_4 at Normal Pressure", *International Journal of Minerals, Metallurgy and Materials*, **20** (5), 493 (2013).

[15] H. Ishikawa, K. Oohira, T. Nakajima, T. Akiyama, "Combustion Synthesis of SrTiO3 using Different Raw Materials", *J Alloy Compd.*, **454** (1-2), 384–388 (2008).

[16] C.R. Bowen, B. Derby, "Self-propagating High-temperature Synthesis of Ceramic Materials", *Br. Ceram. Trans.*, **96** (1), 25 (1997).

[17] P.W.M. Jacobs, T.A.R. Kureishy, "Kinetics of Thermal Decomposition of Sodium azide", *J. Chem. Soc.*, 4718-4723 (1964).

[18] Y. Yu, R.H. Hewins, C.M.O.D. Alexander, J. Wang, "Experimental Study of Evaporation and Isotopic Mass Fractionation of Potassium in Silicate Melts", *Geochim. Cosmochim. Acta.*, **67** (4), 773-786 (2003).

TEM ANALYSIS OF DIFFUSION-BONDED SILICON CARBIDE CERAMICS JOINED USING METALLIC INTERLAYERS

T. Ozaki[1], Y. Hasegawa[1], H. Tsuda[2], S. Mori[2], M. C. Halbig[3], R. Asthana[4], and M. Singh[5]

[1] Technology Research Institute of Osaka Prefecture, Osaka, Japan

[2] Graduate School of Engineering, Osaka Prefecture University, Osaka, Japan

[3] NASA Glenn Research Center, Cleveland, Ohio, USA

[4] University of Wisconsin-Stout, Menomonie, WI, USA

[5] Ohio Aerospace Institute, NASA Glenn Research Center, Cleveland, Ohio, USA

ABSTRACT

SiC fiber-bonded ceramics (SA-Tyrannohex™: SA-THX) diffusion-bonded with Ti/Cu metallic interlayers were investigated. Thin samples of the ceramics were prepared with a focused ion beam (FIB) and the interfacial microstructure of the prepared samples was studied by transmission electron microscopy (TEM) and scanning TEM (STEM). In addition to conventional microstructure observation, for detailed analysis of reaction compounds in diffusion-bonded area, we performed STEM-EDS measurements and selected area electron diffraction (SAD) experiments. The TEM and STEM experiments revealed the diffusion-bonded area was composed of only one reaction layer, which was characterized by TiC precipitates in Cu-Si compound matrix. This reaction layer was in good contact with the SA-THX substrates, and it is concluded that the joint structure led to the excellent bonding strength.

INTRODUCTION

Silicon carbide (SiC) composite materials are attractive materials for applications in high-temperature and extreme environments because of their excellent mechanical properties, oxidation resistance, and thermal stability. In particular, one robust ceramic SA-Tyrannohex™ (SA-THX), which has a structure of highly ordered, close-packed, hexagonal columnar fibers of crystalline β-SiC bonded with thin layers of interfacial carbon[1,2], is a promising material because of its good thermomechanical performance, high strength sustained up to 1600°C, and high fracture toughness (1200 J·m^{-2} at RT)[3]. Hence, SA-THX and related SiC-based ceramics are currently being developed and tested for a wide variety of applications in aerospace and energy[4,5]. However, the geometrical limitations of SiC ceramics prevent the fabrication of large or complex components *via* hot pressing, CVD, machining, or net-shape processing. To fabricate these components from brittle ceramics, simpler units must be joined and integrated. Various joining methods have been developed, such as reaction bonding[6-8] and brazing[9,10]. Diffusion bonding techniques have also been used and hold much promise[11,12].

We have applied diffusion-bonded to a variety of SiC parts with various metallic interlayers[13-16]. We also obtained good diffusion bonding in SA-THX through the use of a 10-µm-thick Ti interlayer and fibers parallel to the interlayer [15].

Recently, to reduce the temperature of diffusion bonding processes, we attempted to use Ti/Mo and Ti/Cu foils as metallic interlayers[17,18]. In the case of Ti/Mo our process achieved metallurgically sound joints, but the results of a Knoop test suggested that the joint area contained some weak areas and microscale cracking was observed[17]. We also reported that the $Ti_5Si_3C_x$ phase generated by diffusion bonding had a large coefficient of thermal expansion (CTE), which contributed to the microscale cracking[16,18]. Conversely for the case of Ti/Cu no defects, such as microscale cracking, were observed around the diffusion-bonded area. Furthermore, the Knoop hardness of the joint area measured for Ti/Cu stability showed a higher strength than that based on Ti/Mo. The area diffusion bonded with Ti/Cu showed a different joint microstructure from that of Ti/Mo[17]. Therefore, to reveal the bonding mechanism, which exhibits a high strength in Ti/Cu interlayer, we investigated the microstructure around diffusion-bonded area.

For the microstructure observations by transmission electron microscopy (TEM) and scanning TEM (STEM), we prepared thin samples of the diffusion bond with focused ion beam (FIB) milling. In addition to conventional microstructure observations, we performed detailed analysis of the reaction compounds in the diffusion-bonded area with STEM- energy-dispersive X-ray spectroscopy (EDS) and selected area electron diffraction (SAD). On the basis of the results obtained by TEM and STEM analysis, the process of formation of the joint microstructure was discussed.

EXPERIMENTAL

SA-Tyrannohex™ (SA-THX) SiC fiber-bonded ceramic was obtained from Ube Industries (Ube, Japan). The material was composed of SA-Tyranno fiber™ bundles in an eight-harness satin weave, with fibers oriented in parallel and perpendicular directions. Ti foil (10 μm) and Cu foil (5 μm) were obtained from Goodfellow Corporation (Glen Burnie, MD, USA). Before joining, all materials were ultrasonically cleaned in acetone for 10 min. Joints were diffusion bonded at 1200 °C for 4 hours under a pressure of 30 MPa in vacuum. For joints involving the Ti/Cu bilayers, three sheets of Cu foil were sandwiched by two sheets of Ti foil as illustrated in Fig. 6(a). Detailed fabrication procedures are reported elsewhere[17].

STEM and STEM-EDS measurements were performed with a Hitachi HD-2700. TEM and SAD were performed with a JEOL JEM-2000FX. All samples for TEM and STEM were prepared with a FIB (Hitachi FB-2200), which allowed us to examine precisely selected, clean, and diffusion-bonded areas with low-damage. Detailed conditions of the FIB process are described in the next chapter.

RESULTS AND DISCUSSION
Preparation of Thin Samples from Diffusion Bonding

We fabricated thin samples for TEM/STEM observation with a FIB micro-sampling technique. Fig. 1 shows scanning ion microscope (SIM) images of the diffusion-bonded area obtained from the FIB. The reacted layers and their boundaries are clearly shown in Fig. 1(a). No cracks or void were observed around the reacted layer. Similar microstructures were imaged by SEM [17]. We selected a thin sample from the joint area which included the reaction layer and a part of the SA-THX substrate (Fig. 1(b)), and processed the sample to be thin enough for TEM observations.

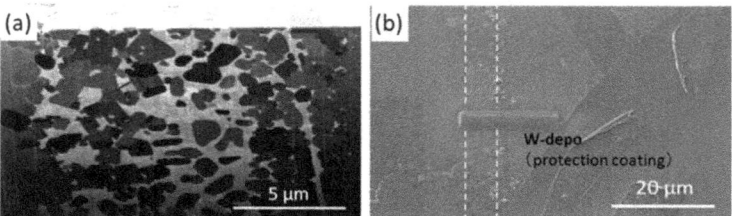

Figure 1. Scanning ion microscope (SIM) images of the diffusion bond obtained with the FIB. (a)Cross-sectional image and (b) image of the area where the thin sample was fabricated.

STEM Imaging

Fig. 2 shows STEM images of the diffusion bond acquired in secondary electron (SE), bright-field (BF), and high angle annular dark-field (HAADF) STEM modes, respectively. The samples were thin enough to observe the microstructure of the entire diffusion-bonded area and the quality of the bonding appeared to be good. In the HAADF image, some dark contrast in SA-THX was attributed to residual carbon or fiber boundaries [18]; apart from the marks induced in the FIB process, no notable defects were observed. Some features should be noted in the diffusion bonded area. Several reaction layers were observed in the diffusion bonded Ti/Mo interlayer [17,18]; however, in the case of Ti/Cu, only one reaction layer was apparent in the diffusion bond. A part of the reaction compound leaked into the boundaries of the SA-THX, but no elemental diffusion from the reaction layer side into the SiC grains was observed. This result suggested that the diffusion bonding process with Ti/Cu interlayer proceeded mainly via a liquid state rather than diffusion in solid. Furthermore, the reaction layer was composed of grains of various sizes and a monolithic matrix, which filled the grains without gap. The reaction layer had few cracks or voids and formed a clean interface with the SA-THX substrate.

Figure 2. STEM images of diffusion bond: (a) SE STEM mode, (b) BF STEM mode, and (c) HAADF STEM mode.

STEM-EDS Mapping

To investigate the elemental composition of the grains and the matrix in the reaction layer, we performed STEM-EDS mapping. As shown in Fig. 3, the elemental distribution of Si, Ti, and Cu could be clearly divided into two areas of the grains and the matrix. The grains appeared to contain more Ti and C whereas the matrix contained more Cu and Si. On the basis of these results, the reaction layer was likely composed of TiC grains and a matrix of a Cu-Si compound. Although the elemental distribution of C appeared not to differ much, carbon located in the nearby SA-THX substrate might have affected the EDS measurement or the matrix may have contained some carbon. Some previous research had reported that a certain amount of SiC can dissolve into melted copper and fine glassy carbon precipitates in Cu-Si solid solution[19].

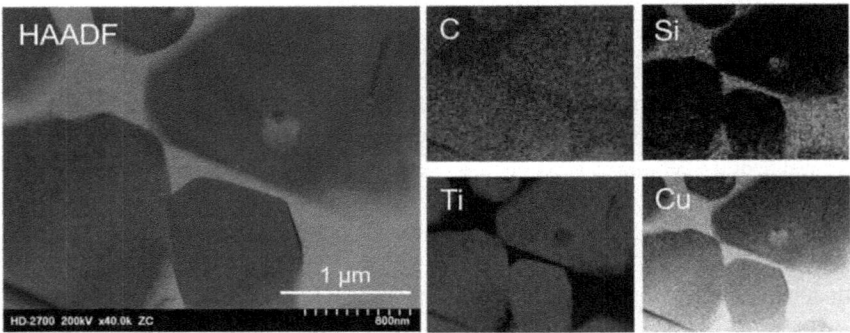

Figure 3. HAADF-STEM image of the reaction layer and elemental mapping images obtained by STEM-EDS.

TEM Imaging and SAD analysis

For more detailed analysis of the reaction compound, we investigated its crystal structure by TEM. We acquired SAD patterns from regions of the grains and matrix in the reaction layer. Fig. 4 shows the TEM images and SAD patterns acquired from one grain, indicated by a circle. Each SAD pattern featured a reciprocal lattice pattern from a simple NaCl-type structure, which was consistent with the standard TiC structure.

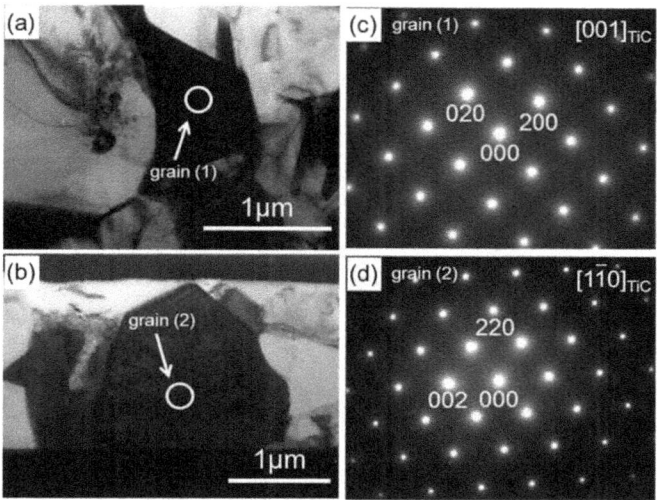

Figure 4. (a, b) TEM micrographs and (b, c) SAD patterns obtained from grains (1) and (2).

Although the crystal structure of the grains in the reaction layer was easily characterized it was more difficult to characterize the crystal structure of the matrix. As shown in Fig. 5, the SAD patterns acquired from the matrix region were complex. The results of the STEM-EDS suggested the matrix featured a Cu-Si system, and Cu_3Si is a potential candidate material. It has been reported that Cu_3Si has a long-period structure which appears as superlattice reflections in its SAD patterns[20,21]. The SAD patterns of the matrix were similar to those of Cu_3Si; however, further studies are required to clarify the matrix structure.

Figure 5. (a) TEM micrograph and (b, c) SAD patterns obtained from Cu-Si matrix region.

Formation of the Joint Microstructure

TEM and STEM analysis revealed the reaction layer of the diffusion bonding was composed of TiC precipitate in a Cu-Si compound matrix. The results described above also indicate the formation mechanism of the joint microstructure formed from the Ti/Cu interlayer.

At the maximum temperature in the diffusion bonding process, the Cu interlayer melts, and the titanium interlayer and SiC (SA-THX) substrate partially dissolve in the molten metal. In the cooling process, TiC grains precipitate from the molten metal, and as the temperature decreases the grains gradually grow by absorbing Ti and C atoms from the molten metal. At the end of the process, the molten metal completely solidifies becoming a monolithic Cu-Si compound. Thus the microstructure of TiC grains embedded in a monolithic Cu-Si compound was formed in this way.

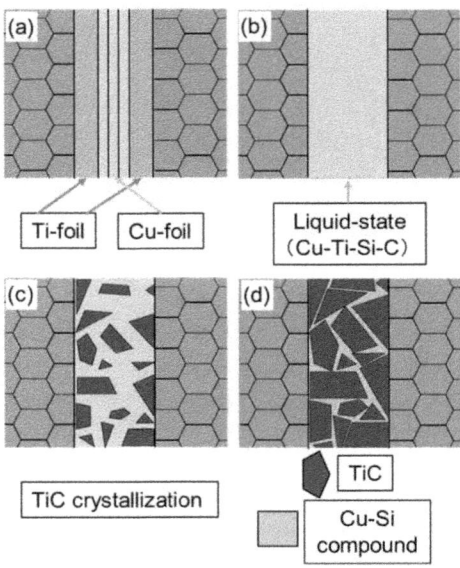

Figure 6. Change in the microstructure of interface of SA-THX/Ti/Cu/Ti/SA-THX (a) before processing, (b) at maximum temperature (1200°C), (c) in cooling, and (d) after processing

Considering the CTE value of the reaction compound, it is favorable for TiC grains to constitute most of interlayer because TiC has a CTE value relatively close to that of SiC. Furthermore, because TiC has a cubic crystal structure, TiC will expand and contract isotropically. Although the exact CTE value of the Cu-Si compound is not known, Cu metal generally exhibits relatively large CTE values. However, because Cu metal also exhibits a low Young modulus, the soft Cu-based compound formed in the matrix of the joint area may be advantageous for the diffusion bonding quality.

CONCLUSIONS

SA-THX was diffusion bonded with Ti-Cu foil interlayers. The diffusion bond regions were examined by TEM and STEM imaging for samples prepared by FIB. The results are summarized as follows.

(1) We selected thin samples from the bonded area of diffusion bonded SA-THX and processed these using a FIB micro-sampling technique. The prepared samples were sufficiently thin and showed low enough damage to allow detailed evaluation by TEM and STEM.

(2) The microstructure of the diffusion bonded area was observed by STEM and TEM. The composition and crystal structures of the reaction compound were investigated by STEM-EDS and SAD methods. The reaction layer of the diffusion bonding was composed of TiC precipitates in a Cu-Si compound matrix.

ACKNOWLEDGEMENTS

This work was funded by JSPS KAKENHI Grant Number JP16K06802.

REFERENCES

[1]T. Ishikawa, S. Kajii, K. Matsunaga, T. Hogami, Y. Kohtoku, and T. Nagasawa, A Tough, Thermally Conductive Silicon Carbide Composite with High Strength up to 1600°C in Air, *Science*, **282**, 1265–1297 (1998).

[2]T. Ishikawa, Y. Kohtoku, K. Kumagawa, T. Yamamura, and T. Nagasawa, High-strength Alkali-Resistant Sintered SiC Fibre Stable to 2200°C, *Nature*, **391**, 773–775 (1998).

[3]T. Ohji, M. Singh, Engineered Ceramics: Current Status and Future Prospects, *Wiley*, **16**, Integration Challenges in Alternative and Renewable Energy Systems, 323 (2016).

[4]P. J. Lamicq, G. A. Bernhart, M. M. Dauchier, and J. G. Mace, SiC/SiC Composite Ceramics, *Am.Ceram. Soc. Bull.*, **65**(2), 336–338 (1986).

[5]M. Halbig, M. Jaskowiak, J. Kiser, and D. Zhu, Evaluation of Ceramic Matrix Composite Technology for Aircraft Turbine Engine Applications, *proceedings of the 51st AIAA Aerospace Sciences Meeting including the New Horizons Forum and Aerospace Exposition* (2013).

[6]M. Singh, A Reaction Forming Method for Joining of Silicon Carbide-based Ceramics, *Scr. Mater.*, **37**(8), 1151–1154 (1997).

[7]M. Singh, Joining of Sintered Silicon Carbide Ceramics for High Temperature Applications, *J. Mater. Sci. Lett.*, **17**(6), 459–461 (1998).

[8]M. Singh, Microstructure and Mechanical Properties of Reaction Formed Joints in Reaction Bonded Silicon Carbide Ceramics, *J. Mater. Sci.*, **33**, 1–7 (1998).

[9]V. Trehan, J. E. Indacochea, and M. Singh, Silicon carbide brazing and joint characterization, *J. Mech. Behav. Mater.*, **10**(5–6), 341–352 (1999).

[10]M. G. Nicholas, Joining Processes: Introduction to Brazing and Diffusion Bonding, *Kluwer Academic Publishers*, Dordrecht, (1998).

[11]B. Gottselig, E. Gyarmati, A. Naoumidis, and H. Nickel, Joining of Ceramics Demonstrated by the Example of SiC/Ti, *J. Eur. Ceram. Soc.*, **6**, 153–160 (1990).

[12]M. Naka, J. C. Feng, and J. C. Schuster, Phase Reaction and Diffusion Path of the SiC/Ti System, *Metall. Mater. Trans. A*, **28A**, 1385–1390 (1997).

[13]H. Tsuda, S. Mori, M. C. Halbig, and M. Singh, TEM observation of the Ti Interlayer between SiC Substrates during Diffusion Bonding, *Proceedings of ICACC 2012*, (2012).

[14]M. C. Halbig, M. Singh, and H. Tsuda, Integration Technology for Silicon Carbide-Based Ceramics for Micro-Electro-Mechanical Systems-Lean Direct Injector Fuel Injector Applications, *Int. J. Appl. Ceram. Tec.*, **9**, 677–687 (2012).

[15]H. Tsuda, S. Mori, M. C. Halbig, and M. Singh, Interfacial Characterization of Diffusion Bonded Monolithic and Fiber Bonded Silicon Carbide Ceramics, *Proceedings of ICACC 2013*, (2013).

[16]H. Tsuda, S. Mori, M. C. Halbig, M. Singh, and R. Asthana, Diffusion Bonding and Interfacial Characterization of Sintered Fiber Bonded Silicon Carbide Ceramics Using Boron-Molybdenum Interlayers, *Proceedings of ICACC 2014*, (2014).

[17]M. C. Halbig, M. Singh, and R. Asthana, Diffusion Bonding of SiC Fiber-Bonded Ceramics using Ti/Mo and Ti/Cu Interlayers. *Ceramics International*, **41**, 2140–2149, (2015).

[18] T. Ozaki, Y. Hasegawa, H. Tsuda, S. Mori, M. C. Halbig, M. Singh and R. Asthana, TEM Analysis of Interfaces in Diffusion-Bonded Silicon Carbide Ceramics Joined Using Metallic Interlayers, *Ceramic Engineering and Science Proceedings (CESP), Proceedings of the 40th International Conference on Advanced Ceramics and Composites*, (2016).

[19]K. Suganuma, K. Nogi, Interface Structure Formed by Characteristic Reaction between α-SiC Single Crystal and Liquid Cu, *J. Japan Inst. Met. Mater.*, **59**(12), 1292-1298 (1995).

[20]M. Heuer,T. Buonassisi, A. A. Istratov, and M. D. Pickett, Transition metal interaction and Ni-Fe-Cu-Si phases in silicon, *J. Appl. Phys.* **101**, 123510 (2007).

[21]C.-Y. Wen, F. Spaepen, In Situ Electron Microscopy of the Phases of Cu_3Si, *Phil. Mag.*, **87**(35), 5581-5599 (2007).

NUMERICAL ANALYSIS OF INHOMOGENEOUS BEHAVIOR IN FRICTION STIR PROCESSING BY USING A NEW COUPLED METHOD OF MPS AND FEM

Hisashi Serizawa
Joining and Welding Research Institute, Osaka University
11-1 Mihogaoka, Ibaraki, Osaka 567-0047, Japan

Fumikazu Miyasaka
Graduate School of Engineering, Osaka University
2-1 Yamadaoka, Suita, Osaka 565-0871, Japan

ABSTRACT
    Friction stir processing (FSP) has been developed as one of the surface modification techniques based on the basic principles of friction stir welding (FSW). In FSP, the heat is generated due to the friction between the tool and the base material and it produces the residual stress as same as that in the welding process, where the transient temperature distributions shows the inhomogeneous behavior. In addition, the plastic flow might affect the residual stress. Therefore, in order to reveal the heterogeneous behavior in FSP, a new coupled method of moving particle semi-implicit (MPS) and finite element method (FEM) was developed and it was applied for simulating both thermal and mechanical transient behavior in FSW butt joint of two plates simply and precisely. The computational results suggested that the inhomogeneous transient temperature distributions could be precisely predicted by using this coupled method. Also, it was revealed that the longitudinal plastic strain near the tools can be predicted by assuming the boundary temperature which is the same as the annealing temperature in the welding.

INTRODUCTION
    Based on the basic principles of friction stir welding (FSW), friction stir processing (FSP) has been developed as one of the surface modification techniques and various surface composites produced by FSP, which are for example Al-SiC, Mg-CNT (carbon nanotube), Mg-SiC and so on, were reported[1-3]. Most of these researches were conducted experimentally by varying FSP conditions and evaluating the microstructural changes. Although the residual stress is one of the most important factors for estimating the mechanical properties, the residual stress produced in FSP has not been studied precisely because the influence of plastic flow in FSP on the residual stress is unclear. In addition, the heat is generated during FSP due to the friction between the tool and the base material.
    In order to reveal the thermal and mechanical behavior near the tool of FSW, many numerical studies have been conducted by employing finite element method (FEM)[4-7], finite different method (FDM)[8], computational fluid dynamics (CFD)[9], arbitrary Lagrangian-Eulerian (ALE) formulation[10,11] and so on. In addition, the particle methods such as the moving particle semi-implicit (MPS) method[12] and the smoothed-particle hydrodynamics (SPH) method[13] have been applied for simulating the flow behavior in FSW and the plastic flow can be demonstrated qualitatively[14]. However, the complex processes were required in the former methods and it is unreasonable to calculate a whole model of FSW joint or FSP surface composites by the particle methods due to the computational time.
    Because the inherent strain is the origin of welding distortion and residual stress based on the welding mechanics, Terasaki *et al.* studied about the inherent strain along the joining line produced through FSW experimentally[15]. The authors numerically examined the inherent strain in FSW by using FEM[7]. However, the effect of plastic flow on the inherent strain has not been studied. In addition, a combined method using both MPS and FEM has been developed, and the

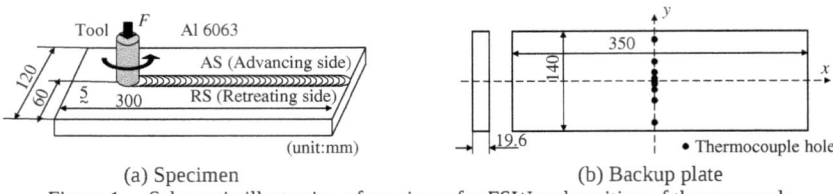

(a) Specimen   (b) Backup plate

Figure 1. Schematic illustration of specimen for FSW and position of thermocouple.

inherent strain can be predicted except for those near the joint interface, which is caused by a simplification in the definition of heat source and a neglect of the plastic flow[16].

In this research, in order to overcome these problems, a new coupled method of MPS and FEM is developed and the applicability of this coupled method is examined through the butt joining process of Al plates by FSW.   In concrete terms, the heterogeneous heat density distribution near FSW tool were obtained precisely through MPS analyses of the large deformation surrounding the rotational tool of FSW.   The temperature distributions in the whole model except for the area computed by MPS were calculated by FEM with the inhomogeneous heat source density, which could be computed by superimposing the heat generation distributions of MPS on the nodes of FEM.   Then, the elastic-plastic finite element analysis was conducted using the transient temperature obtained by this coupled method.   In addition, in order to include the influence of plastic flow, the concept of boundary temperature which is as same as the annealing temperature in the welding.

METHOD FOR ANALYSIS
Modeling for Analysis
The model for analysis is the butt joint of aluminum alloy plates (A6063-T5) whose size is 300 mm in length, 60 mm in width and 5 mm in thickness according to the experiment conducted by Terasaki et al. as shown in Figure 1 [15].   FSW tool has a cylindrical shoulder of 15.5 mm in diameter and a probe of 5.5 mm in diameter and the rotational and traveling speed of tool are 1750 rpm and 3.3 mm/s, respectively.   The compressive load for tool is 3450 N.   In the experiment, a backup plate was attached to the back of aluminum alloy plates using the clamp and the temperature changes at the bottom of aluminum alloy plates were measured by the thermocouples, which were directly attached to the back of aluminum alloy plates by drilling the holes in the backup plate as shown in Figure 1.

MPS Method
In this research, the plastic flow of aluminum alloy near the FSW tool was set to be approximately described by flow of highly viscous fluid.   Therefore, the governing equations for plastic flow motion are Navier-Stokes equation.   In addition, the viscosity of aluminum alloy during plastic flow is assumed to be described as follows because the distribution of viscosity depends on temperature and equivalent strain rate generally[17].

$$\eta = \frac{\sigma(\dot{\varepsilon}, T)}{3 \cdot \dot{\varepsilon}} \tag{1}$$

Where η, $\dot{\varepsilon}$, T and σ are viscosity, equivalent strain rate, temperature and equivalent flow stress, respectively.

Table 1. Material constants in Eq. (6) and (7) for A6063.

α [1/MPa]	ln A [1/s]	n	Q [kJ/mol]
0.04	22.5	5.385	141

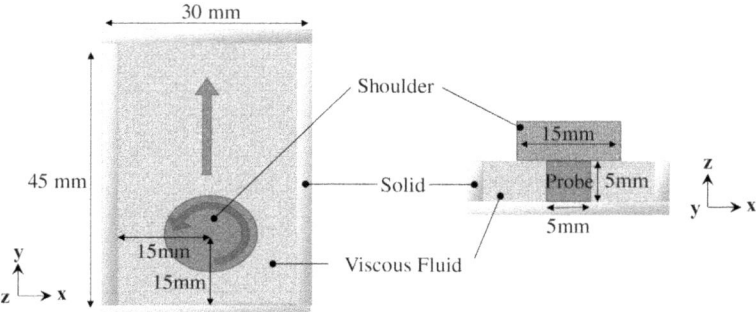

Figure 2. Schematic illustration of model for MPS method.

The equivalent strain rate is generally obtained by the velocity distribution. On the other hand, the value of equivalent flow stress at any temperature and any equivalent strain rate depends on the type of materials. In this study, the equivalent flow stress is approximated using the following equation, which was proposed by Sheppard et al.[18].

$$\sigma = \frac{1}{\alpha} ln\left[\left(\frac{Z}{A}\right)^{\frac{1}{n}} + \sqrt{\left(\frac{Z}{A}\right)^{\frac{2}{n}} + 1}\right] \qquad (2)$$

$$Z = \dot{\varepsilon} \cdot exp\left(\frac{Q}{RT}\right) \qquad (3)$$

Where, R is gas constant. α, A, n, Q are material constants, and their values were set as shown Table 1 according to the report of Sheppard et al..

During FSW, heat is generated due to plastic deformation. In this paper, the heat generation per unit volume is assumed to be that 90 % of the work for plastic deformation changes into heat[14].

Modeling for MPS Method

In order to reveal the temperature distributions near FSW tool precisely, a small volume near the FSW tool was employed for MPS. Figure 2 shows a schematic illustration of the model for MPS. The size of aluminum alloy plate studied was 45 mm in length, 30 mm in width and 5 mm in thickness. Although the diameters of tool shoulder and probe used in the experiment were 15.5 and 5.5 mm, respectively, they were assumed to be respectively 15 and 5 mm which are dependent on the initial distance between particles. As shown in Figure 2, the aluminum alloy plate was described as the highly viscous fluid and its edge was set as the rigid solid. Also, the FSW tool was modeled as the rigid solid. The shapes of shoulder and probe were set as a simple cylindrical though the shape of shoulder has little taper and the probe has thread in usual

Table 2. Material constants and boundary conditions for MPS analysis.

Time interval	[s]	0.0001
Initial minimum particle distance	[mm]	0.25
Number of particles		641564
Density of aluminum alloy	[kg/m^3]	2690
Specific heat of aluminum alloy	[J/kg·K]	900
Thermal conductivity of aluminum alloy	[W/m·K]	173
Density of tool	[kg/m^3]	7850
Specific heat of tool	[J/kg·K]	460
Thermal conductivity of tool	[W/m·K]	31
Traveling speed of tool	[mm/min]	200
Rotation speed of tool	[rpm]	1750
Tilt angle of tool	[degree]	0
Pitch of thread on probe	[mm]	0.7
Heat transfer coefficient on air	[W·m^{-2}·K^{-1}]	50
Room temperature	[°C]	20

experiment.   Therefore, the vertical velocity which may correspond to the effect of thread was added on the lateral face particles of probe.   In actual FSW process, there is the plug-in process of tool before traveling the tool in the plate.   However, in this research, the tool was set into the plate at the initial state, where the temperature of whole model was equal to room temperature and the initial velocity was zero, because only the heat generation behavior in traveling the FSW tool is important.   The conditions for MPS method are summarized into Table 2.

Plastic Flow of FSW
        In order to define the plastic behavior in MPS, the equivalent stress is computed from the equivalent strain rate $\dot{\varepsilon}$ and is compared with the yield stress of aluminum alloy.   From this comparison, it is revealed that the plastic behavior seems to start after the temperature of aluminum alloy decreases to a boundary temperature which is computed to 400 °C.

Combination of MPS and FEM
        From our previous studies using both MPS and FEM for FSW process, it was revealed that the transient temperature distributions except for area near the tool could be predicted using a homogeneous volumetric heat source[7,16].   In this research, since a precise and heterogeneous heat density distribution can be obtained as the result of MPS computations, a non-uniform heat input distribution of FEM is calculated from the heterogeneous heat density distribution through the integration method.   Figure 3 shows the finite element mesh divisions employed.   Since the backup plate was attached to the bottom of the aluminum alloy plates joined in the experiment, the heat transfer coefficient of back face was set to 0.0002 J/(s•mm^2•K) while those of other surfaces were set to 0.00002 J/(s•mm^2•K) according to our previous result[7].   Then, in order to improve the precision of temperature distributions near the tool, the result of MPS method was superposed on that of FEM.   Finally, by using the temperature distributions improved, the inherent strain was computed through the elastic-plastic finite element analysis.   Although the plates were clamped tightly in the experiment during FSW, only the rigid body motion was fixed in the mechanical analysis as shown in Figure 3 since the deformation of plate would be small and the jig was released after FSW.   In both thermal and mechanical analyses using FEM, the physical properties should be defined to be temperature dependent[7].   In addition, in order to demonstrate the

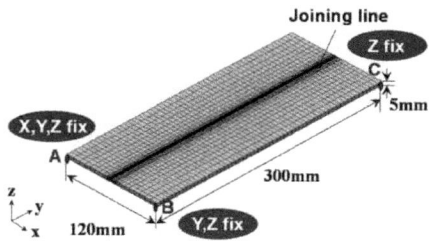

Figure 3.   Finite element model for finite element analysis.

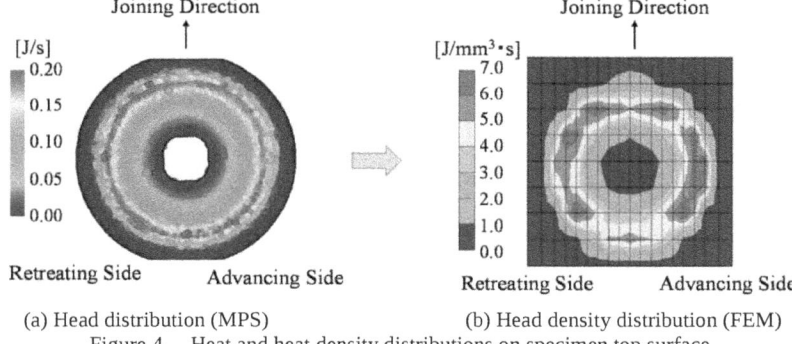

(a) Head distribution (MPS)   (b) Head density distribution (FEM)
Figure 4.   Heat and heat density distributions on specimen top surface.

influence of plastic flow during FSW, the equivalent plastic strain at the temperature higher than the boundary temperature is assumed to be zero as same as the annealing process in the welding.

RESULTS AND DISCUSSION
Results of MPS Method
        According to our previous studies using MPS method for FSW process, it was revealed that the heat generation area through the stirring of FSW tool can be defined as a circular truncated cone and the average and maximum temperatures of this heat generation area becomes to be almost constant after 2.0 s [16].   Figure 4(a) shows the heat generation distribution on the specimen surface computed after 2.0 s and the difference of heat generation between retreating and advancing sides is obviously demonstrated.   Because the computational grids for MPS are much finer than those for FEM, the heat density distribution was computed by superimposing the heat generation distribution of MPS on the nodes of FEM (Figure 4(b)) and it was employed as the heat input distribution for FEM.
        By moving the heterogeneous heat density distribution along the joining line, the transient temperature distributions were obtained through the thermal analysis of FEM.   Because the temperature distribution near the tool at 2.0 s computed by MPS seems to be quasi-stationary, the temperature at the finite element nodes within 5.0 mm from the center of tool is replaced by the temperature estimated from MPS.   The maximum temperature distributions on the central top and bottom surface transverse to the joining line are summarized in Figure 5, where the experimental

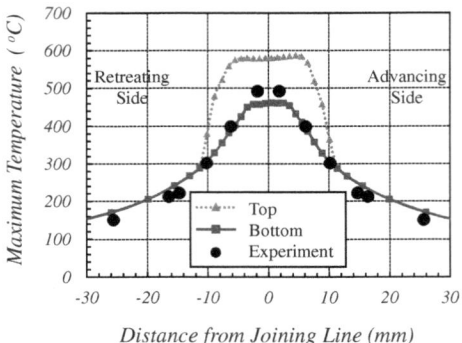

Figure 5. Maximum temperature distributions on central top and bottom surface transverse to joining line.

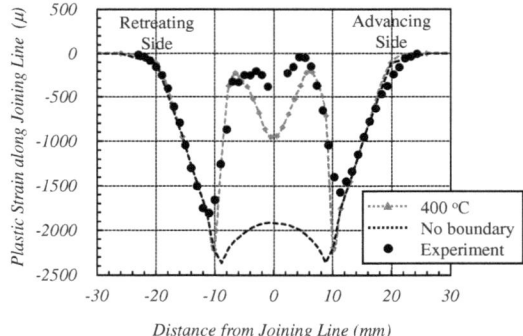

Figure 6. Boundary temperature effect of plastic strain distributions along joining line at central view transverse to joining line .

results on the bottom surface are also plotted in this figure.   From this figure, it is revealed that the coupled result of MPS and FEM shows a smooth curve and the computed result has a very good agreement with the experiment.   Therefore, it can be concluded that this coupled method seems to be a good tool for estimating the transient temperature distributions during FSW simply and precisely.

Mechanical Analysis of FEM
        In order to predict the inherent strain generated at FSW, the elastic-plastic finite element analysis was conducted using the transient temperature distributions obtained by the coupled method of MPS and FEM.   The average of longitudinal plastic (inherent) strains computed through the plate thickness are summarized in Figure 6, because the experimental result of Terasaki *et al.* is also the average of the plate thickness.   Where the equivalent plastic strain at the

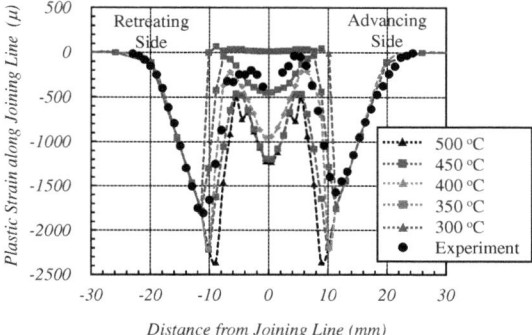

Figure 7. Influence of boundary temperature on plastic strain distributions along joining line at central view transverse to joining line .

temperature higher than 400°C is assumed to be zero. In addition, the calculation result without assuming the boundary temperature is plotted in this figure. From this figure, it is revealed that the numerical result assuming the boundary temperature has a fairly good agreement with the experimental result and the concept of boundary temperature seems to be reasonable for demonstrating the influence of plastic flow on the inherent strain.

Then, in order to examine the effect of boundary temperature on the inherent strain distributions, the longitudinal inherent strains are computed by varying the boundary temperature from 300 to 500 °C and the results are summarized in Figure 7. These results suggest that the absolute value of the longitudinal inherent strain near the tool decreased with decreasing the boundary temperature. In addition, it is found that when the boundary temperature is set to be 350 °C, the numerical result has a good agreement with the experiment. Because the plastic flow might enhance the recrystallization in the stir zone and the recovery in thermos-mechanical affected zone, the recrystallization and the recovery might reduce the equivalent plastic strain generated by the plastic flow and then the boundary temperature would be decreased apparently.

CONCLUSIONS

In order to estimate the residual stress produced in FSP, a new coupled method of MPS and FEM was developed and it was applied for simulating both thermal and mechanical transient behavior in FSW butt joint of two plates simply and precisely. The conclusions can be summarized as follows.
(1)   The inhomogeneous heat density distribution for thermal analysis of FEM can be obtained by superimposing the heterogeneous heat generation distribution of MPS on the nodes of FEM.
(2)   The maximum temperature distributions on the central bottom surface transverse to the joining line obtained by the coupled method shows a smooth curve and this result has a very good agreement with the experimental result. Therefore, this coupled method seems to be a good tool for estimating the inhomogeneous transient temperature distributions during FSW simply and precisely.
(3)   From the elastic-plastic finite element analysis using the temperature distributions obtained by the coupled method of MPS and FEM, the longitudinal plastic strain has a very good agreement with the experimental result by employing the concept to the boundary temperature.

REFERENCES

[1] R.S. Mishra, Z.Y. Ma and I. Charit, Friction Sir Processing : A Novel Technique for Fabrication of Surface Composite, *Materials Science and Engineering A*, **341**, 307-310 (2003).

[2] Y. Morisada, H. Fujii, T. Nagaoka and M. Fukusumi, MWCNTs/AZ31 Surface Composites Fabricated by Friction Stir Processing, *Materials Science and Engineering A*, **419**, 344-348 (2006).

[3] Y. Morisada, H. Fujii, T. Nagaoka and M. Fukusumi, Effect of Friction Stir Processing with SiC Particles on Microstructure and Hardness of AZ31, *Materials Science and Engineering A*, **433**, 50-54 (2006).

[4] S.R. Rajesh, H.S. Bang, W.S. Chang, H.J. Kim, H.S. Bang, C.I. Oh and J.S. Chu, Numerical Determination of Residual Stress in Friction Stir Weld Using 3D-Analytical Model of Stir Zone, *Journal of Materials Processing Technology*, **187-188**, 224-226 (2007).

[5] P. Ulysse, Three-Dimensional Modeling of the Friction Stir-Welding Process, *International Journal of Machine Tools & Manufacture*, **42**, 1549-1557 (2002).

[6] G. Buffa, J. Hua, R. Shivpuri and L Fratini, A Continuum Based FEM Model for Friction Stir Welding – Model Development, *Materials Science and Engineering A*, **419**, 389-396 (2006).

[7] H. Serizawa, J. Shimazaki and H. Murakawa, Numerical Study of Factors for Generating Inherent Strain in Friction Stir Welding, *Trends in Welding Research 2012, Proceedings of the 9th International Conference*, 922-929 (2013).

[8] M Song and R Kovacevic, Numerical and Experimental Study of the Heat Transfer Process in Friction Stir Welding, *Proceedings of the Institution of Mechanical Engineers, Part B: Journal of Engineering Manufacture*, **217** (1), 73-85 (2003).

[9] H. Atharifar, D. Lin and R. Kovacevic, Numerical and Experimental Investigations on the Loads Carried by the Tool During Friction Stir Welding, *Journal of Materials Engineering and Performance*, **18** (4), 339-350 (2009).

[10] H Schmidt and J Hattel, A Local Model for the Thermomechanical Conditions in Friction Stir Welding, *Modelling and Simulation in Materials Science and Engineering*, **13**, 77-93 (2005).

[11] S Guerdoux and L Fourment, A 3D Numerical Simulation of Different Phases of Friction Stir Welding, *Modelling and Simulation in Materials Science and Engineering*, **17**, 075001 (32pp) (2009).

[12] S. Koshizuka and Y. Oka, Moving-Particle Semi-implicit Method for Fragmentation of Incompressible Fluid, *Nuclear Science and Engineering*, **123**, 421-434 (1996).

[13] J.J. Monaghan, An Introduction to SPH, *Computer Physics Communications*, **48**, 89-96 (1988).

[14] G. Yoshikawa, F. Miyasaka, Y. Hirata, Y. Katayama and T. Fuse, Development of Numerical Simulation Model for FSW employing Particle Method, *Science and Technology of Welding and Joining*, **17** (4), 255-263 (2012).

[15] T. Terasaki and T. Akiyama, Mechanical Behavior of Joints in FSW : Residual Stress, Inherent Strain and Heat Input Generated By Friction Stir Welding, *Welding in the World*, **47** (11/12) 24-31 (2003).

[16] H. Serizawa and F. Miyasaka, New Combined Method of MPS and FEM for Simulating Friction Stir Processing, *Ceramic Engineering and Science Proceedings*, **36** [2], 27-36 (2015).

[17] O.C. Zienkiewicz, P.C. Jain, and E. Onate, Flow of Solids During Forming and Extrusion: Some Aspects of Numerical Solutions, *International Journal of Solid and Structures*, **14** (1), 15-38 (1978).

[18] T. Sheppard and A. Jackson, Constitutive Equations for Use in Prediction of Flow Stress During Extrusion of Aluminum Alloys, *Materials Science and Technology*, **13** (3), 203-209 (1997).

Advanced Materials and Innovative Processing Ideas for the Industrial Root Technology

STUDY OF SHIELDING METHOD TO REDUCE LEAKAGE MAGNETIC FIELDS OF AN OPENING IN A MAGNETICALLY SHIELDED ROOM

H. SUGIYAMA[1] and K. KAMATA[2]

[1] Department of Mechanical and Electronic Control Systems Engineering, National Institute of Technology, Kagoshima College, KAGOSHIMA, JAPAN

[2] Department of Electronic Control Engineering, National Institute of Technology, Kagoshima College, KAGOSHIMA, JAPAN

ABSTRACT

Magnetic noise is generated by several common sources, including electrical transmission lines and the movement of magnetic materials through Earth's magnetic field. Several types of precision instruments may be adversely affected by magnetic noise, requiring them to be placed in a magnetically shielded room (MSR) to reduce magnetic noise. An MSR typically consists of multiple shielding layers of ferromagnetic material, with openings in the walls and the floor to allow for ventilation, wiring, and plumbing connections. However, magnetic noise can leak into the MSR through those openings, and anything inside can be influenced by the leaked magnetic flux density. Generally, a duct or a partition plate made of a ferromagnetic material is used to reduce magnetic flux density entering through room openings. In this study, we focus on the partition plate installed in the duct, and examine the effects of partition plate length and the number of plate divisions. The shape of the partition plate with high shielding effect was studied by 3-dimensional magnetic field analysis using the finite element method. As a result, we found that if the width of the partition plate cell is 50 mm or less, the leakage magnetic flux density can be efficiently reduced by setting the length of the partition plate to about 0.5 times the width of each cell.

INTRODUCTION

There are several sources of unwanted magnetic noise, such as electrical transmission lines, induction motors, and the movement of ferromagnetic material through the Earth's magnetic field. Several types of precision instruments may be adversely affected by magnetic noise. For example, the patterning precision of an electron beam lithography (EBL) system decreases when a source of magnetic noise is present[1]. Precision instruments susceptible to magnetic fields must be operated in a magnetically shielded room (MSR) to reduce the effects of magnetic noise[2, 3]. These rooms enable sensitive tools such as EBL systems and instruments for biomagnetic measurement to maintain high precision. A typical MSR consists of multiple shielding layers, with some openings installed in the walls and floor to allow ventilation, wiring, and plumbing connections[4]. However, magnetic noise can leak into the MSR from those openings, and anything inside can be influenced by the leaked magnetic flux density. The leaked magnetic flux density is changed by value and direction of it outside of the MSR. Generally, a duct made of a magnetic material is used to reduce the leaked magnetic flux density entering from the opening. Many studies have examined in detail connection methods and the length of the duct installed in the MSR consisting of multiple shielding layers[5]. It is also known that the magnetic resistance of the duct opening is reduced by dividing the inside of the duct opening by a partition plate made of a ferromagnetic material[5]. However, there do not appear to be published studies that explore the effects of the length and number of partition plates on magnetic shielding performance.

In the case of the electron beam patterning apparatus, since the size of an MSR increases with the size of the apparatus being shielded, it is inevitable to use multilayer construction to secure performance[6]. Moreover, many openings are necessary in an MSR to allow for room ventilation[3].

Conventional studies have shown that the optimum length of an MSR duct is 50% of the opening width, and the inside of the duct should be divided to reduce the magnetic flux density leakage into the MSR[5]. The purpose of the present study is to examine the most suitable shape of the partition plate to install in the opening of an MSR. A 3-dimensional magnetic field analysis was carried out using a finite element method[2]. This analysis determined the dependence of the magnetic flux density leaking from an opening on the length of the partition plate and the number of partitions present.

The validity of the analytical result was verified by experiment. Since a full scale MSR experiment requires a large space and may be quite expensive, the experimental verification addressed only the duct portion of the MSR. The most effective design of the partition plate that can reduce magnetic noise was determined in this study.

ANALYTICAL MODEL AND ANALYSIS METHOD
Analytical model

The analytical model consisted of a MSR design with a 3-layer structure, as shown in Figure 1. The depth (x), width (y) and height (z) of the MSR were 2,400 mm, 3,400 mm, and 3,000 mm, respectively. The distance between the inner and outer layers was 100 mm. The shielding material was μ metal with a thickness of 2 mm. Because the purpose was to examine the characteristics of the magnetic flux density leaked from an opening, the relative permeability was assumed to be 10,000 without considering nonlinearities[7]. In practice, many ventilation openings are necessary for any MSR[3]. The purpose of this study is to determine the optimal design of the partition plate designed to fill these openings. Therefore, in order to simplify the model, an opening with dimensions of 300 mm × 300 mm was positioned in the upper and lower parts of the model MSR's wall. The partition plates in the opening and the duct used μ metal with a thickness of 1 mm. The length of the duct was set at one half the width of the opening, or 150 mm.

Four different partition plate lengths were examined, as shown in Figure 2. The lengths were selected to be fractions of the duct length, i.e., 100% (150 mm), 75% (112.5 mm), 50% (75 mm), and 25% (37.5 mm) of the length of the duct. An air gap of 10 mm was provided between

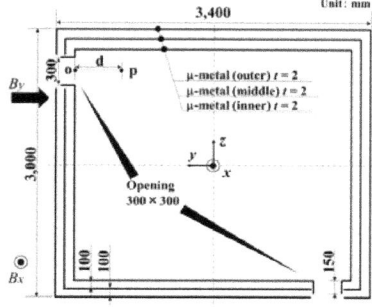

Figure 1. The model of the MSR.

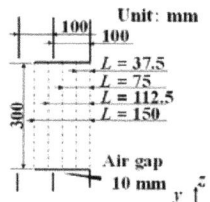

Figure 2. Partition plate lengths examined in this study.

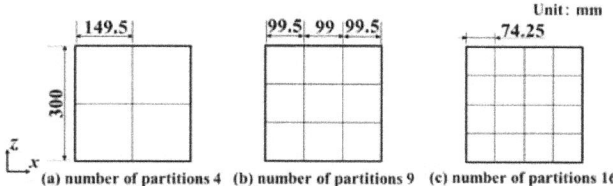

(a) number of partitions 4   (b) number of partitions 9   (c) number of partitions 16

Figure 3. The number of divisions of the partition plates.

the duct and the outer layer to prevent direct. Figure 3 shows the different ways in which the partition plates installed in the MSR duct were subdivided. Models (a), (b), and (c) were respectively divided into 4, 9, and 16 partitions, as shown in Figure 3.

Analysis method

The MSR model was analyzed using a 3-dimensional magnetic field analysis method incorporating finite element analysis. The basic equation for three-dimensional linear static magnetic field analysis is

$$\text{rot}(\text{vrot}\mathbf{A}) = \mathbf{J_0} \quad (1)$$

where, $\mathbf{A}$ is the magnetic vector potential, ν is magnetic reluctivity, and $\mathbf{J_0}$ is magnetization current density.

Since the source of the magnetic noise exists outside the MSR, the analysis area was approximately 5 times the size of the outermost dimensions. This model was divided into approximately 950,000 sections for numerical calculation. The model used a fine numerical mesh in the regions where the change of the magnetic flux density was expected to be large, such as the corner part of the MSR and near the opening. Conversely, the mesh was coarser in the parts of the modeled space where the change was expected to be relatively small, such as outside the MSR and in the room interior. Magnetic noise outside the MSR was assumed to consist of an extremely low frequency magnetic field of less than 1 Hz, such as those caused by the movement of trains[8] and automobiles[9]. A uniform magnetic field of 5 μT (DC) was applied to the opening in the horizontal direction (x) or the vertical direction (y). Also, since the DC magnetic field was applied, the shielding effect of eddy currents was not considered. Under this condition, the x, y, and z components of the magnetic flux density inside the MSR were calculated and evaluated as the absolute value of flux density.

ANALYSIS RESULTS
Shielding performance in the horizontal direction (x)

Figure 4, 5, and 6 show the comparison of the magnetic flux density along the x axis from an opening for partition plates with 4, 9, and 16 partitions and various lengths. In addition to the four models with different partition plate lengths as shown in Figure 2 and Figure 3, Figure 4 also shows the cases where the ducts were made without connecting the partition plates ($L = 0$ mm). The distance d from the wall of the inner layer of the MSR is plotted along the horizontal axis, as shown in Figure 1. Moreover, the leaked magnetic flux density is plotted along the vertical axis. Normally, the magnetic flux density on the central axis should be determined and compared. However, in this study, since the number of divisions by the partition plate was varied, the arrangement of the partition plate on the plane of the opening (the xz plane) was different as shown in Figure 3. Therefore, it is inappropriate to use the magnetic flux density on the center axis of the opening for comparison. In this study, the magnetic flux densities in all the divided elements in the

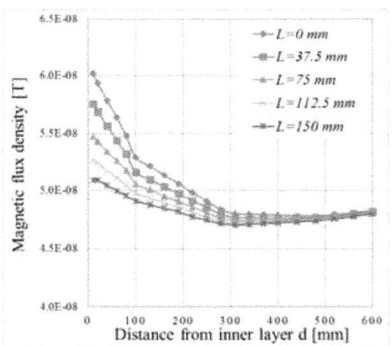
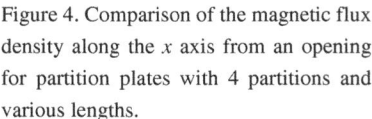

Figure 4. Comparison of the magnetic flux density along the x axis from an opening for partition plates with 4 partitions and various lengths.

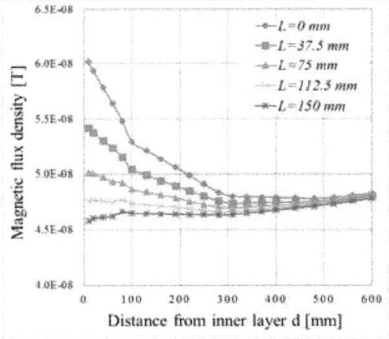

Figure 5. Comparison of the magnetic flux density along the x axis from an opening for partition plates with 9 partitions and various lengths.

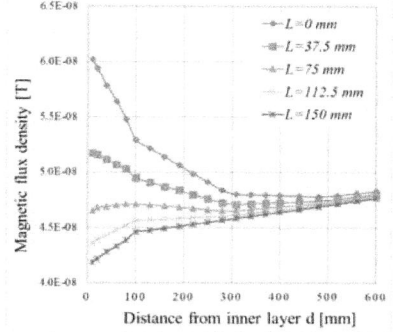

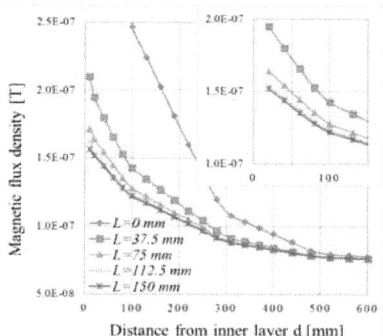

Figure 6. Comparison of the magnetic flux density along the x axis from an opening for partition plates with 16 partitions and various lengths.

Figure 7. Magnetic flux density with distance along the y axis for plates with 4 partitions.

opening plane (the xz plane) were calculated and the average value was determined.

The change of magnetic flux density with distance from the opening differed for different numbers of plate divisions. The magnetic flux density in the vicinity of the opening decreased with plates having a greater number of divisions; flux density also decreased with increasing partition plate length. For modelled plates with 16 partitions, the flux density also increased with distance from the opening. On the other hand, in the models using partition plates with 4 partitions (Figure 4), the magnetic flux density in the vicinity of the opening increased with shorter partition plates. Moreover, the leakage flux tended to decrease with distance from the opening. The difference of each magnetic flux density became larger near the opening ($d = 10$ mm). However, the difference of each magnetic flux density became smaller with distance from the opening.

Shielding performance in the vertical direction (y)

Figure 7 shows the change in magnetic flux density with distance from an opening on the y axis. Results are shown for plates using 4 partitions over the range of partition lengths. An enlarged view of the data near the opening ($d = 0 \sim 150$ mm) is shown as an inset on the graph, highlighting the shielding performance in the vertical direction. The magnetic flux density decreased with distance from the opening, as shown in Figure 7. For distances less than 280 mm, the magnetic flux density reached a minimum when $L = 150$ mm. However, for $d = 280$ mm or more, the model plate with $L = 112.5$ mm provided the most flux attenuation, and the difference of magnetic flux density between this model and the model with $L = 150$ mm was about 0.2 nT. If the length of the partition plate was 112.5 mm or more, the analysis suggests that the magnetic flux density cannot be reduced efficiently. Therefore, a partition plate of $L = 112.5$ mm could effectively reduce magnetic flux density if the duct is divided into 4 partitions.

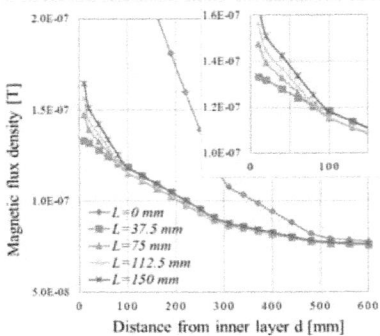

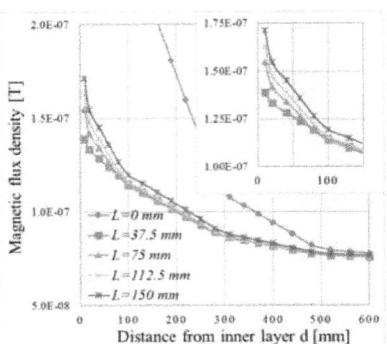

Figure 8. Magnetic flux density with distance along the y axis for plates with 9 partitions.

Figure 9. Magnetic flux density with distance along the y axis for plates with 16 partitions.

Figure 8 shows a similar comparison of model results as Figure 7, this time using plates with 9 partitions. The magnetic flux density in the vicinity of the opening ($d = 10$ mm) was smaller with shorter partition plate lengths, as shown in Figure 8. For distances within 80 mm of the opening, the magnetic flux density was smallest for $L = 37.5$ mm. However, when $d = 80$ mm or more, the model with $L = 75$ mm provided the smallest flux density, and the difference of magnetic flux density between this model and the model with $L = 37.5$ mm was about 2 nT. Therefore, it is considered that the length of the partition plate should be between 37.5 mm and 75 mm.

Figure 9 shows the magnetic flux density with distance from an opening along the y axis, for plates with 16 partitions and a range of lengths. For distances less than 130 mm, the magnetic flux density was the smallest when $L = 37.5$ mm. However, when $d = 130$ mm or more, the modelled flux density with $L = 75$ mm was the smallest, and the difference of magnetic flux density between this model and the model with $L = 37.5$ mm was about 0.4 nT. Moreover, the difference between $L = 75$ mm and $L = 37.5$ mm was quite small. Therefore, a partition plate with $L = 37.5$ mm gave optimum shielding performance when dividing the duct into 16 partitions.

Shielding performance by plates with 9 partitions in the vertical direction (y)

According to the modelled result of plates with nine divisions, the length of the partition plate should lie in the range $37.5 < L < 75$ mm. The results for plate length $L = 37.5$ mm were optimum when the duct was divided into 16 partitions. The width of each cell on a partition plate with 16 partitions is 74.25 mm, so for this case, the length of the partition plate was about 50% of the cell width. Therefore, it is considered that the optimal length of the partition plate is 50% of the cell width for plates divided into 16 partitions. When dividing the duct into 9 partitions, the model results suggest that $L = 50$ mm is the optimum length of the partition plate, because the width of the divided cell is about 100 mm.

Figure 10 shows a comparison similar to that of Figure 9, using a plate with 9 partitions. Because the model results for 9-partition plates shown above suggest that the length of the partition plate should be between 37.5 mm and 75 mm, Figure 10 shows only the results for $L = 37.5$, 50, and 75 mm.

At distances less than 80 mm, the magnetic flux density was smallest when $L = 37.5$ mm. The difference of magnetic flux density between this model and the model with $L = 50$ mm was about 2.2 nT. On the other hand, the difference of magnetic flux density between this model and the model with $L = 75$ mm was about 3.7 nT. When $d = 80$ mm or more, the model with $L = 75$ mm yielded the smallest leakage flux density, and the difference of flux density between this model and the model with $L = 50$ mm was about 0.7 nT. On the other hand, the difference of magnetic flux density between this model and the model with $L = 37.5$ mm was about 1.5 nT. These results suggest that the leakage flux density can be efficiently reduced when the length of the partition plate is 50 mm.

CONSIDERATION OF ANALYSIS RESULTS
Plates with 4 and 16 partitions

According to the model results obtained in the vertical direction (y), when the duct was divided into 4 partitions, the shielding efficiency was high in the model with the partition plate length of 112.5 mm. In the case of plates with 9 and 16 divisions, the shielding efficiency was high for models with a plate length approximately 50% of the width of each partition cell. To analyze the factors contributing to these results, Figures 11 and 12 show the modelled **B** field vector distribution for plates with 4 and 16 divisions, for the case of applying a uniform magnetic field in the vertical direction (y).

Figure 11 shows the magnetic flux density distribution inside the duct opening as a vector field. The diagrams in Figs. 11(a) through 11(d) give the vector distributions for modelled plates

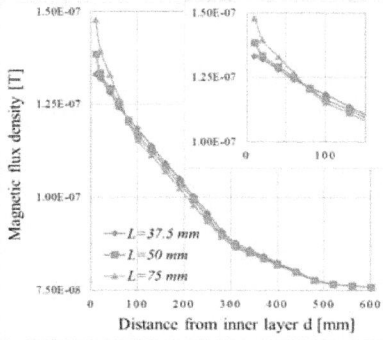

Figure 10. Magnetic flux density on the y axis for plates with 9 partitions.

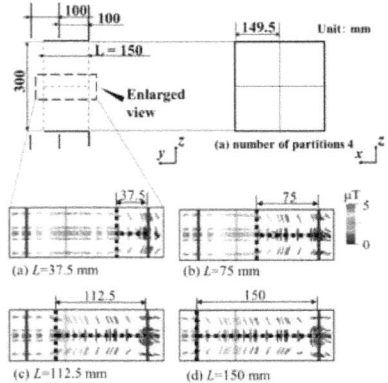

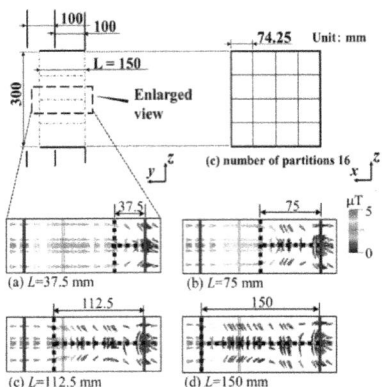

Figure 11. Distribution of magnetic flux density on the *y* axis: comparison of the vector distributions for 4-partition plates of four lengths.

Figure 12. Distribution of magnetic flux density on the *y* axis: comparison of the vector distributions for 16-partition plates of four lengths.

with 4 partitions at four lengths from the opening. The vector distribution diagrams show an enlarged view at the opening center in the *yz*-plane; gray bold lines show both ends of the duct and black dotted lines show the partition plates. The leakage magnetic field appeared to collect in the direction of the partition plate for longer plate lengths. In the models with $L = 37.5$ and 75 mm, the partition plates were too short to collect the magnetic field effectively, compared to the model results with $L = 112.5$ mm. In the model with $L = 150$ mm, the magnetic field around the partition plate at the MSR side duct end was not collected by the partition plate, and the magnetic field was spread inside the MSR. This is considered to be due to the phenomenon that the magnetic field is locally strengthened by the magnetic material at the boundary between the magnetic material and air. The longer partition plates gathered more leakage magnetic flux, however the amount of spreading from the end of the partition plate increased. Therefore, if the length of the partition plate was 150 mm or more, the leakage magnetic field was found to increase. In the model with $L = 112.5$ mm, the magnetic field seems to be spreading inside the MSR, however it was smaller than that for the model with $L = 150$ mm. Considering the analysis results, the vector distribution diagram, and overall material costs, a plate length of 112.5 mm was found to be the most suitable for plates divided into four partitions.

Figure 12 shows a vector distribution diagram similar to that of Figure 11, for a plate with 16 partitions. In the model with length $L = 112.5$ and 150 mm, the magnetic field spread from parts of the partition plates. In the models with $L = 75$, 112.5, and 150 mm, the magnetic field around the partition plates at the MSR side duct end was not collected by the partition plate, and the

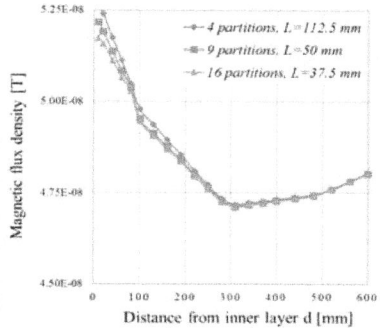

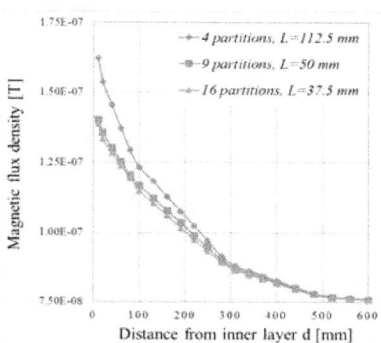

Figure 13. Comparison of the best magnetic flux attenuation performance along the x axis for each number of partitions.

Figure 14. Comparison of the best magnetic flux attenuation performance on the y axis for each number of partitions.

magnetic field was spread inside the MSR. On the other hand, in the model with $L = 37.5$ mm, the magnetic field was collected along all parts of the partition plate. Moreover, the leakage magnetic field did not spread, even at the end of the partition plate. Considering these results, the vector distribution diagram, and material costs, the length of 37.5 mm was judged to be most suitable magnetic shielding among plates divided into 16 partitions.

Comparison of shielding performance by plates with 4 to 16 divisions

In this study, the duct was divided into 4, 9, and 16 cells by the partition plate and the results analyzed. Models having the highest shielding performance are compared in Figures 13 and 14. Figure 13 shows the magnetic flux attenuation at the MSR wall opening in the horizontal direction (x). The responses of the highest performing plates with 4, 9, and 16 divisions are shown in the figure. The magnetic flux density was largest for plates of four partitions and $L = 112.5$ mm, as shown in Figure 13. On the other hand, the magnetic flux density became smallest for plates with 16 partitions and $L = 37.5$ mm.

Figure 14 shows a comparison similar to that of Figure 13, only for magnetic field attenuation performance in the vertical direction (y). The figures shows that magnetic shielding improved when the number of duct partitions was increased. Almost no difference was observed between plates with 9 partitions and 16 partitions. The magnetic flux density in the vertical direction (y) became smaller when the number of duct partitions was increased.

Therefore, considering the material and processing costs for producing magnetic shielding, the plate with length of 50 mm and 9 partitions appeared to be the most efficient approach to reducing the leaked magnetic flux density.

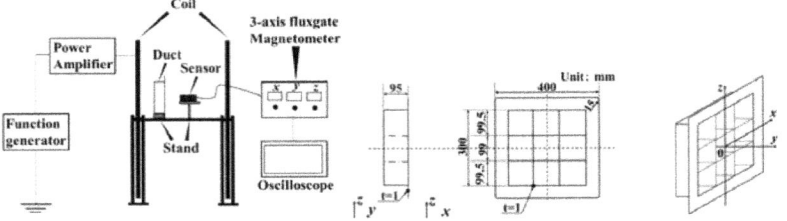

Figure 15. The measurement
system used in the experiment.

Figure 16. Duct dimensions.

Figure 17. Measurement
coordinate scheme.

EXPERIMENTAL METHOD

The validity of the analytical results above were verified by experiment. Since a full scale MSR experiment requires a large space and may be quite expensive to carry out, the experimental verification addressed only the duct portion of the MSR.

Figure 15 shows the configuration of the measuring device used in the experiment[10]. Two circular coils (Helmholtz coils) with a diameter of 1,500 mm and 14 turns were used to generate a uniform magnetic field assumed to represent the field from a distant source. A flat plate was installed at the center of the two circular coils, and a duct was placed on the flat plate to measure the magnetic field. AC magnetic noise (in the y axis direction) of 5 μT was generated with the circular coil using a function generator and power amplifier. Figure 16 shows the dimensions of the duct, a 1 mm thick silicon steel plate with a relative permeability of 3,000. Figure 17 shows the coordinate scheme of the measurements. The magnetic noise was measured along the x and y axes from the center axis of the duct. Measurements were taken along the y axis from -120 mm to 330 mm and along the x axis from the origin to 230 mm using the coordinate system defined in Figure 17. A 3-axis fluxgate magnetometer was used for the evaluation sensor. The magnetic noise detected by the magnetometer was monitored with an oscilloscope, the output waveform of which was recorded to obtain the shielding effect by the duct.

EXPERIMENTAL RESULT

Figure 18 shows a comparison of the magnetic flux density on the x axis predicted by the analysis and measured by experiment. Since the flux density was measured on the x axis, it was impossible to measure the region overlapping the duct. As the figure shows, the analysis results and the experimental results were similar. Since the magnetic field inside the cell divided by the partition plate was smaller than 5 μT, magnetic noise was reduced. Moreover, since the material of the duct concentrates the magnetic noise, the measurements showed that the magnetic flux density was larger within the duct than at the other measurement points.

Figure 19 shows a comparison of the magnetic flux density on the y axis predicted by the

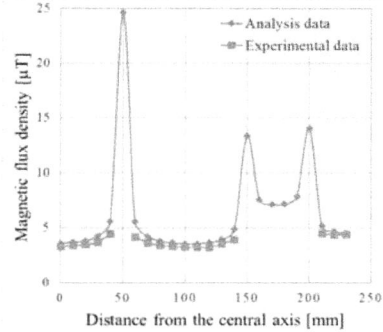
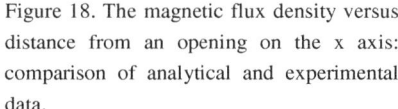
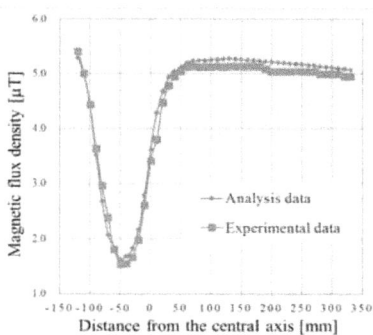

Figure 18. The magnetic flux density versus distance from an opening on the x axis: comparison of analytical and experimental data.

Figure 19. The magnetic flux density versus distance from an opening on the y axis: comparison of analytical and experimental data.

analysis and measured by experiment. The analysis data and the experimental data were similar. Both showed that the magnetic flux density inside the duct decreased toward the center of the duct and then increased. Magnetic flux density between 50 and 200 mm exceeded 5 µT, a value larger than the applied magnetic field. In typical use, the ducts would be connected to the MSR. Stray magnetic noise is intercepted by the duct and directed away from it, thereby reducing the magnetic noise. Therefore, there was a point in the duct where the magnetic flux density did not decrease below 5 µT. Based on these results, the analytical predictions appear to be valid.

CONCLUSIONS

The purpose of this study was to examine the most suitable shape for a partition plate to be installed in the opening of an MSR consisting of multiple shielding layers. In the opening with dimensions of 300 mm × 300 mm, the dependencies of the leakage magnetic field distribution on the length of the partition plate installed in the duct and on the number of plate divisions were analyzed using a three-dimensional magnetic field analysis and a finite element method. The study yielded the following results.

1. The magnetic flux density could be reduced by using a partition plate. Moreover, the shielding performance improved with increased number of divisions.

2. The shielding performance along the horizontal direction (x) near the opening was better with longer partition plates for any number of plate divisions. In addition, the magnetic flux density slightly increased from near the opening to the inside of the MSR as the number of plate divisions increased.

3. For shielding the magnetic flux in the vertical direction (y), it was found that the length of

the partition plate must be increased if the width of the cell divided by the partition plate is greater than 100 mm. On the other hand, the magnetic flux density could be efficiently reduced by setting the length of the partition plate to about 50% of the cell width when the width of the cell is 100 mm or less.

4. Considering not only the shielding performance but also the material and processing costs, the model in which the duct was divided into 9 cells by a partition plate 50 mm in length was able to reduce the leakage magnetic field most efficiently.

REFERENCE

[1] K. Yamazaki, K. Kato, K. Ono, H. Saegusa, K. Tokunaga, Y. Iida, S. Yamamoto, K. Ashiho, K. Fujiwara, N. Takahashi, "Analysis of magnetic disturbance due to buildings", *IEEE Trans. Magn.*, Vol.39, No.5, PP.3226-3228 (2003).

[2] K. Yamazaki, K. Kato, K. Fujiwara, "Effective combination of magnetic and conductive layers of magnetically shielded room", *IEEE Trans. Magn.*, Vol.36, No.5, PP.3649-3651 (2000).

[3] K. Yamazaki, "Magnetically shielded Cleanroom", Clean Technology, Vol.5, No.12, PP.76-79 (1995) (in Japanese).

[4] A. Mager, "The Berlin magnetically shielded room", Biomagnetism, PP.51-78, Walter de Gruyer &amp,Co. (1981).

[5] H. Fujita, Y. Hatsukade, K. Yamazaki, S. Tanaka, "Study of a Leakage Magnetic Field due to an Opening in a Multilayer type Magnetically Shielded Room using Finite Element Method", *J. Magn. Soc. Jpn.*, Vol.31, No.5, PP.416-420 (2007) (in Japanese).

[6] VAINO O. KELHA, JUSSI M. PUKKI, RISTO S. PELTONEN, AUVO J. PENTTINEN, RISTO J.ILMONIEMI, AND JAROM J. HEINO, "Design, Construction, and Performance of a Large Volume Magnetic Shield", *IEEE Trans. Magn.*, Vol.MAG-18, No.1, PP.260-270, (January 1982).

[7] T. Nakata, N. Takahashi, "Denkikougaku no yuugenyousohou ver.2", Morikita Shuppan, (1986) (in Japanese).

[8] K. Yamazaki, K. Kato, K. Kobayashi, Y. Uchikawa, Y. Kumagai, A. Haga, K. Fujiwara, "Characteristics and prediction of magnetic noise due to DC electric railcars for biomagnetic measurements", *IEEE Trans.FM*, Vol.37, No.4, PP.2884-2887 (2001) (in Japanese).

[9] K. Kamata, K. Yunokuchi, K. Yamazaki, K. Kato, T. Ueda, A. Haga, "Magnetic Field Fluctuation Due to Movement of Automobile", *IEEJ Trans.FM*, Vol.125, No.2, PP.92-98(2005) (in Japanese).

[10] JEITA : "Method for evaluation of shielding factor of magnetically shielded rooms for environmental magnetic noises at very low frequencies of 1 Hz or less", JEITA standard EM-4502, (2012) (in Japanese).

Materials for
Extreme Environments

LOW-TEMPERATURE SYNTHESIS OF HAFNIUM DIBORIDE POWDER VIA MAGNESIOTHERMIC REDUCTION IN MOLTEN SALT

Ke Bao[a], Joseph Massey[a], Juntong Huang[b], and Shaowei Zhang[a,]
[a] College of Engineering, Mathematics and Physical Sciences, University of Exeter, Exeter EX4 4QF, UK
[b] School of Materials Science and Engineering, Nanchang Hangkong University, Nanchang 330063, PR China

ABSTRACT

Hafnium diboride (HfB_2) powder was synthesized in molten NaCl, KCl or $MgCl_2$ at a relatively low temperature using HfO_2 and B_2O_3 powders as the main starting materials. The effects of processing parameters such as salt type, initial batch composition, and firing temperature/time on the synthesis were investigated, based on which the main mechanisms dominating the synthesis process proposed. Compared with NaCl and KCl, $MgCl_2$ facilitated the synthesis more effectively. Upon using appropriately excessive amounts of Mg and B_2O_3 to compensate for their evaporation losses at reaction temperatures, phase-pure HfB_2 superfine particles of 100-200 nm were synthesized after 4 h at 1000°C or 6 h at 950°C. The "dissolution-precipitation" mechanism was found to be more dominant in the overall molten salt synthesis process than the "template-growth" mechanism.

INTRODUCTION

Hafnium diboride (HfB_2) is a representative member in the so-called ultra-high-temperature ceramics (UHTCs) family. It has a high melting point (3380°C), high hardness/Young's modulus (28 GPa (Hv)/480 GPa), high thermal/electrical conductivity (104 W/m·K/9.1×10^6 S/m), high thermal neutron cross-section (^{10}B isotope: 3980 barn), and good oxidation and corrosion resistance.[1-4] Because of these, HfB_2 has been extensively investigated as a potential candidate material for several demanding applications, such as in hypersonic flights, atmospheric re-entry vehicles, rocket propulsion systems, plasma-arc electrodes, radiation shielding, and control rods of pressurized water reactors.[5-7]

Various techniques/methodologies have been developed to date to synthesize HfB_2 powders with various morphologies and sizes, including direct elemental reaction,[8-10] borothermic or carbo/borothermic reduction of sol-gel-derived precursors[11-14] or mechanically activated or mixed powders,[15-21] metallothermic reduction,[22] hydrothermal synthesis,[23] and high-energy ball milling (HEBM)-assisted annealing.[24] Unfortunately, these techniques suffered from one or more of the following problems: 1) use of expensive and/or hazardous raw materials (e.g., elemental boron, hafnium and $NaBH_4$), 2) requirement of specialty equipment/vessels, 3) requirement of high operation temperature and/or long processing time, 4) heavy agglomeration and contamination in final product, and 5) high production cost.

In addition to these techniques/methodologies, a molten salt synthesis (MSS) technique originally developed for complex oxides preparation has been modified and attempted recently to synthesize some metal boride materials. For example, Portehault et al.[25] synthesized several nanosized borides in a eutectic LiCl/KCl salt at relatively low temperatures using expensive and hazardous $NaBH_4$ as both boron source and reducing agent. Ran et al.[26] synthesized nanosized NbB_2 powder in NaCl/KCl using expensive elemental B as both boron source and reducing agent. The present authors synthesized submicron sized ZrB_2 and TiB_2 powders in molten chloride salts via magnesiothermic reduction of relatively cheap oxide-based raw materials.[27, 28]

In this work, such an MSS technique was further extended to synthesize high quality submicron sized HfB_2 powder at much lowered temperatures, also from relatively cheap oxide-based precursors. The effects of key processing factors such as salt type, starting batch composition, and heating temperature/time on the synthesis process were studied, based on which, synthesis conditions were optimized. In addition, based on the results, the dominant reaction mechanisms were discussed.

EXPERIMENTAL PROCEDURES

(1) Sample preparation
Raw materials used were mainly Sigma-Aldrich (Gillingham, UK) HfO_2 (98%), B_2O_3 (99.98%) and Mg ($\geq$99%) powders, and salts used were NaCl (>99%), KCl (>99%) and anhydrous $MgCl_2$ ($\geq$98%). HfO_2, Mg and B_2O_3 in the stoichiometric (corresponding to the overall Reaction (1)) or nonstoichiometric ratios (with excessive amounts of Mg and B_2O_3) were mixed, and further combined with 5 times weight of NaCl, KCl or $MgCl_2$, in a 250 ml mortar and pestle. The resultant batch powder was placed in a lid covered alumina crucible (L82 mm × W40 mm × H24 mm) and heated in Ar atmosphere in an alumina tube (I/D 60 mm) furnace at 3°C/min to a target temperature between 750 and 1000°C and held for 4-8 h, and then furnace-cooled to room temperature.
To remove the residual salts and byproduct MgO, the reacted mass was subjected to repeated-washing with hot distilled water and subsequent 2 h acid leaching with 1 M HCl. The remaining product powder was collected centrifugally and further rinsed with deionized water several times (until no Cl$^-$ was detected by $AgNO_3$ in the centrifugal liquid) before oven-drying overnight at 80°C.

$$HfO_2 + B_2O_3 + 5Mg = HfB_2 + 5MgO \qquad (1)$$

(2) Sample characterization
Powder X-ray diffraction (XRD) analysis (X-ray diffractometer, D8 Advance, Bruker, Germany) was used to identify crystalline phases in samples. XRD spectra were taken at 40 mA and 40 kV using CuKα radiation (λ = 1.5418 Å), at a scan rate of 2° (2θ)/min with a step size of 0.03°. ICDD cards used for identification are MgO (65-476), HfB_2 (38-1398), HfO_2 (34-104), $Mg_3B_2O_6$ (38-1475), and Hf (38-1478). A scanning electron microscope (SEM, Nova 600; FEI, Hillsboro, OR, USA) and a transmission electron microscope (TEM, JEM 2100; JEOL, Tokyo, Japan) were used to examine microstructures and morphologies of samples, and the linked energy-dispersive X-ray spectroscopy (EDS) was used to semi-quantitatively analyze elemental compositions of phases in samples.

RESULTS AND DISCUSSION

(1) HfB_2 formation in different salts
Figure 1 presents XRD results of samples of stoichiometric composition, after 4 h heating in different salts at 850°C (here and in the cases of Figs. 2-6 below, samples had only been subjected to water-washing). When NaCl was used (Figure 1a), both HfB_2 and MgO were identified, indicating the occurrence of Reaction (1). However, large amounts of unreacted HfO_2 still remained, along with some intermediate $Mg_3B_2O_6$ (whose formation will be discussed in Section 6 below), suggesting the low extents of magnesiothermic reduction and HfB_2 formation, and limited accelerating effect from NaCl in this case.
Upon replacing NaCl with KCl, HfB_2 and MgO peaks became higher, whereas those of HfO_2 and $Mg_3B_2O_6$ became lower (Figure 1b), indicating enhanced reaction extents and better accelerating effect of KCl than NaCl. On the other hand, upon replacing KCl with $MgCl_2$, HfB_2

and MgO peaks further increased whereas those of HfO_2 and $Mg_3B_2O_6$ further decreased (Figure 1c), indicating further enhanced reaction extents and an even better accelerating effect of $MgCl_2$. These results suggested that $MgCl_2$ was the most effective among the three salts in facilitating the magnesiothermic reduction as well as HfB_2 formation. This was similar to that found by our recent MSS work on TiB_2,[28] and could be explained based on the similar reasons: higher solubility levels of Mg and MgO in $MgCl_2$, than in KCl or NaCl. The higher solubility of Mg in $MgCl_2$ made the magnesiothemic reduction and subsequent HfB_2 formation in it proceed more rapidly than in the other two salts, and the higher solubility of MgO in $MgCl_2$ led to more effective removal of byproduct MgO from the barrier layer initially formed on the remaining unreacted solid reactants in it than in other two salts, avoiding otherwise significantly delayed/inhibited magnesiothermic reduction and HfB_2 formation (see Section 6 below).

(2) HfB_2 formation at different temperature

Figure 2 shows the effects of heating temperature on phase evolution in samples of stoichiometric composition after 4 h heating in $MgCl_2$. At 750°C, despite the evident formation of HfB_2, large amounts of HfO_2 still existed, along with some intermediate $Mg_3B_2O_6$ (Figure 2a). Upon increasing the temperature to 850 (Figure 2b/Figure 1c) and 950°C (Figure 2c), HfB_2 and MgO peaks increased while those of HfO_2 and $Mg_3B_2O_6$ decreased, revealing enhanced reaction extents. Nevertheless, upon further increasing the temperature to 1000°C (Figure 2d), HfO_2 and $Mg_3B_2O_6$ peaks started to increase adversely, indicating reduced extents of magnesiothermic reduction and HfB_2 formation. This was attributable to the evaporation loss of Mg at a relatively high temperature,[29-31] as similarly observed and discussed in MSS of ZrB_2 and TiB_2.[27, 28] Therefore, the effect of using excessive Mg on the HfB_2 formation was further investigated, as described and discussed next.

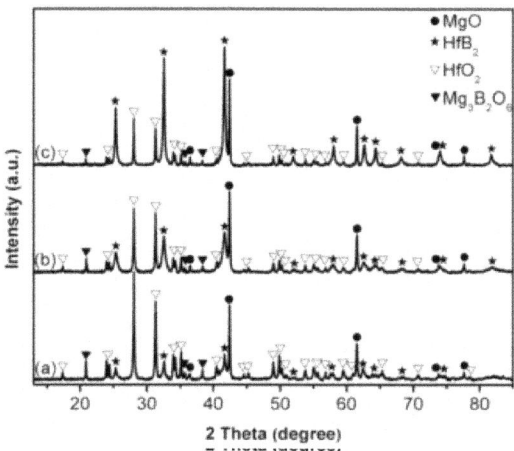

Figure 1. XRD patterns of samples of stoichiometric composition, after 4 h heating at 850°C in: a) NaCl, b) KCl, and c) $MgCl_2$, respectively.

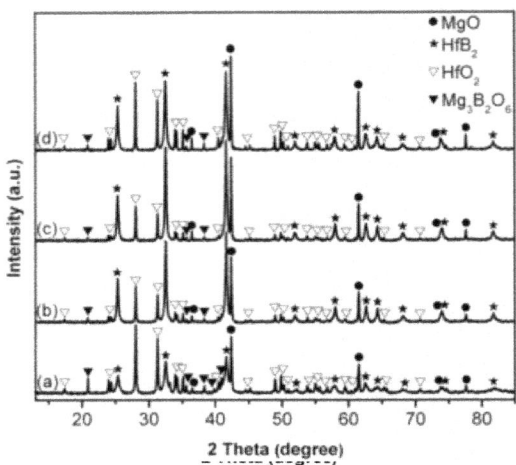

Figure 2. XRD patterns of samples of stoichiometric composition, after 4 h heating in MgCl$_2$ at: a) 750, b) 850, c) 950, and d) 1000°C, respectively.

(3) HfB$_2$ formation using excessive Mg and B$_2$O$_3$

　　Figure 3 illustrates the effect of using excessive Mg on phase evolution in samples resultant from 4 h heating in MgCl$_2$ at 1000°C. When 20 wt% excessive Mg was used (Figure 3b), HfO$_2$ and Mg$_3$B$_2$O$_6$ decreased substantially, compared to in the case of using the stoichiometric amount (Figure 3a), suggesting that use of excessive Mg led to enhanced HfO$_2$ and B$_2$O$_3$ reduction and HfB$_2$ formation. When 80 wt% excessive Mg was used (Figure 3c), Mg$_3$B$_2$O$_6$ disappeared and both HfB$_2$ and MgO increased. However, much residual HfO$_2$ still remained, due to lack of B$_2$O$_3$ caused by its evaporation loss.[12, 20] This was verified by the results shown in Figure 4.

　　As shown in Figure 4, along with 80 wt% excessive Mg, when 10 wt% excessive B$_2$O$_3$ was used, both HfB$_2$ and MgO increased, and only minor HfO$_2$ remained, and when 20 wt% excessive B$_2$O$_3$ was used, HfO$_2$ disappeared completely, and only HfB$_2$ and the byproduct MgO were formed, suggesting that it was necessary to use appropriately excessive B$_2$O$_3$, along with excessive Mg, to complete the magnesiothermic reduction and HfB$_2$ formation.

(4) Further optimization of synthesis conditions

　　Given in Figure 5 are XRD results of samples with respectively 80 and 20 wt% excessive Mg and B$_2$O$_3$, after heating in MgCl$_2$ at 950°C for different times. Increasing reaction time from 4 to 6 h (Figs. 5a&b) led to much increased HfB$_2$ and MgO, decreased HfO$_2$ and complete disappearance of Mg$_3$B$_2$O$_6$. Unfortunately, further extending the time to 8 h (Figure 5c) did not result in any further improvement in the reaction extents, and some HfO$_2$ still remained, which was related to the evaporation loss of B$_2$O$_3$ with prolonged firing. This was verified by XRD results shown in Figure 6 which reveals that in addition to 80 wt% excessive Mg, using 20 to 60 wt% excessive B$_2$O$_3$, led to significantly reduced HfO$_2$ and its final disappearance, and the final completion of the HfB$_2$ formation reaction (Figure 6c).

　　According to Figs. 4c&6c and the discussion above, the HfB$_2$ formation reaction could be completed in MgCl$_2$ after 4 h at 1000°C using 80 wt% excessive Mg and 20 wt % excessive B$_2$O$_3$ or 6 h at 950°C using 80 wt% excessive Mg and 60 wt% excessive B$_2$O$_3$. In both cases, only HfB₂ was formed along with the byproduct MgO. So after leaching out the MgO with HCl, phase-pure HfB₂ powders were obtained (Figure 7).

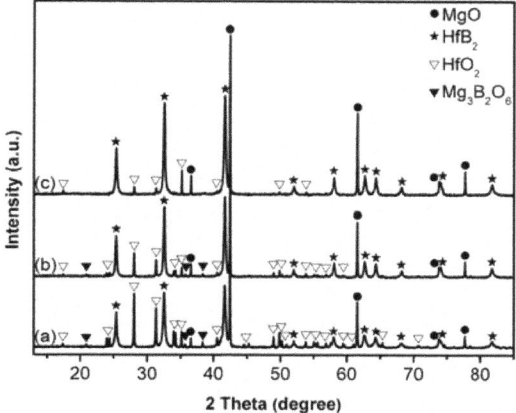

Figure 3. XRD patterns of samples after 4 h heating at 1000°C in MgCl₂ using: a) 0, b) 20, and c) 80 wt% excessive Mg, respectively.

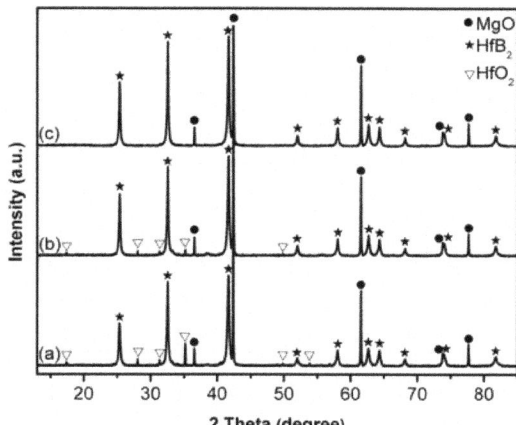

Figure 4. XRD patterns of samples resultant from 4 h heating in MgCl₂ at 1000°C, using 80 wt% excessive Mg and respectively: a) 0, b) 10, and c) 20 wt% excessive B₂O₃.

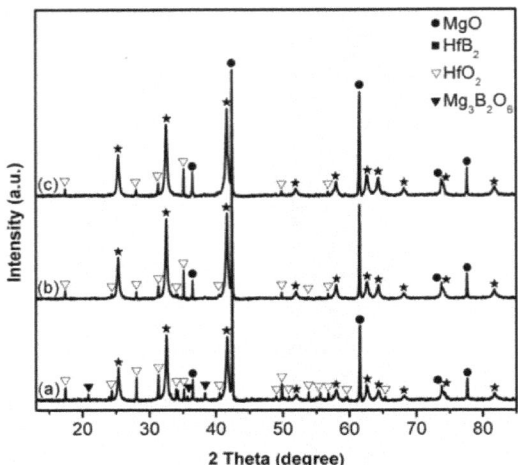

Figure 5. XRD patterns of samples resultant from heating the batch powders containing respectively 80 and 20 wt% excessive Mg and B_2O_3 in $MgCl_2$ at 950°C for a) 4, b) 6, and c) 8 h, respectively.

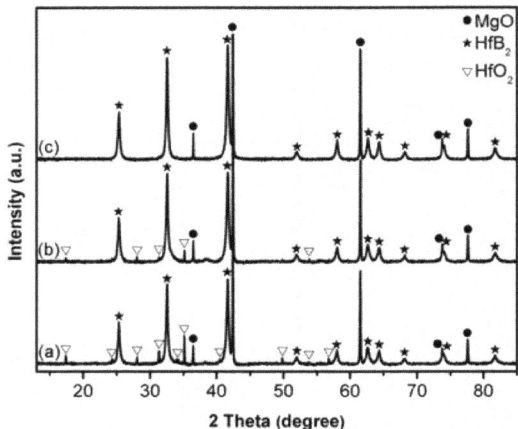

Figure 6. XRD patterns of samples resultant from 6 h heating in $MgCl_2$ at 950°C, using 80 wt% excessive Mg and a) 20, b) 40, and c) 60 wt% excessive B_2O_3, respectively.

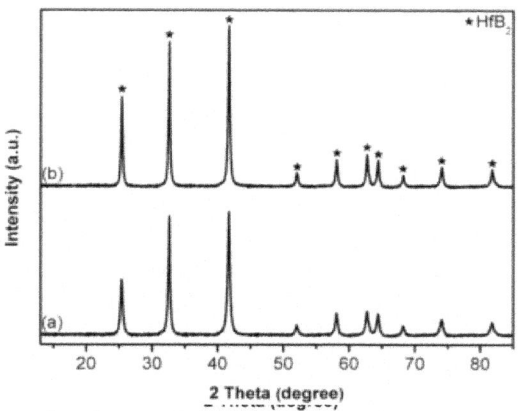

Figure 7. XRD patterns of product samples resultant from, a) 6 h heating at 950°C using respectively 80 and 60 wt% excessive Mg and B_2O_3, and b) 4 h at 1000°C using respectively 80 and 20 wt% excessive Mg and B_2O_3 (after water washing and subsequent acid leaching).

(5) Microstructure of HfB_2 product powder prepared under optimal conditions

Figure 8 displays together SEM and TEM images and EDS results of the HfB_2 product powders whose XRD patterns are given in Figure 7. Spheroidal HfB_2 particles with an average size of about 100 nm were resulted from 6 h heating at 950°C (Figure 8a), whereas angular particles with a larger average size of 100-200 nm resulted from 4 h heating at 1000°C (Figure 8b). Such morphology and size differences can be seen more clearly from their corresponding TEM images (Figs. 8c&d). These results indicated that synthesis temperature had greater effect than reaction time, on morphology and size of HfB_2 particle. To prepare HfB_2 particles with fine sizes, a lower synthesis temperature (in the present case, 950°C) should be preferred. As for the purity level of product powder, EDS only detected Hf and B along with tiny O contamination (the Cu and C peaks were from the carbon film/Cu grid used for TEM) in both cases, further confirming the formation of phase-pure HfB_2, as already confirmed earlier by XRD (Figure 7).

(6) Reaction mechanisms

Reaction (1) was used to indicate the overall reaction process. However, several individual reaction steps had been most likely involved, which, based on Figs. 1-8, could be considered as follows.

$$B_2O_3 \text{ (melt)} + 3Mg \text{ (dissolved in the salt)} = 2B + 3MgO \qquad (2)$$

$$HfO_2 + 2Mg \text{ (dissolved in the salt)} = Hf + 2MgO \qquad (3)$$

The firing temperature used to investigate the effect of salt type on synthesis was 850°C (Figure 1) which is above the melting points of the three chloride salts. Hence, at a test temperature, the chloride salt used would initially melt, forming a desired liquid medium. In this liquid medium, Mg partially dissolved[32] and diffused through it to the two reactants, B_2O_3 and HfO_2, and subsequently reduced them correspondingly to B and Hf (Reaction (2) and (3)). In the case of Reaction (2), the reactant B_2O_3 was in a liquid state (melted due to its low melting point: ~450°C[17]), thus Reaction (2) was essentially a liquid-liquid reaction. In this case, B resulted would not retain morphologies and sizes of the original B_2O_3 or Mg. In the case of Reaction (3), however, the reactant HfO_2 could be in a solid and/or a liquid state (depending on its solubility in the molten chloride salt used). Unfortunately, the exact solubility values of HfO_2 in the three molten salts are not available from literature. Nevertheless, our recent work on MSS of Hf from magnesiothermic reduction of HfO_2 (unpublished) found that HfO_2 did exhibit some solubility in the molten chloride salts. Therefore, for Reaction (3), two parallel reactions might occur simultaneously, as indicated by Reactions (4) and (5).

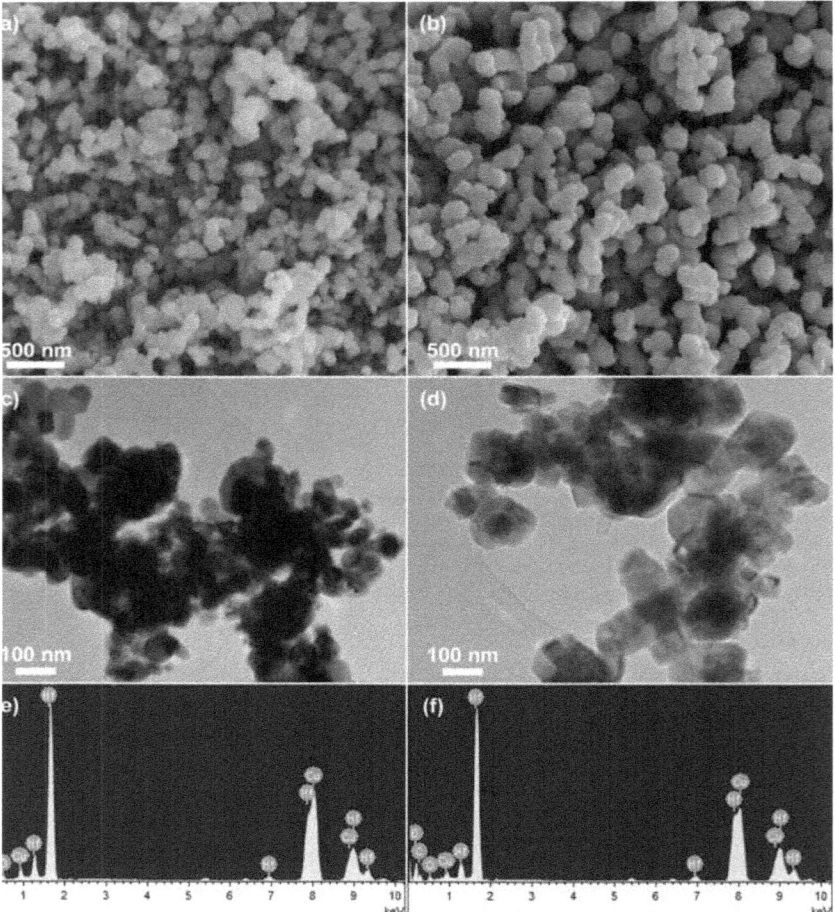

Figure 8. SEM/TEM images and EDS of the HfB_2 product powders whose XRD patterns are shown in Figure 7: (a, c and e) 6 h at 950°C and (b, d and f) 4 h at 1000°C.

$$\text{HfO}_2 \text{ (solid)} + 2\text{Mg (dissolved in the salt)} = \text{Hf} + 2\text{MgO} \qquad (4)$$

$$\text{HfO}_2 \text{ (dissolved in the salt)} + 2\text{Mg (dissolved in the salt)} = \text{Hf} + 2\text{MgO} \qquad (5)$$

In the case of Reaction (4), after initial reaction, a protective barrier layer composed of Hf and MgO would be formed on unreduced solid HfO_2, causing some delay in the further reaction. Fortunately, MgO also has "high" solubility in chloride salts, in particular in MgCl_2,[33, 34] so MgO in the initially formed barrier layer would be removed via its dissolution in the salt, making the barrier layer less continuous and less "protective", and avoiding otherwise significant delay in Reaction (4). Nevertheless, our recent work on MSS of Hf mentioned above revealed that morphologies and sizes of Hf particles prepared under the same firing conditions, from magnesiothermic reduction of HfO_2 in chloride salts (e.g., MgCl_2) appeared to be quite different from those of raw material HfO_2 (Figure 9), implying that in the present work, Reaction (5) controlled by the "dissolution-precipitation" mechanism was much more dominant than Reaction (4) controlled by the "template-growth" mechanism.

On the other hand, upon the formation of MgO from the magnesiothermic reduction (Reactions (2)-(3)), it would also react with some unreduced B_2O_3 to form intermediate $\text{Mg}_3\text{B}_2\text{O}_6$ (Reaction (6), Figs. 1-3&5) which is very difficult to eliminate by acid leaching.[35, 36] Interestingly, due to the presence of a molten salt medium, this intermediate phase could be well dispersed in the salt, and thus readily reduced to B and acid-leachable MgO, according to Reaction (7). This, in addition to other advantages such as low synthesis temperature and fine size/better dispersion of final product powder, is an "unexpected" advantage of the MSS technique over other conventional magnesiothermic reduction techniques reported previously (see "Introduction" section above).

$$\text{B}_2\text{O}_3 \text{ (melt)} + 3\text{MgO} = \text{Mg}_3\text{B}_2\text{O}_6 \qquad (6)$$

$$\text{Mg}_3\text{B}_2\text{O}_6 + 3\text{Mg (dissolved in the salt)} = 6\text{MgO} + 2\text{B} \qquad (7)$$

Similarly to the case of Reaction (4), Reaction (7) also would not be delayed much by the barrier layer possibly formed on $\text{Mg}_3\text{B}_2\text{O}_6$, as the MgO in the barrier layer also would be removed via its dissolution in the molten salts.

Our recent studies on MSS of ZrB_2 and TiB_2 powders,[27, 28] and B_4C and HfC coatings on carbon fibers (unpublished), also confirmed that both B and Hf had at least some solubility in the chloride salts (though the actual solubility seemed to be very small). So B resultant from Reactions (2) and (7) and Hf resultant from Reaction (3) (i.e. Reactions (4) and (5)) would partially dissolve in the molten salt medium, diffuse through it and finally react with each other to form HfB_2 via the "dissolution-precipitation" mechanism (Reaction (8)).

$$\text{Hf (dissolved in the salt)} + 2\text{B (dissolved in the salt)} = \text{HfB}_2 \qquad (8)$$

Apart from this, based on the findings from the work on borothermic reduction of HfO_2,[17] reduction of HfO_2 by B at the test temperatures was thermodynamically favorable. Therefore, B formed from Reactions (2) and (7) also would partially dissolve in the salt and react directly with some HfO_2 dissolved in the salt and unreduced solid HfO_2 to form additional HfB_2 according to Reactions (9) and (10) controlled respectively by the "dissolution-precipitation" mechanism and "template-growth" mechanism.

$$3\text{HfO}_2 \text{ (dissolved)} + 10\text{B} = 3\text{HfB}_2 + 2\text{B}_2\text{O}_3 \qquad (9)$$

$$3HfO_2 \text{ (solid)} + 10B = 3HfB_2 + 2B_2O_3 \qquad (10)$$

Based on the individual reaction steps indicated by Reactions (2)-(10) and discussed above, it can be considered that the "dissolution-precipitation" mechanism was more dominant in the whole molten salt synthesis process, than the "template-growth" mechanism.

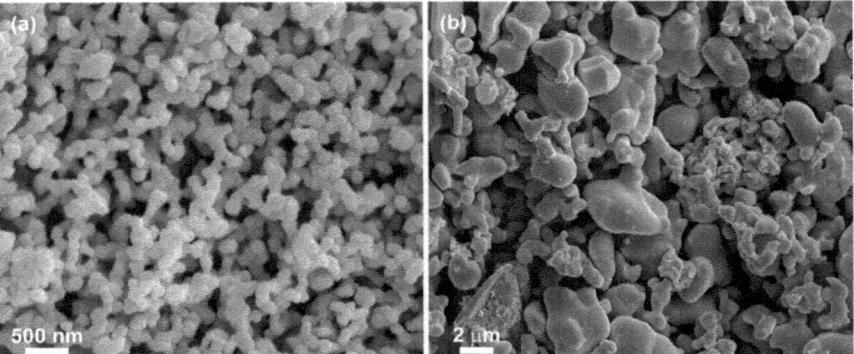

Figure 9. SEM images of a) as-received HfO_2 powder and b) Hf prepared via magnesiothermic reduction of HfO_2 (Reaction (3)) in $MgCl_2$ at 950°C for 6 h.

CONCLUSIONS

Submicron sized HfB_2 powder was successfully synthesized in chloride salts via magnesiothermic reduction of HfO_2 and B_2O_3. The effects of salt type, initial batch composition, and heating temperature/time, on the synthesis process were examined. Compared to NaCl and KCl, $MgCl_2$ accelerated the overall reaction process more effectively. Phase-pure HfB_2 powder of 100-200 nm could be prepared in $MgCl_2$ after 6 h heating at 950°C using respectively 80 and 60 wt% excessive Mg and B_2O_3 or 4 h heating at 1000°C using respectively 80 and 20 wt% excessive Mg and B_2O_3. These synthesis conditions were much milder than those required by many other techniques reported previously. The "dissolution-precipitation" mechanism was found to be more dominant in the overall MSS process, than the "template-growth" mechanism.

REFERENCES

[1] W. G. Fahrenholtz, G. E. Hilmas, I. G. Talmy, and J. A. Zaykoski, "Refractory Diborides of Zirconium and Hafnium," Journal of the American Ceramic Society, 90[5] 1347-64 (2007).
[2] E. Wuchina, M. Opeka, S. Causey, K. Buesking, J. Spain, A. Cull, J. Routbort, and F. Guitierrez-Mora, "Designing for ultrahigh-temperature applications: The mechanical and thermal properties of HfB_2, HfC x , HfN x and αHf(N)," Journal of Materials Science, 39[19] 5939-49 (2004).
[3] M. M. Opeka, I. G. Talmy, and J. A. Zaykoski, "Oxidation-based materials selection for 2000°C + hypersonic aerosurfaces: Theoretical considerations and historical experience," Journal of Materials Science, 39[19] 5887-904 (2004).
[4] M. M. Nasseri, "The Investigation of Neutron Interactions with HfB_2 – A Simulation Study," Transactions of the Indian Ceramic Society, 74[3] 177-80 (2015).

[5] M. M. Opeka, I. G. Talmy, E. J. Wuchina, J. A. Zaykoski, and S. J. Causey, "Mechanical, Thermal, and Oxidation Properties of Refractory Hafnium and zirconium Compounds," Journal of the European Ceramic Society, 19[13–14] 2405-14 (1999).

[6] R. Savino, M. De Stefano Fumo, L. Silvestroni, and D. Sciti, "Arc-jet testing on HfB_2 and HfC-based ultra-high temperature ceramic materials," Journal of the European Ceramic Society, 28[9] 1899-907 (2008).

[7] M. M. Nasseri, "The behavior of HfB_2 at neutron irradiation: a simulation study," Radiation Effects and Defects in Solids, 171[3-4] 252-58 (2016).

[8] I. P. Borovinskaya, A. G. Merzhanov, N. P. Novikov, and A. K. Filonenko, "Gasless combustion of mixtures of powdered transition metals with boron," Combustion, Explosion and Shock Waves, 10[1] 2-10 (1974).

[9] Y. D. Blum, J. Marschall, D. Hui, B. Adair, and M. Vestel, "Hafnium Reactivity with Boron and Carbon Sources Under Non-Self-Propagating High-Temperature Synthesis Conditions," Journal of the American Ceramic Society, 91[5] 1481-88 (2008).

[10] S. E. Kravchenko, A. G. Burlakova, I. I. Korobov, Y. M. Shul'ga, N. N. Dremova, L. S. Volkova, G. V. Kalinnikov, S. P. Shilkin, and R. A. Andrievskii, "Preparation of hafnium diboride nanopowders in an anhydrous $Na_2B_4O_7$ ionic melt," Inorg Mater, 51[4] 380-83 (2015).

[11] H. Wang, S.-H. Lee, H.-D. Kim, and H.-C. Oh, "Synthesis of Ultrafine Hafnium Diboride Powders Using Solution-Based Processing and Spark Plasma Sintering," International Journal of Applied Ceramic Technology, 11[2] 359-63 (2014).

[12] S. Venugopal, E. E. Boakye, A. Paul, K. Keller, P. Mogilevsky, B. Vaidhyanathan, J. G. P. Binner, A. Katz, and P. M. Brown, "Sol–Gel Synthesis and Formation Mechanism of Ultrahigh Temperature Ceramic: HfB2," Journal of the American Ceramic Society, 97[1] 92-99 (2014).

[13] L. S. Walker and E. L. Corral, "Structural Influence on the Thermal Conversion of Self-Catalyzed HfB2/ZrB2 Sol–Gel Precursors by Rapid Ultrasonication of Oxychloride Hydrates," Journal of the American Ceramic Society, 97[2] 399-406 (2014).

[14] C. Yan, R. Liu, C. Zhang, and Y. Cao, "Synthesis of zirconium, hafnium and their ternary borides by a polymer complex route," Journal of Sol-Gel Science and Technology, 76[3] 686-92 (2015).

[15] N. Akçamlı, D. Ağaoğulları, Ö. Balcı, M. L. Öveçoğlu, and İ. Duman, "Synthesis of HfB_2 powders by mechanically activated borothermal reduction of $HfCl_4$," Ceramics International, 42[3] 3797-807 (2016).

[16] P. Peshev and G. Bliznakov, "On the borothermic preparation of titanium, zirconium and hafnium diborides," Journal of the Less Common Metals, 14[1] 23-32 (1968).

[17] W.-M. Guo, Z.-G. Yang, and G.-J. Zhang, "Synthesis of submicrometer HfB_2 powder and its densification," Materials Letters, 83 52-55 (2012).

[18] H. Wang, S.-H. Lee, and H.-D. Kim, "Nano-Hafnium Diboride Powders Synthesized Using a Spark Plasma Sintering Apparatus," Journal of the American Ceramic Society, 95[5] 1493-96 (2012).

[19] Z. Wang, X. Liu, B. Xu, and Z. Wu, "Fabrication and properties of HfB_2 ceramics based on micron and submicron HfB_2 powders synthesized via carbo/borothermal reduction of HfO2 with B4C and carbon," International Journal of Refractory Metals and Hard Materials, 51 130-36 (2015).

[20] D.-W. Ni, G.-J. Zhang, Y.-M. Kan, and P.-L. Wang, "Synthesis of Monodispersed Fine Hafnium Diboride Powders Using Carbo/Borothermal Reduction of Hafnium Dioxide," Journal of the American Ceramic Society, 91[8] 2709-12 (2008).

[21] J. K. Sonber, T. S. R. C. Murthy, C. Subramanian, S. Kumar, R. K. Fotedar, and A. K. Suri, "Investigations on synthesis of HfB_2 and development of a new composite with $TiSi_2$," International Journal of Refractory Metals and Hard Materials, 28[2] 201-10 (2010).

[22] T. Ohkubo, T. Ono, K. Nishiyama, S. Niwa, H. Sakai, M. Koishi, and M. Abe, "Preparation of ZrB_2 and HfB_2 by Metallothermic Reduction of ZrO_2, $ZrSiO_4$, and HfO_2," Journal of the Japan Society of Powder and Powder Metallurgy, 52[9] 664-69 (2005).

[23] L. Chen, Y. Gu, L. Shi, Z. Yang, J. Ma, and Y. Qian, "Synthesis and oxidation of nanocrystalline HfB_2," Journal of Alloys and Compounds, 368[1–2] 353-56 (2004).

[24] E. Barraud, S. Bégin-Colin, and G. L. Caër, "Nanorods of HfB_2 from mechanically-activated $HfCl_4$ and B-based powder mixtures," Journal of Alloys and Compounds, 398[1–2] 208-18 (2005).

[25] D. Portehault, S. Devi, P. Beaunier, C. Gervais, C. Giordano, C. Sanchez, and M. Antonietti, "A General Solution Route toward Metal Boride Nanocrystals," Angewandte Chemie International Edition, 50[14] 3262-65 (2011).

[26] S. Ran, H. Sun, Y. n. Wei, D. Wang, N. Zhou, and Q. Huang, "Low-Temperature Synthesis of Nanocrystalline NbB_2 Powders by Borothermal Reduction in Molten Salt," Journal of the American Ceramic Society, 97[11] 3384-87 (2014).

[27] S. Zhang, M. Khangkhamano, H. Zhang, and H. A. Yeprem, "Novel Synthesis of ZrB_2 Powder Via Molten-Salt-Mediated Magnesiothermic Reduction," Journal of the American Ceramic Society, 97[6] 1686-88 (2014).

[28] K. Bao, Y. Wen, M. Khangkhamano, and S. Zhang, "Low-Temperature Preparation of Titanium Diboride Fine Powder via Magnesiothermic Reduction in Molten Salt," Journal of the American Ceramic Society (2017) (DOI: 10.1111/jace.14649).

[29] N. J. Welham, "Formation of Nanometric TiB_2 from TiO_2," Journal of the American Ceramic Society, 83[5] 1290-92 (2000).

[30] A. Nekahi and S. Firoozi, "Effect of KCl, NaCl and $CaCl_2$ mixture on volume combustion synthesis of TiB_2 nanoparticles," Materials Research Bulletin, 46[9] 1377-83 (2011).

[31] E. Bilgi, H. E. Çamurlu, B. Akgün, Y. Topkaya, and N. Sevinç, "Formation of TiB_2 by volume combustion and mechanochemical process," Materials Research Bulletin, 43[4] 873-81 (2008).

[32] J. Wypartowicz, T. Østvold, and H. A. Øye, "The solubility of magnesium metal and the recombination reaction in the industrial magnesium electrolysis," Electrochimica Acta, 25[2] 151-56 (1980).

[33] R. L. Martin and J. B. West, "Solubility of magnesium oxide in molten salts," Journal of Inorganic and Nuclear Chemistry, 24[1] 105-11 (1962).

[34] M. Ito and K. Morita, "The Solubility of MgO in Molten $MgCl_2$-$CaCl_2$ Salt," Materials Transactions, JIM, 45[8] 2712-18 (2004).

[35] U. Demircan, B. Derin, and O. Yücel, "Effect of HCl concentration on TiB_2 separation from a self-propagating high-temperature synthesis (SHS) product," Materials Research Bulletin, 42[2] 312-18 (2007).

[36] S. Yazici and B. Derin, "Effects of process parameters on tungsten boride production from WO_3 by self propagating high temperature synthesis," Materials Science and Engineering: B, 178[1] 89-93 (2013).

TRIBOLOGY STUDY OF NOVEL Ti$_3$SiC$_2$ MATRIX COMPOSITES REINFORCED WITH CERAMICS (Al$_2$O$_3$, BN, B$_4$C) PARTICULATES

J. Nelson, M. Olson, and S. Gupta*

Department of Mechanical Engineering, University of North Dakota
Grand Forks, ND, USA
Corresponding Address: surojit.gupta@engr.und.edu

ABSTRACT

This paper reports the synthesis of Ti$_3$SiC$_2$ matrix composites by incorporating (1 and 6 vol%) Al$_2$O$_3$, (1 and 5 vol%) BN, and (1 and 5 vol%) B$_4$C ceramics particulate additives in the Ti$_3$SiC$_2$ matrix. All the composites were fabricated by pressureless sintering by using ~1 wt% Ni as a sintering agent at 1550 °C for 2h. SEM and XRD studies showed that Al$_2$O$_3$ is relatively inert in the Ti$_3$SiC$_2$ matrix whereas BN and B$_4$C reacted significantly with the Ti$_3$SiC$_2$ matrix to form TiB$_2$. Detailed tribological studies showed that Ti$_3$SiC$_2$-1wt%Ni (baseline) samples showed dual type tribological behavior where the friction coefficient (μ) was low (μ ~0.2) during stage 1, thereafter μ increased sharply and transitioned into stage 2 (μ ~0.8). The addition of Al$_2$O$_3$ as an additive had no effect on the tribological behavior, but the addition of B$_4$C and BN was able to enhance the tribological behavior by increasing the transition distance (TD). For example, the TD of Ti$_3$SiC$_2$-1wt% Ni was ~10 m and TD increased to ~90 m and ~150 m in 312Si-5%B4C and 312Si-5%BN, respectively.

INTRODUCTION

The M$_{n+1}$AX$_n$ (MAX) phases (over 70+ phases) are thermodynamically stable layered hexagonal (space group D$^4_{6h}$–P6$_3$/mmc) with two formula units per cell [1-4]. These phases possess a M$_{n+1}$AX$_n$ chemistry, where n is 1, 2, or 3, M is an early transition metal element, A is an A-group element and X is C or N. Barsoum et al. [5] pointed out in the first paper that Ti$_3$SiC$_2$ due to its layered structure may have good solid-lubricant qualities [5]. Later, Myhra et al. [6] demonstrated that by using a lateral force microscope with a Si$_3$N$_4$ tip that the friction coefficients, μ, of the basal planes were ultra-low (2–5 × 10^{-3}), but the μ's of non-basal planes were much higher. Interestingly, the tribological behavior of polycrystalline samples showed different results. For example, El-Raghy et al. [7] showed that both coarse-grained, CG (~25–50 μm) and fine-grained, FG (~4 μm) polycrystalline Ti$_3$SiC$_2$ samples during testing by using a pin-on-disc method against a 9.5 mm diameter 440C steel ball, a load of 5 N, with a sliding speed of 0.1 m/s showed low μ, initially, which increased linearly from 0.15 to 0.4 and then attained a steady state value of 0.8. Due to third body abrasion, the average sliding wear rates were high, 1.34 × 10^{-3} mm^3/N m and 4.25 × 10^{-3} mm^3/N m, for the FG and CG samples, respectively. Souchet et al. [8] studied the dual type tribology of FG (~4 μm) and CG (~25–50 μm) Ti$_3$SiC$_2$ samples against steel and Si$_3$N$_4$ balls by using a reciprocating type tribometer in details. Souchet et al. [8] concluded that the transition between the two regimes occurred at different times, and depended on various factors such as grain size, type of pin, and normal load applied. Several investigators have observed similar dual type tribological in Ti$_3$SiC$_2$ based ceramics [9]. Clearly, due to high wear rate, pristine Ti$_3$SiC$_2$ ceramics cannot be used as a coating material for high performance tribological applications. The aim of this paper is to study the effect of different ceramics particulates like Al_2O_3, B_4C, and BN on the tribological behavior of Ti_3SiC_2. These composites will be designated as MAXCERs.

EXPERIMENTAL

Ti_3SiC_2 powder (-325 mesh, Kanthal, Hallstahammar, Sweden) and calculated concentrations of ceramic powders (Alpha Al_2O_3 powders (nanopowder, Inframat Advanced Materials, Manchester, CT) or BN powders ((~1 µm, Sigma Aldrich, St. Louis, MO)) or B_4C powders (<10 µm, Sigma Aldrich, St. Louis, MO) were dry ball milled (8000 M mixer Mill, SPEX SamplePrep, Metuchen, NJ) for 5 minutes. All the powders were then poured in a die and were cold-pressed at ~232 MPa (the cycle was repeated twice) in a ~12.7 mm die (EQ-Die-12D-B, MTI Corporation, Richmond, CA), and sintered at 1550 °C for ~120 min in a tube furnace by flowing Ar though it. All the MAXCER composites were designed by adding 1 vol% (312Si-1%Al_2O_3 or 312Si-1%BN or 312Si-1%B_4C), 5 vol% (312Si-5%BN or 312Si-5%B_4C), and 6 vol% (312Si-6%Al_2O_3) of ceramics particulates in the Ti_3SiC_2 matrix. In all compositions, ~1 wt% Ni of the Ti_3SiC_2 content was added as a sintering aid. In addition, Ti_3SiC_2 with ~1 wt% Ni (Ti_3SiC_2-1wt% Ni) was also fabricated for comparison as a baseline sample.

Rule of mixture was used to calculate the theoretical density of all the composite samples by using the theoretical density of Ti_3SiC_2 and ceramics particulates. The experimental density was also determined from the mass and dimensions of each sample. The relative density was then calculated by normalizing the experimental density with theoretical density. In Ti_3SiC_2-BN and Ti_3SiC_2-B_4C composites, it is difficult to calculate the exact theoretical density due to the interfacial reaction and formation of different phases like TiB_2. Due to this reason, the density calculations are based on starting compositions of the precursor powders and should be used for *qualitative comparison only*. All the composites were polished (R_a < 1 µm) and then tested by a Vicker's micro-hardness indentor (Mitutoyo HM-112, Mitutoyo Corporation, Aurora, IL) by loading the samples at ~9.8 N for 12 s, and an average of five readings for each composite system is reported in the text.

The phase analysis was performed by XRD (SmartLab, Rigaku, Japan) at a scan rate of 0.05 °/min from 20° to 70°. The tribological behavior of the samples were tested by using a block (tab)-on-disc tribometer (CSM Instruments SA, Peseux, Switzerland) at 5 N, ~31 cm/s linear speed, and ~5 mm track radius against stainless steel balls. In this paper, the transition distance (TD) is defined as the distance after which the tribological behavior changes from Type I (low µ) to Type II (high µ). An average of three TD from each composition is reported in the text (Fig. 7d).

JEOL JSM-6490LV Scanning Electron Microscope (SEM JEOL USA, Inc., Peabody, Massachusetts.) in Secondary electron (SE) and Backscattered Electrons (BSE) mode was used to study the samples. Chemical analysis was done with X-ray analysis by using a Thermo Nanotrace Energy Dispersive X-ray detector with NSS-300e acquisition engine. For identifying the chemistry of a region, a combination of BSE and X-ray analysis is used to determine the tribochemistry of tribocouples. For example, if a region is determined to be chemically uniform at the micron scale then it will be identified with two asterisks as *microconstituent* to emphasize that these areas are not necessarily single phases. In addition, it is difficult to get the exact quantification of C by X-ray analysis, and the presence of C in these tribofilms was shown by adding {C_x} in the microconstituent composition [9].

RESULTS AND DISCUSSION

Figure 1 shows the microstructure of the Ti_3SiC_2-Al_2O_3 composites. Al_2O_3 particulates are well dispersed in the Ti_3SiC_2 matrix. Figures 2 and 3 show the Ti_3SiC_2-B_4C and Ti_3SiC_2-BN composites, respectively. Both the microstructures are very heterogeneous and it is difficult to discern individual particles. Figure 4 summarizes the XRD profile of all the compositions. In the Ti_3SiC_2-Al_2O_3 composites (Fig. 4a) – alumina was observed in the XRD peaks which indicate that Al_2O_3 particulates are stable in the Ti_3SiC_2 microstructure. Comparably, Figs. 4b and 4c show the formation of TiB_2 which indicate that B_4C and BN are not stable in the Ti_3SiC_2 matrix and result in the formation of TiB_2.

Figure 5a shows the plot of relative density versus the additions of ceramics particulates. As described in the experimental section, it is difficult to calculate the exact theoretical density due to the interfacial reaction and formation of different phases like TiB_2 (Fig. 4), thus the plot of Fig. 5a is for qualitative comparison only which is based on the calculations based on starting powders. The Ti_3SiC_2-1wt% Ni samples were predominantly single phase and had a relative density of ~96%. The addition 1 vol% of ceramic particulates had no deleterious effect on the relative density, and the relative density retained similar values. However, the addition of higher vol% (5-6 vol%) of ceramics particulate had a negative effect on the densification, and the resultant samples were porous as compared to Ti_3SiC_2-1wt% Ni (Fig. 5a). Comparatively, the hardness of Ti_3SiC_2-1wt% Ni was ~3 GPa, it gradually increased to ~4 GPa in 312Si-5%BN or 312Si-5%B4C, and it similar value of ~3 GPa in all the compositions of Ti_3SiC_2-Al_2O_3 composites (Fig. 5b).

Figures 6-7 plot the µ versus distance plot of different Ti_3SiC_2 based composites. Figure 6a shows the µ versus distance profile of Ti3SiC2-1wt%Ni composition. As observed in previous studies, the Ti_3SiC_2-1wt%Ni composition displayed a dual type behavior where the µ changes from low value of ~0.2 (stage 1) to high value of ~0.8 (stage 2). The addition of ~1 vol% Al_2O_3 had no significant effect on the TD (Fig. 6b), however the addition of ~1 vol% B_4C or BN enhanced the TD slightly. The addition of ~6 vol% Al_2O_3 had no effect on the TD (Fig. 7a), comparatively, the addition of ~5 vol% B_4C or BN had

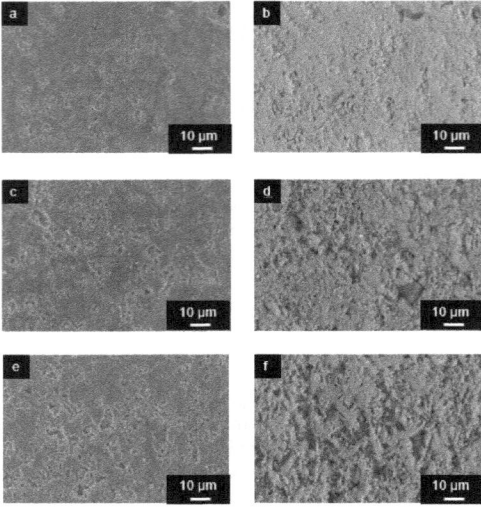

Figure 1: FESEM SE micrographs of, (a) sintered Ti_3SiC_2, (b) BSE image of the same region, (c) 312Si-1%Al_2O_3, (d) BSE image of the same region, (e) 312Si-6%Al_2O_3, and (f) BSE image of the same region.

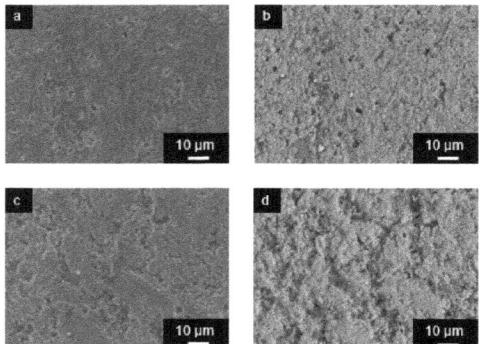

Figure 2: FESEM SE micrographs of, (a) 312Si-1%B_4C, (b) BSE image of the same region, (c) 312Si-5%B_4C, and (d) BSE image of the same region.

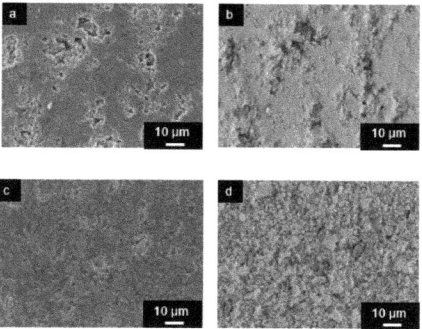

Figure 3: FESEM SE micrographs of, (a) 312Si-1%BN, (b) BSE image of the same region, (c) 312Si-5%BN, and (d) BSE image of the same region.

a significant effect on the TD (Figs. 7c-d). Figure 7d summarizes the variation of TD as a function of ceramics particulate content. The TD of Ti_3SiC_2-1wt% Ni was ~10 m. It retained similar values in Ti_3SiC_2-Al_2O_3 composites. Comparatively, TD increased to ~90 m and ~150 m in 312Si-5%B_4C and 312Si-5%BN, respectively. This study shows that the microstructure where Ti_3SiC_2 has reacted with the ceramic particulates showed longer TD. Based on these results, further studies are warranted whether interpenetrating matrix formed by adding reactive ceramics particulates can further enhance the tribological performance. Figure 8 shows a BSE SEM image of stainless steel and 312Si-5%BN wear surfaces. Both the surfaces are covered with third body debris which is very typical of Type II tribological behavior [9]. More fundamental studies should be focused on minimizing the third body formation.

CONCLUSIONS

Ti_3SiC_2 matrix composites were fabricated by pressureless sintering by using ~1 wt% Ni as a sintering agent. Ti_3SiC_2 matrix composites were designed by adding ~1 and ~6 vol% Al_2O_3, 1 and 5 vol% BN, and 1 and 5 vol% B_4C ceramics as particulate additives. SEM and XRD studies showed that Al_2O_3 is relatively inert in the Ti_3SiC_2 matrix whereas BN and B_4C reacted significantly with the Ti_3SiC_2 matrix to form TiB_2. Detailed tribological studies showed that Ti_3SiC_2-1wt%Ni samples showed dual type tribological behavior. The addition of Al_2O_3 as a particulate additive had no effect on the TD, but the addition of B_4C and BN particulates were able to enhance the TD. Detailed SEM studies also showed that both the stainless steel and 312Si-5%BN wear surface was covered with third body debris.

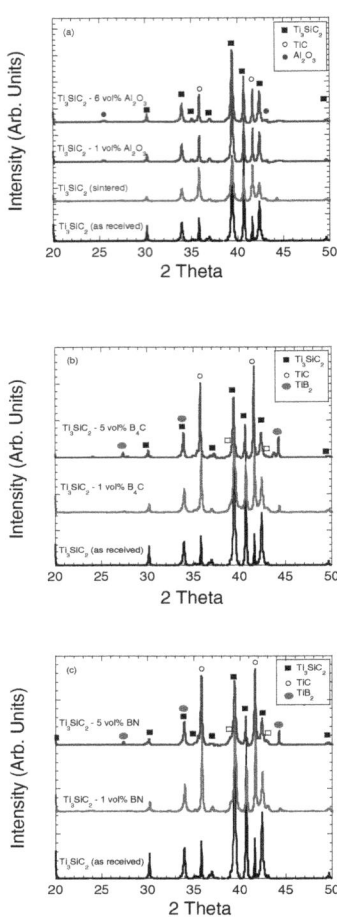

Figure 4: XRD patterns of, (a) Ti_3SiC_2-Al_2O_3 composites, (b) Ti_3SiC_2- B_4C composites, and (c) Ti_3SiC_2-BN composites.

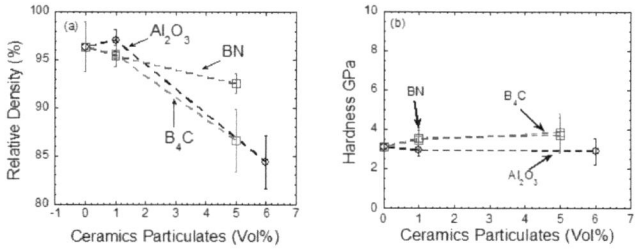

Figure 5 : Plot of, (a) relative density, and (b) hardness versus the vol (%) additions of ceramics particulates.

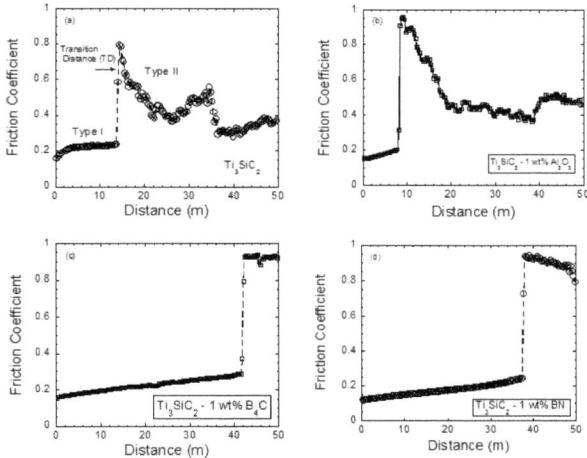

Figure 6: Plot of friction coefficient versus distance of, (a) Ti_3SiC_2-1wt%Ni, (b) 312Si-1%Al_2O_3 ,(c) 312Si-1%B_4C and (d) 312Si-1%BN.

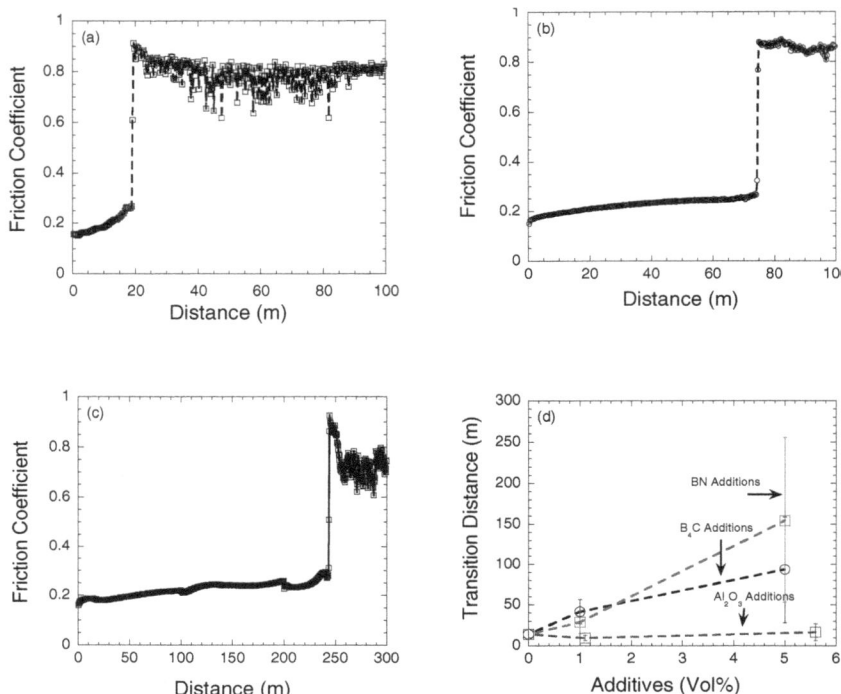

Figure 7: Plot of friction coefficient (μ) versus distance of, (a) 312Si-6%Al$_2$O$_3$, (b) 312Si-5%B$_4$C, (c) 312Si-5%BN, and (d) Transition Distance (TD) versus additives (vol%).

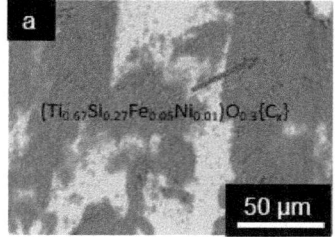

Figure 8: BSE SEM micrographs of, (a) stainless, and (b) 312Si-5%BN surface after tribological testing.

ACKNOWLEDGEMENTS

One of the authors (SG) would like to acknowledge the University of North Dakota start up funding, NASA EPSCoR under the NASA grant number NNX13AB20A, and NSF EPSCoR for support. Authors would like to thank Kanthal Inc. for supplying the Ti_3SiC_2 powders. NDSU Electron Microscopy Center core facility is also acknowledged for the microscopy. This material is also based upon work supported by the National Science Foundation under Grant No. 0619098, and 1229417. Any opinions, findings, and conclusions or recommendations expressed in this material are those of the author(s) and do not necessarily reflect the views of the National Science Foundation. Kanthal Inc. is acknowledged for the supply of Ti_3SiC_2 powders.

REFERENCES
1. "MAX Phases: Properties of Machinable Ternary Carbides and Nitrides", M.W. Barsoum, John Wiley & Sons (2013).
2. "Elastic and Mechanical Properties of the MAX Phases", M.W. Barsoum and M. Radovic Annu. Rev. Mater. Res. **41**, 195-227 (2011).
3. "Synthesis and characterization of a remarkable ceramic: Ti_3SiC_2", M.W. Barsoum, T. El-Raghy, J. Am. Ceram. Soc. **79**, 1953–1956 (1996).
4. M.W. Barsoum, The $M_{n+1}AX_n$ phases: a new class of solids; thermodynamically stable nanolaminates, Prog. Solid State Chem. **28**, 201–281 (2000).
5. "Synthesis and characterization of a remarkable ceramic: Ti_3SiC_2", M.W. Barsoum, T. El-Raghy, J. Am. Ceram. Soc. **79**, 1953–1956 (1996).
6. "Ti_3SiC_2 – a layered ceramic exhibiting ultra– low friction", S. Myhra, J.W.B. Summers, E.H. Kisi, Mater. Lett. Mater. Lett. **39**, 6–11 (1999).
7. "Effect of grain size on friction and wear behavior of Ti_3SiC_2", T. El-Raghy, P. Blau, M.W. Barsoum, Wear **238** (2), 125–130 (2000).
8. A. Souchet, J. Fontaine, M. Belin, T. Le Mogne, J.-L. Loubet, M.W. Barsoum, Tribological duality of Ti3SiC2, Tribol. Lett. **18**, 341–352 (2005).
9. "On the tribology of the MAX phases and their composites during dry sliding: A review", S. Gupta and M.W. Barsoum, Wear **271**, 1878– 1894 (2011).

Advanced Materials for Sustainable Nuclear Fission and Fusion Energy

INTERFACIAL REACTION AND MECHANICAL PROPERTIES OF SiC BONDED WITH ZIRCALOY-4 USING Ni, Zr /Al DOUBLE INTERLAYERS

Xin Geng [1,2], Guangwu Wen[1,2,*] and Xiaoxiao Huang[1]

[1]School of Materials Science and Engineering, Harbin Institute of Technology, Harbin 150001, People's Republic of China; [2] School of Materials Science and Engineering, Shandong University of Technology, Zibo 255091, People's Republic of China.

ABSTRACT

Hydrogen release, generated by the reaction between Zr-based alloy claddings and water steam, raises concerns over safety (explosion) of light water reactors (LWRs) during a loss of coolant accident. This study offers a potential method through solid-state reaction to fabricate SiC/zircaloy-4 joints in order to enhance the accident tolerance of LWR fuel claddings. Within experiments, SiC/zircaloy-4 diffusion couples are subject to high temperature heat treatment. Results from various characterisation techniques indicated that reaction occurs. In addition, Vickers indentation test was performed to evaluate the adhesion properties of the SiC/ zircaloy-4 diffusion couples.

INTRODUCTION

The high mechanical strength, high melting point (1852°C), low neutron-absorption cross-section and excellent corrosion-resistance properties of zirconium-based alloys under normal operating conditions have successfully make them wide application as fuel claddings and pressure tubes in light water reactors (LWRs) for more than 50 years [1]. However, in the severe Fukushima accident, Zr-based alloys corroded rapidly in high temperature steam with extensive hydrogen generation, when coolant system stopped working. It caused an explosion in the reactor building and fission products leaking [2]. Corrosion tests indicated that 17% of Zr cladding had been consumed at 1200°C in a steam environment only for 400 s [2]. After that, the LWR industry developed accident tolerant fuel (ATF), which new cladding materials has a better oxidation resistance property by hot steam and better mechanical strength under a loss of coolant accident (LOCA) condition in comparison with commercial available Zr-based alloys [3].

Silicon carbide (SiC), due to its high melting point, good chemical stability, no appreciable creep at high temperatures, and a low neutron absorption cross section [4], has proven as an effective coating in tristructural isotropic (TRISO) fuel particles irradiated in high temperature gas reactors (HTGRs) [4]. Moreover, ref [5] indicated that SiC have a better oxidation resistance properties by hot steam than Zr-based alloys. Silicon carbide composite (SiC_f/SiC_m) cladding is proposed as the long-term solution, but many practical issues should be resolved, including manufacturing technology, costs, brittleness of SiC and licensing [2]. Therefore, SiC coated Zr-based alloys is considered as a mid-term solution, and the fabrication is still in the very initial stage. The fabrication temperature of SiC coated Zr-based alloys is required to no more than Zr-based alloys phase transformation point (α-Zr-base alloys transforms to β-Zr-based alloys at ~ 866°C [6]) or the temperature of grain growth of Zr-based alloys (generally final annealing at ~600 °C to avoid microstructural change from the initial state) [7]. However, stoichiometric SiC with a high degree of crystalline and low percentages of oxygen and excess carbon, usually be synthesized higher than 1500 °C [8]. Diffusion bonding method and double metal interlayers for joining SiC and zircaloy-4 alloy are used in this work to extricate from this dilemma: high temperature is required for bonding between silicon carbide and metal interlayers (Zr or Ni), and low temperature is annealed to bond the metal interlayer (Zr/Al) and zircaloy-4 alloys, and for avoiding microstructure and phase change in zircaloy-4 alloys.

The objective of the present work is to propose a new way to joining between SiC and Zr-based alloys by diffusion bonding. Combination double metal interlayers (Ni or zircaloy-4 alloy/Al) were employed to join SiC to Zr-based alloys. Two types of reaction modes have been identified in the solid reaction between SiC and metal: Mode I is formation of carbon and silicides and Mode II is formation of carbides and silicides [9]. Ni and Zr (or zircaloy-4 alloys), which represents the above two modes respectively, were used as an interlayer to bond the silicon carbide. The results indicate SiC was well bonded with the Zr interlayer annealed at 600-1400°C. Moreover, an Al foil is used to bond Zr (zircaloy-4 alloys) with zircaloy-4 alloy substrates at lower annealing temperature (600 °C). The assembling design of the SiC-zircaloy-4 alloys diffusion couples is shown in Fig.1. Interfacial reaction products and its morphology were identified by XRD, SEM and EDS. The Vickers indentation tests were employed to evaluate the mechanical properties of the bonding interface (reaction zone).

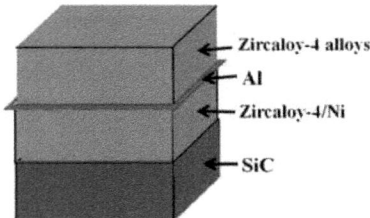

Fig.1 a schematic graph for assembling of α-SiC/Ni or Zircaloy-4/Al/Zircaloy-4 joints.

EXPERIMENTAL

1) Raw materials

Commercially available Zr powder (purity ≥99.8%, particle size 1-3 μm, Shanghai Chaowei, China), Ni powder (purity >99.5%, particle size < 5 μm, Aladdin, China) and β-SiC powder (purity 99.8%, particle size ~1 μm, Alfa Aesar, China) were employed to prepare composites. Commercial Ni foils (>99.95% purity, 0.05 mm thick, Qingyuan Metal Materials Co., Ltd.) and Al foils (>99.95% purity, 0.02 mm thick, Qingyuan Metal Materials Co., Ltd.) were used as diffusion bonding interlayers. The commercially available Zircaloy-4 alloys (Western Energy Material Technologies Co. , Ltd. , China; the composition of Zircaloy-4 alloys is shown in Table 1.) and α-SiC bulk (a pressureless sintered SiC without sintering additives, supplied by Harbin Institute of Technology, China) were utilized as starting materials. The α-SiC bulk and Zircaloy-4 alloys were cut into 5 mm×5 mm squares with the thickness of 1mm and 1.4 mm, respectively. Note that the joining surfaces were polished by standard metallographic polishing procedures.

Afterward, they were rinsed and cleaned by acetone and ethanol in an ultrasonic cleaner and then dried in hot air.

Table 1. Chemical composition of Zircaloy-4 alloys.

ID	Sn (%)	Fe (%)	Cr (%)	Al (%)	B (%)
Zircaloy-4	1.37	0.20	0.10	0.0028	<0.00005

2) Ni-SiC composites and Ni/SiC joints

Powders (50 wt.% Ni powder +50 wt.% β-SiC powder) were manually mixed by pestle and mortar and cold-pressed into pellets of 0.5g each with the diameter of 10 mm. Both Ni/SiC composites and Ni/α-SiC joints were heat-treated at temperatures between 600 and 1000 °C with a heating and cooling rate of 300 °C/h and 480°C/h under a high purity argon flow.

3) Zr/SiC composites, Zircaloy-4 /α-SiC joints and zircaloy-4/Al/zircaloy-4 joints

Powders (50 wt.% Zr powder + 50 wt.% β-SiC powder) were manually mixed by pestle and mortar and cold-pressed into pellets of 0.5g each with a diameter of 10 mm. Both Zr/SiC and Zircaloy-4 alloy/α- SiC joints were then subjected to a tube furnace (Thermcraft incorporated, N.C. U.S.A). (Note that zircaloy-4 substrate was also used as the Zr foil in this paper). Heat treatment was conducted at temperatures in the range from 600 to 1400 °C for 4 hours with a heating and cooling rate of 300 °C/h and 480°C/h under a high purity argon flow. Secondly, the zircaloy-4 alloy/Al/ zircaloy-4 alloy joints were conducted at 600 °C for 10 h in an Ar atmosphere. In order to achieve a good contact between the polished surfaces, a small compressive force was applied on all diffusion couples. The diffusion bonded couples were sectioned perpendicular to the cross-section and then mounted in resin. All the embedded samples were polished by standard metallographic polishing procedures down to 0.5 µm diamond paste, then cleaned ultrasonically in soap water and ethanol, and subsequently dried in hot air.

The phase compositions were identified by X-ray diffraction (D&A25ADVANCE, BRUKER, Germany) using Cu K_α radiation and the measurements were performed from 20 to 80°. The interfacial microstructure evolution of annealed specimens were examined using a scanning electron microscope (SEM) (SUPRATM55, ZEISS, UK) coupled with an energy-dispersive X-ray spectrometer (EDS), with an accelerating voltage of 20 kV. Chemical compositions of the interfacial reaction zone were quantitatively analyzed by EDS. The hardness of the α-SiC/Zircaloy-4 alloys joints in cross section were measured by Vickers microhardness tester (HV-1000). The load was 0.098 N and the dwelling time was 15 s for all the measurements. At least 5 indents were made on each region. The interval between indents was more than 100 µm in order to avoid the influence from adjacent indents.

RESULTS AND DISCUSSION

1) Ni/β-SiC composites and Ni/α-SiC diffusion couples

The interlayer Ni represents type (I) solid reaction in metal/SiC system. For the Ni/β-SiC composites, XRD is used to investigate the interfacial reaction between Ni and SiC (shown in Fig.2). All XRD peaks shown in Fig.2(a) and (b) correspond to phases of Ni and β-SiC, respectively. It concludes that the raw materials are pure without other crystallographic impurities. As Fig.2(c) shows, XRD peaks of Ni_3Si phase begin to appear, and both 3C-SiC and Ni peaks are also involved as main phases. Thus, interfacial reaction occurs between SiC and Ni at 600 °C. When the annealing temperature goes up to 1000 °C, Fig.2(d) indicates that SiC and Ni₂Si become as the main phases and XRD peaks of Ni and Ni₃Si disappear.

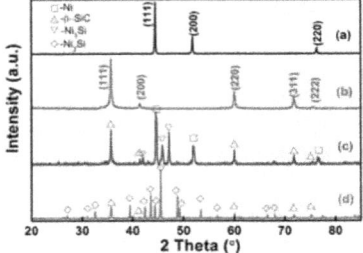

Fig.2 XRD patterns of (a) pristine Ni powder; (b) pristine SiC powder; and (50wt.%) Ni/β-SiC composites heat-treated at various temperatures: (b) 600 °C and (d) 1000 °C.

Ni/6H-SiC diffusion couples are annealed at 600 °C for 10h and 1000 °C for 0.5h. However, Ni foils are peeled off from the SiC bulk after the heat treatment process. Fig.3 shows the optical image of polished surface of the pristine SiC and the fracture surface of the Ni/6H-SiC diffusion couples in the SiC side. In comparison with Fig.3(a), the fracture surface of annealed Ni/SiC diffusion couples on the SiC side are rough (Fig.3(b)and (c)). It concludes that interfacial reaction between Ni and 6H-SiC occurred and the interfacial reaction zone is the weakest site.

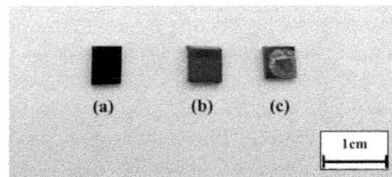

Fig.3 An optical image of (a) pristine polished SiC; and fracture surface of the Ni/6H-SiC diffusion couples after heat-treatment at various temperatures: (b) 600 °C for 10h; and (c) 1000 °C for 0.5h.

SEM images of the diffusion couple 6H-SiC/Ni annealed at 1000°C in cross section are shown in Fig.4. Cracks are found inside the 6H-SiC bulk near to the interfacial reaction zone (Fig.4(a)), since the coefficient of thermal expansion (CTEs) among Ni, SiC and interfacial reaction products are different, which causes the residual stresses at Ni/6H-SiC interface during the cooling process. Moreover, in ceramic/metal systems, the maximum stress generally occurred in the ceramic side near the interface [10-13]. The interfacial reaction between 6H-SiC/Ni leads to the formation of periodic layers (Fig. 4(b)).   EDX results, corresponding to locations of point 1 to 3 in the inset image of Fig.4 (b), are listed in Table 2. Periodic bands consist of Ni₂Si and graphite. In addition, brittle silicide (Ni₂Si) and the precipitation of separate graphite layers reduce the mechanical properties of the joints [10, 14]. Therefore, cracks are easily formed inside the interfacial reaction layer (indicated by the red arrow in Fig.4(b)).

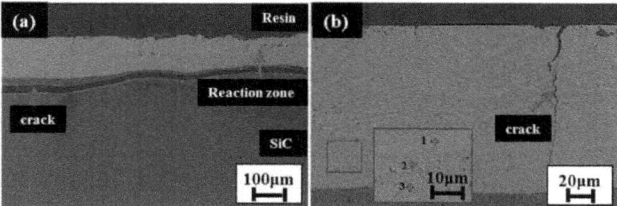

Fig.4. Cross-section SEM micrographs of Ni/SiC diffusion couples annealed at 1000°C for 0.5h : (a) general view; and (b) interfacial reaction zone, the inset image is an enlarged view of the red rectangular area, and points 1-3 corresponds to the chemical composition analysis position by EDS.

Table 2. Composition measurement of the Ni/6H-SiC diffusion couple annealed at 1000 °C for 0.5 h by EDS corresponding points in Fig.4(b).

ID number	Analyzed phase	Measured composition (at.%)		
		Ni	Si	C
1	Ni_2Si+graphite	15.90	6.56	77.55
2	Ni_2Si+graphite	14.91	6.24	78.84
3	N_2Si	64.78	35.22	0

In summary, the reaction between SiC and Ni occurred at the temperature higher than 600 °C. Periodic layers including Ni_2Si and graphite are formed between SiC and Ni annealed at 1000 °C. The 6H-SiC/Ni joints are failure after heat treatment, it is due to the residual stresses caused by the differences in CTEs and lower mechanical properties of the interfacial reaction products (brittle silicide and graphite). Therefore, the metal interlayer, which the reaction with SiC belongs to type (I), is not suitable for this study. The candidate inserted materials between SiC and Zircaloy-4 alloys should meet the following requirements: (1) avoiding formation of the periodic bands and separate graphite layers at the interface, which means meets the type (II) reaction mode by formation of carbides and silicides; (2) preventing the intense interfacial reaction between SiC and inserted material, (3) reducing the residual stress in the joint, and (4) without affecting neutron-absorption cross-section properties. The combination of Zr (Zircaloy-4 alloys) and Al interlayers are considered as the good candidates to join SiC with Zircaloy-4 alloys through two step diffusion bonding process: (1) Zr is a strong carbide former. According to Zr-Si-C ternary phase diagram section at 1200°C and 1300°C [14-17], it reacts with SiC forming not only silicide (Zr_3Si, Zr_2Si, $ZrSi$, $ZrSi2$ etc.), but also ZrC and ternary phase $Zr_5Si_3C_x$. All those reaction products have higher melting points (at least 1620°C) [6]. (2) Moreover, the reaction between Zr and SiC is mild and the reaction rate is about two orders of magnitude lower than that between Ni and SiC [14]. The thickness of the interfacial reaction zone in the SiC/Zr joint after annealing at 1200°C for 8h is nearly 10μm. (3) the zircaloy-4 alloy (4.44×10⁻⁶ K⁻¹) has a close coefficient of thermal expansion (CTE) to SiC (4.02×10⁻⁶ K⁻¹), which helps to reduce the residual stress in the joint [18, 19]. (4) Although the high temperature heat treatments change the microstructure or cause the phase transformation $(\alpha \rightarrow \beta)$ of Zircaloy-4 alloys, it still has better neutron-absorption cross-section properties than other metals. Although the mechanical properties of the metal/SiC joints are affected by the microstructure and chemical compositions of the interfacial reaction zone [16], the interfacial microstructure evolution and reaction mechanism between Zr and SiC depending on annealing temperatures are limit [15, 16]. In the following part, the Al foil is used as a interlayer to join the Zr interlayer with Zr-based alloys and then the 6H-SiC/zircaloy-4 alloys diffusion couples annealing at 600-1400°C are used to investigate the interfacial reaction.

2) Zr/β-SiC composites, α-SiC/Zircaloy-4 joints and Zircaloy-4/Al/Zircaloy-4 joints

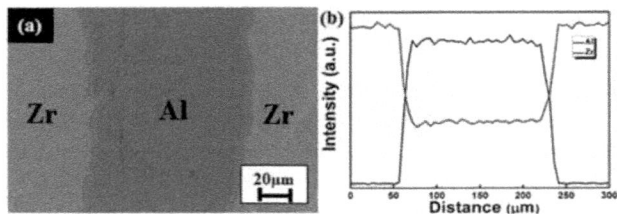

Fig.5. A SEM micrographs and EDS linescans taken from the cross section of zircaloy-4/Al/ zircaloy-4 joints annealed at 600°C for 10 hours: (a) overview; and (b) Al and Zr composition profiles across the zircaloy-4/Al/ zircaloy-4 joint corresponding to Fig.5(a).

It is confirmed that Zircaloy-4 alloys are well bonded by an Al interlayer annealing at 600°C for 10 hours (shown in Fig.5). No intermetallic compounds of Al-Zr are formed. Therefore, the α-SiC/Zr/Al/Zircaloy-4 alloys joints only depend on the bonding behaviors and reaction mechanism between SiC and the Zr (Zircaloy-4 alloys) interlayer. XRD was used to investigate the phase evolution after annealing the β-SiC/Zr composites at 600-1400 °C (shown in Fig.6.). Although the interfacial reaction rate in Zr/SiC composites is higher than that in zircaloy-4 alloys/SiC diffusion couples at the same temperature, it is due to the higher contact area ratio between Zr powder and SiC powder than that between the SiC bulk and zircaloy-4 alloys bulk. It is expected that the reaction products of heat treated Zr/SiC composites will be helpful in understanding the interfacial reaction evolution in the α-SiC/Zircaloy-4 joints. As Fig.6(c) and (d) show, the XRD spectra collected from the SiC/Zr composites annealed at 600 °C and 800 °C indicate that β-SiC and α-Zr are identified as main phases in the reference of XRD patterns of pure β-SiC powder (Fig.6(a)) and pure α-Zr powder (Fig.6(b)). Furthermore, a XRD peak corresponding to Zr₅Si₄ also appear, but the amount of this phase is little. Moreover, XRD peaks corresponding to α-Zr phase in Fig.6(c) and (d) shift to lower 2θ angles with an expand lattice compared to the pure α-Zr powder (Fig.6(b)), since about 0.3 at.% Si and 3 at.% C can dissolve in α-Zr and make the lattice of α-Zr swell, according to phase diagrams of Zr-Si and Zr-C[6, 20]. Thus, it infers that tiny reaction and interdiffusion occurs between Zr and SiC at 600-800 °C. When the annealing temperature goes up to 1000 °C (shown in Fig. 6(e)), XRD peaks of Zr_2Si and ZrC begin to appear, and the peaks of SiC are also involved as main phases. As the temperature increases to 1200 °C, ZrC, ZrSi and SiC become as the main phases (Fig. 6(f)). As the temperature reaches to 1400 °C, only ZrC and SiC phases are identified and XRD peaks of intermetallic compounds of Zr-Si disappear (Fig. 6(g)). Moreover, XRD peaks of Zr disappear in Zr/SiC composites annealing higher than 1000 °C (Fig.6(e)-(g)). It suggests that Zr powders have been completely transformed to the intermetallic compounds of Zr-Si and ZrC at a temperature higher than 1000 °C. Note that no XRD peaks corresponding to the ternary phase of $Zr_5Si_3C_x$ are detected in Zr/SiC composites annealing at 600 to 1400°C for 4 hours, since the Zr-SiC may not reach to the thermodynamic equilibrium state.

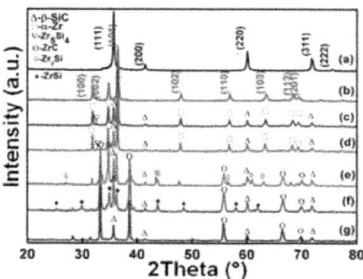

Fig.6. XRD patterns of (a) pristine β-SiC powder; (b) pristine α-Zr powder; and (50wt.%) Zr/β-SiC composites heat-treated at various temperatures for 4h: (c) 600 °C; (d) 800 °C; (e) 1000 °C; (f) 1200 °C; and (g) 1400 °C.

The α-SiC/Zircaloy-4 joints are bonded after annealing at 600-1400°C for 4h, and the interfacial reaction between zircaloy-4 alloys and SiC at various temperatures is investigated in detail. Cross section SEM images and EDS linescans of the zircaloy-4 alloy/α- SiC joints annealed at 600-800 °C are shown in Fig.7. An obvious two-phase microstructure is shown: the darker phase (left) is attributed to SiC, and the brighter phase (right) is the zircaloy-4 alloy. The interface between SiC and zircaloy-4 alloys is planar (shown in Fig.7(a)and (c)). Fig.7(c) indicates that no apparent interfacial reaction occurs between SiC and Zr at 600 °C, but the interdiffusion of C and Si from SiC and Zr from zircaloy-4 alloys happens and the thickness of the interdiffusion zone is about 4µm. When the annealing temperature increases to 800 °C, A thin reaction layer with a thickness of ~300 nm is formed between SiC and Zr (shown in Fig.7(e)). A tiny step in the composition profiles of Si and Zr also confirmed the formation of the reaction layer. Moreover, the thickness of the interdiffusion zone is about 5~6 µm. According to the carbon-zirconium (C-Zr) and silicon-zirconium (Si-Zr) phase diagrams [6, 20], dissolved carbon (~3at.%) and silicon (~0.3at.%) in zircaloy-4 alloys (α-Zr phase) slightly affect the phase transformation temperature (α-Zr → β-Zr) from 863 to 870°C. No phase transformation occurred in zircaloy-4 alloys annealed at 600-800°C.

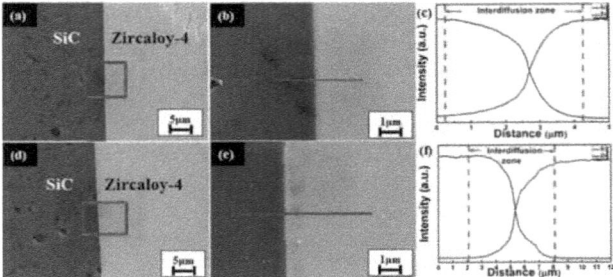

Fig.7. SEM micrographs in BSE mode and EDS linescans taken from zircaloy-4 alloy/α- SiC joints in cross section annealed at various temperatures for 4 hours: (a)(b)(c) 600 °C and (d)(e)(f) 800°C; Fig.7(b)and (e) are an enlarged view of the red rectangular area in Fig.7(a) and (d).

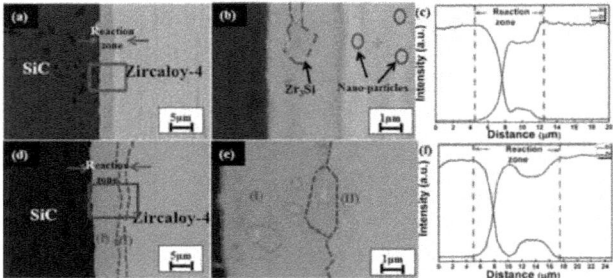

Fig.8. SEM micrographs in BSE mode and EDS linescans taken from zircaloy-4 alloy/α-SiC joints in cross section annealed at various temperatures for 4 hours: (a)(b)(c) 1000 °C and (d)(e)(f) 1200°C; Fig.8(b)and (e) are an enlarged view of the red or blue rectangular area in Fig.8(a) and (d).

Fig.8 shows SEM micrographs and EDS linescan along zircaloy-4 alloy/α-SiC joints annealed at 1000-1200 °C for 4 hours. A multi-phase structure has formed in the interfacial reaction zone between SiC and Zircaloy-4 alloys annealed at temperature higher than 1000 °C. Fig.7(b) indicates that the interfacial reaction zone contains mainly two phases: gray phase (grains are outlined by the red dash lines) and dark nanoparticles precipitated in the Zr matrix (highlighted by the red circles). The chemical compositions analysis of these two phases (point 1 and 2) is summarized in Table 3. The gray grains are attributes to Zr_3Si (point 1) and the dark nanoparticles are possible ZrC_{1-x}, which is in agreement with Wang's results [15]. The EDS linescan result indicate the thickness of the reaction layer grows to about 8-8.7μm (shown in Fig.8(c)). When the annealing temperature goes to 1200 °C, the reaction zone divides into two zones by contrast as outlined by the red dashed lines: the darker layer (zone (I), near to SiC) and the brighter layer ((zone (II)), near to zircaloy-4 alloys). Reaction zone (I) contains two phases: ZrC (grains are highlighted by the blue dash lines) and Zr_2Si (grains are outlined by the red dash lines). Reaction zone (II) contains the mixture of ZrC_{1-x} and Zr. The chemical composition analyses corresponding to the locations (point 3 to 5) are summarized in Table 3. The thickness of the reaction layer grows to about 12-14μm measured by the EDS linescan (shown in Fig.8(f)).

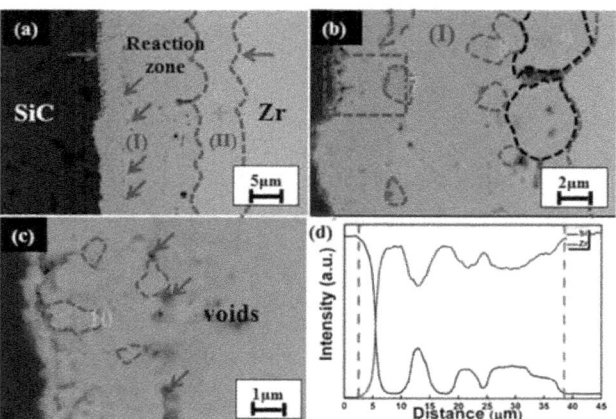

Fig.9. SEM micrographs in BSE mode and EDS linescans taken from zircaloy-4 alloy/α- SiC joints in cross section annealed at 1400 °C for 4 hours: (a) overview; (b) an enlarged view of reaction zone (I) in Fig.8(b); (c) an enlarged view of the purple rectangular area in Fig.8(b); and (d) chemical composition profiles measured by EDS.

Fig.9 indicates the interfacial reaction zone between zircaloy-4 alloys and SiC annealed at 1400°C for 4 hours. The interfacial reaction zone can be divided by two zones by contrast (highlighted by the red dash lines). Zone (I) consists of ZrC and complex intermetallic and zone (II) contains Zr_3Si. The chemical composition of the reaction products is listed in Table 3 (point 6-10). The interfacial reaction zone between SiC and zircaloy-4 alloys annealing at 1400 °C for 4 hours leads periodic layers as: $SiC/ZrC-ZrSi-ZrC-Zr_2Si-Zr_3Si/zircaloy$-4 alloys. The first layer near the SiC bulk is ZrC with the grain size of 500nm-1μm (as outlined by the blue dashed lines in Fig.9(c)) and the second layer near ZrC layer is ZrSi with the grain size of 1-2μm (as outlined by the red dashed lines in Fig.9(b)). Moreover, some nanometer-sized voids are formed between ZrC and ZrSi grains (as indicated by purple arrows in Fig.9(a) and (c)). These nanometer-sized voids are ascribable to the Kirkendall effect, because the intrinsic diffusion coefficients of Zr and Si atoms are different. Moreover, bigger ZrC grains with the width of about 2μm are formed the third layer next to ZrSi, and the fourth layer containing the phase of Zr_2Si (highlighted by the black dashed lines) with grain size of about 5μm is adjacent to ZrC. The inner layer contains Zr_3Si

(reaction zone (II)), which is adjacent to the Zr matrix. The thickness of the interfacial reaction zone between SiC and zircaloy-4 alloys grows significantly to ~36μm.

Table 3. Chemical composition measurement in the interfacial reaction zone of the zircaloy-4 /6H-SiC joints annealing at 1000-1400 °C for 4 h by EDS.

Temperature (°C)	Point number	Analyzed phase	Measured composition (at.%)		
			Zr	Si	C
1000	1	Zr_3Si	32.64	8.66	58.70
	2	$ZrC_{1-x}+Zr$	37.46	0	62.54
1200	3	ZrC	28.22	0.47	71.32
	4	Zr_2Si	31.89	14.80	53.32
	5	$ZrC_{1-x}+Zr$	39.58		60.42
1400	6	Zr_3Si	33.48	9.53	56.99
	7	$ZrSi$	25.97	23.11	50.92
	8	ZrC	28.83		71.17
	9	$Zr2Si$	31.06	14.49	54.45
	10	ZrC	24.26		75.74

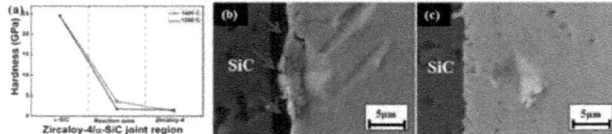

Fig.10(a) Vickers hardness in different regions, and SEM images of Vickers indent impressions in interfacial reaction region of the zircaloy-4 /SiC joints annealing at different temperatures: (b) 1200°C and (c) 1400°C.

Indentation hardness is a useful tool to evaluate the mechanical properties of the joints. Fig.10(a) shows the Vickers hardness (GPa) in different regions along the cross section of the zircaloy-4 /SiC joints annealing at 1200-1400°C. The hardness of the α-SiC substrate with a high value of about 25GPa is insensitive to the annealing temperatures. However, the hardness values of the Zr interlayer (zircaloy-4 alloy) and the interfacial reaction zone are significantly affected by the heat-treatment process. For the zircaloy-4 alloy interlayer, the values were decreased from ~1.53 GPa at 1200°C to ~1.24 GPa at 1400°C; however, The values of hardness in the interfacial reaction zone is improved from ~1.75GPa at 1200°C to ~3.62 GPa at 1400°C, which is affected by the composition and thickness of the reaction zone. High hardness values (1.75~3.62 GPa) observed in the interfacial reaction zone is associated with the formation of ZrC (~25GPa)[21]. In addition, detaching of the interfacial reaction zone from the SiC was observed after the Vickers indent loading on the interfacial reaction zone of a zircaloy-4 /SiC joint annealing at 1200°C (shown in Fig.10(b)), it is due to the brittleness of the forming ZrC phase in the reaction zone. However, when the zircaloy-4/SiC joint is annealing at 1400°C, periodic structure $ZrC/ZrSi/ZrC/Zr_2Si/Zr_3Si$ were formed in the interfacial reaction zone, the formed ZrC grains can be considered as dispersive distribution in the reaction zone. It helps to prevent continuous cracks in single phase and nanosized ZrC grains also inhibit cracks propagation as shown in Fig.10(c).

CONCLUSIONS

In this study, Ni or Zr/Al double interlayers are used to join of SiC to zircaloy-4 alloys, in order to improve the anti-corrosion properties by hot steam. 1)Ni/SiC joints are failed after annealing at 600-1000°C, due to the residual stresses caused by the differences in CTEs and formation of separated graphite layer with low strength. Cracks are generated inside the reaction zone and the SiC bulk. 2) An Al foil is used to bond two zircaloy-4 alloys substrates together at 600°C; no intermetallic compounds of Zr-Al were formed. 3) Annealing temperatures affects significantly on the composition and thickness of the interfacial reaction zone between SiC and Zr. At 600-800°C, interdiffusion of C and Si from SiC and Zr from zircaloy-4 alloys occurs. At 1000°C, Zr_3Si and ZrC_{1-x} are formed. At 1200°C, a multiphase layer consisting of $ZrC-Zr_2Si-ZrC_{1-x}$ is formed in the reaction zone. Interfacial reaction zone with the thickness of ~36μm contains $ZrC-ZrSi-ZrC-Zr_2Si-Zr_3Si$ for the zircaloy-4 alloys/SiC joint after annealing at 1400°C for 4h. Moreover, this interfacial reaction zone has good mechanical properties (Hv=3.62 GPa), since fine ZrC grains with a high value of hardness are formed and well dispersed in the reaction zone between Zr and SiC.

ACKNOWLEDGMENTS

The present work was financially supported by the National Natural Science Foundation of China under Grant No. NSFC 51602075.

REFERENCES

[1] M. Bojinov, V. Karastoyanov, P. Kinnunen, T. Saario. Influence of water chemistry on the corrosion mechanism of a zirconium–niobium alloy in simulated light water reactor coolant conditions, *Corrosion Science* 52 (2010) 54-67.
[2] L. Hallstadius, S. Johnson, E. Lahoda. Cladding for high performance fuel, *Progress in nuclear energy* 57 (2012) 71-76.
[3] B.A. Pint, K.A. Terrani, M.P. Brady, T. Cheng, J.R. Keiser. High temperature oxidation of fuel cladding candidate materials in steam–hydrogen environments, *Journal of Nuclear Materials* 440 (2013) 420-427.
[4] L.L. Snead, T. Nozawa, Y. Katoh, T.-S. Byun, S. Kondo, D.A. Petti. Handbook of SiC properties for fuel performance modeling, *Journal of Nuclear Materials* 371 (2007) 329-377.
[5] K.A. Terrani, B.A. Pint, C.M. Parish, C.M. Silva, L.L. Snead, Y. Katoh. Silicon carbide oxidation in steam up to 2 MPa, *Journal of the American Ceramic Society* 97 (2014) 2331-2352.

[6]   H. Okamoto. The Si-Zr (silicon-zirconium) system, *Bulletin of Alloy Phase Diagrams* 11 (1990) 513-519.

[7]   H.G. Kim, I.H. Kim, Y.I. Jung, D.J. Park, J.Y. Park, Y.H. Koo. Adhesion property and high-temperature oxidation behavior of Cr-coated Zircaloy-4 cladding tube prepared by 3D laser coating, *Journal of Nuclear Materials* 465 (2015) 531-539.

[8]   A.R. Bunsell, A. Piant. A review of the development of three generations of small diameter silicon carbide fibres, *Journal of Materials Science 41* (2006) 823-839.

[9]   J. Park, K. Landry, J. Perepezko. Kinetic control of silicon carbide/metal reactions, *Materials Science and Engineering: A* 259 (1999) 279-286.

[10]   M. Hattali, S. Valette, F. Ropital, G. Stremsdoerfer, N. Mesrati, D. Tréheux. Study of SiC–nickel alloy bonding for high temperature applications, *Journal of the European Ceramic Society* 29 (2009) 813-819.

[11]   K. Suganuma, Y. Miyamoto, M. Koizumi. Joining of ceramics and metals, Annual Review of Materials Science 18 (1988) 47-73.

[12]   H.P. Xiong, W. Mao, Y.H. Xie, W.L. Guo, X.H. Li, Y.Y. Cheng. Brazing of SiC to a wrought nickel-based superalloy using CoFeNi (Si, B) CrTi filler metal, *Materials Letters* 61 (2007) 4662-4665.

[13]   K. Geib, C. Wilson, R. Long, C. Wilmsen. Reaction between SiC and W, Mo, and Ta at elevated temperatures, *Journal of Applied Physics* 68 (1990) 2796-2800.

[14]   K. Bhanumurthy, R. Schmid-Fetzer. Interface reactions between silicon carbide and metals (Ni, Cr, Pd, Zr), *Composites Part A: Applied Science and Manufacturing* 32 (2001) 569-574.

[15]   Y. Wang, A.H. Carim. Ternary Phase Equilibria in the Zr Si C System, *Journal of the American Ceramic Society* 78 (1995) 662-666.

[16]   T. Fukai, M. Naka, J. Schuster. Bonding and Interfacial Structures of SiC/Zr Joint (Materials, Metallurgy & Weldability),   (1996).

[17]   Y. Zhang, F. Di, Z.-Y. He, X.-C. Chen. Progress in joining ceramics to metals, *Journal of iron and steel research*, international 13 (2006) 1-5.

[18]   Z. Zhong, T. Hinoki, H.C. Jung, Y.H. Park, A. Kohyama. Microstructure and mechanical properties of diffusion bonded SiC/steel joint using W/Ni interlayer, *Materials & Design* 31 (2010) 1070-1076.

[19]   P. MacDonald, L. Thompson. MATPRO: a handbook of materials properties for use in the analysis of light water reactor fuel rod behavior. SEE CODE-9502158 Aerojet Nuclear Co., Idaho Falls, Idaho (USA). Idaho National Engineering Lab., 1976.

[20]   R. Sara. The system zirconium-carbon, *Journal of the American Ceramic Society* 48 (1965) 243-247.

[21]   D. Sciti, S. Guicciardi, M. Nygren. Spark plasma sintering and mechanical behaviour of ZrC-based composites, *Scripta Materialia* 59 (2008) 638-641.

Single Crystalline Materials for Electrical, Optical, and Medical Applications

CHARACTERIZATION APPROACHES OF FEMTOSECOND DIRECT LASER WRITING (DLW) MODIFICATIONS INSIDE CUBIC YAG CRYSTALS

W. Gebremichael[1,2], I. Manek-Hönninger[1], S. Rouzet[1], M. Chamoun[1], A. Fargues[2], V. Jubera[5], T. Cardinal[5], Y. Petit[1,5], L. Canioni[1]

[1] Centre Lasers Intenses et Applications (CELIA - University of Bordeaux-CNRS-CEA UMR5107), Talence, France

[2] Amplitude Systèmes, Cité de la Photonique, 11 Avenue de Canteranne, 33600 Pessac

[5] Institut de Chimie de la Matière Condensée de Bordeaux (ICMCB-CNRS-UPR9048), Pessac, France

ABSTRACT

This study is focused on direct laser writing (DLW) of waveguides inside cubic $Y_3Al_5O_{12}$ (YAG) crystals doped with Nd^{3+} and their characterization. We show a mode-field analysis at 633 nm for different double track waveguides and first results on gain measurements at 1064 nm and compare the result with free propagation in the bulk material.

INTRODUCTION

Recent developments in femtosecond DLW allow for realizing a vast range of photonic devices which is a driving force behind industrial interest for rapid prototyping of optical components. Thus, it is necessary to have a better understanding and approach for qualitative and quantitative characterizations in both waveguide and luminescence properties of modified transparent materials. Yttrium aluminium garnet (YAG) is a common crystal widely used as gain medium in solid state lasers. Owing to its crystallographic structure, optical specificities and restrictions of low-symmetry crystal optics are neglected which enables a simple systematic approach to study modifications due to femtosecond DLW inside crystals along the damaged tracks.

Waveguides based on a permanent refractive index modification by femtosecond lasers were first inscribed in glasses [1], and later also in crystalline optical materials like silicon [2]. Doped YAG crystals have been used for waveguide inscription by DLW and waveguide laser demonstration [3-5], even with curved forms of waveguides showing a slope efficiency of 79 % [6]. DLW waveguides and waveguide lasers in rare earth doped YAG ceramics are also reported for different active ions like Neodymium [7-9] and Ytterbium [10] for laser operation at 1 μm, and with Holmium, Erbium [11] and Thulium [12] for lasers in the mid-infrared.

In glasses, the refractive index is typically increased along the damage line, and the guiding of the light takes place within the damage track. However, in crystals, the refractive index is usually lowered, demanding more complex structures for guiding between damage tracks. The simplest form of a waveguide consists of double lines forming a channel where the light is guided between the two writing tracks. A study of the refractive index change mechanisms by DLW in Nd:YAG ceramic waveguides can be found in [13]. Reference [14] gives a very nice review on optical waveguides in crystals describing the writing mechanism and giving an exhaustive overview over the different crystalline materials.

SAMPLE PREPARATION

A rectangular piece with the dimensions of 11.3 mm x 9.0 mm x 0.9 mm of Nd:YAG (doped at 0.8 % wt.) was cut out of a standard laser rod with the original dimensions of 100 mm length and 16.5 mm diameter and was then polished to optical quality. A femtosecond laser beam (see Fig. 1) at 800 nm with an energy of 0.63 mJ delivered by a commercial Ti:Sapphire amplifier system (RegA, Coherent) is focused by a microscope objective with NA = 0.9 focusing at 100 μm mechanically below the crystal surface, and thus leading to a focal point at 184 μm underneath the crystal surface taking into account the refractive index of around 1.8 of the material. Due to the elongated form of the focalized laser beam, the damage tracks are elliptical as can be easily understood from Fig. 1.

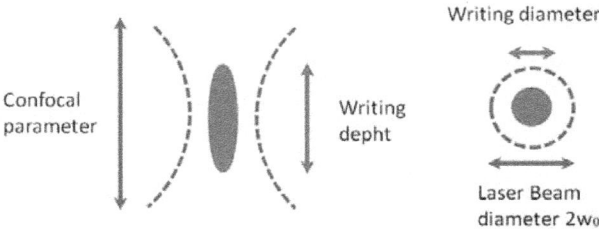

Figure 1. Schematic drawing of the laser beam for DLW on the Nd:YAG crystal.

The sample was translated perpendicularly to the laser beam with a translation speed of v=10 μm/s writing speed. All DLW parameters that were applied in this experiment are summarized in Table 1.

Table I. Laser source, beam parameters and writing parameters for waveguide writing in Nd:YAG.

Laser source parameters		Writing parameters		Beam parameters	
Wavelength	$\lambda = 800$ nm	Speed	$v = 10$ μm/s	Beam waist	$w_0 = 0.97$ μm
Energy	$E = 0.63$ μJ	# of pulses	$N = 48\,570$	Rayleigh range	$z_R = 3.7$ μm
Pulse duration	$\tau = 150$ fs			Irradiance	$I = 226$ TW/cm^2
Repetition rate	$f = 250$ kHz	Objective	$NA = 0.9$	Energy deposition	$E_{dep} \sim 30.6$ mJ

We chose to write 3 different series consisting each of three equal double lines forming each pair of tracks a waveguide channel. We opted for 3 series of double tracks with a distance between the lines of 10 μm, 20 μm and 25 μm, respectively. The distance between two double tracks of the same series was 200 μm which should be sufficently large to avoid cross talking between the different waveguide channels. The distance between two series was chosen to be 350 μm in order to easily distinguish between two different series. The resulting double track waveguides are schematically illustrated in Fig. 2 showing a top view (a) and a side view (b).

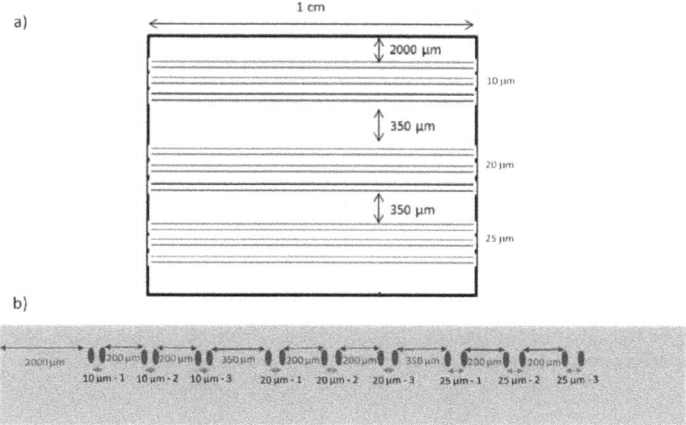

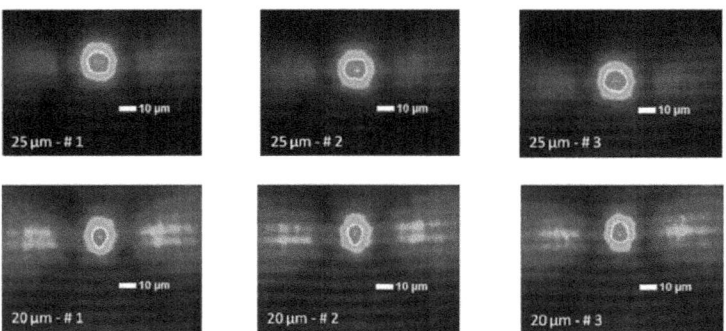

Figure 2. Schematic view seen from the top (a) and from the side (b) of the DLW structured Nd:YAG crystal.

GUIDING PROPERTIES

In order to study the guiding properties of the double tracks, we injected a HeNe laser beam at 633 nm using a microscope objective of x20 with an NA of 0.5 into the waveguides. The transmitted beam was imaged on a CCD camera (Thorlabs, BC106N-VIS/M) using exactly the same type of objective. The objectives and the sample were mounted on a three axes translation stage allowing for accurate injection into the waveguides and good and reproducible alignment.

Figure 3. Images of the beam profile of the guided modes in the 25 μm double tracks (upper row) and in the 20 μm double tracks (lower row) taken with a CCD camera, respectively.

As can be seen on Fig. 3, we obtained good guiding with Gaussian beam profiles for the 25 μm and 20 μm double tracks, however with some small side lobes in the x direction, especially in the case of the 20 μm waveguide channels. We observed no guiding in the 10 μm double tracks due to a too high numerical aperture of the microscope objective. This can be overcome by using a lens for focusing the injected beam.

We studied furthermore the polarization dependence of the waveguides. Therefore, we turned the polarization by a half wave-plate just before injection into the waveguides. The results are depicted in Fig. 4 for the 20 μm and 25 μm double tracks and show a Malus law behavior with a maximum transmission corresponding to s-polarized light. The minimum transmission for p-polarized light is on the order of 20 %. Further work is needed to identify if these 20% truly result from the p-polarized transmission of the waveguide, or if it results from non spatially filtered light not being guided but still collected after free propagation since (i) the waveguides were not so long and (ii) the numerical aperture of the injecting objective was visibly higher than that of the waveguiding structures. Still, our double track waveguides are likely to be stress-induced corresponding to their polarization-selective behavior.

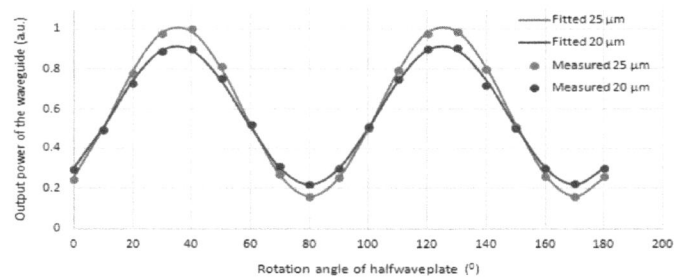

Figure 4. Transmitted output power of the waveguide as a function of the rotation angle of the halfwave-plate.

GAIN MEASUREMENTS

In order to compare the amplification potential within the waveguides with respect to ordinary bulk propoagation, we performed gain measurements. Therefore, we injected a pump beam around 808 nm, corresponding an absorption maximum of Nd:YAG, delivered by a Ti:Sapphire laser (Cameleon, Coherent) into the sample. We carefully mode-matched a signal beam at 1064 nm delivered by a diode-pumped microchip laser (Teemphotonics) and overlapped the two laser beams for single-pass injection. The gain was measured for different pump power and signal power levels for all three waveguide series and for non guided free propagation in the bulk. The injected seed power was 0.07 mW, 0.21 mW, 0.66 mW and 1.08 mW, respectively, whereas the pump power was varied up to 750 mW. Using a lens of focal length f' = 75 mm for injection allowed us to have a better coupling efficiency, and especially to be able to inject into the 10 μm double tracks. In fact, considering the Fresnel losses of 8.8% per face, the measured coupling efficiencies of the pump / seed were 88% / 76%, 73% / 58% and 27% / 19% for the 25 μm, 20 μm and 10 μm waveguide channels, respectively.

Fig. 5 depicts the single-pass gain as a function of the injected pump power for the three different waveguide channels and for bulk propagation. It is clearly visible that the smaller the waveguide the higher the gain. For the 25 μm double tracks the gain is still comparable to the bulk propagation and no advantage of waveguide propagation is visible.

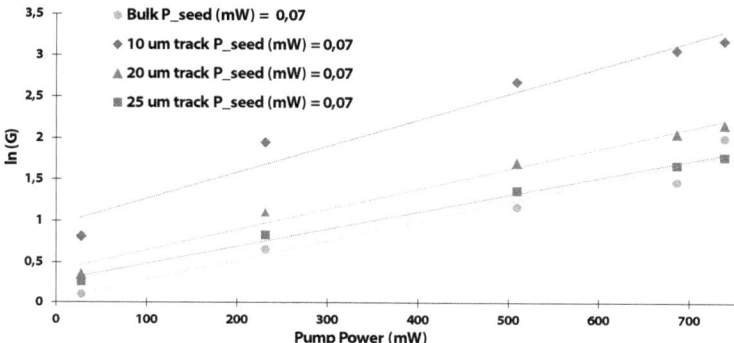

Figure 5. Single-pass gain as a function of injected pump power for the three different waveguide channel series and bulk propagation in comparison; the injected seed power at 1064 nm was $P_{seed} = 0.07$ mW.

CONCLUSION AND PERSPECTIVES

We have successfully written double track waveguides by direct femtosecond laser writing into a Neodymium-doped YAG crystal. The guided modes are Gaussian in both x and y direction. However, some side lobes are visible in the x direction due to leakage but can be filtered out easily. Moreover, the waveguides are polarization sensitive and mainly guide s-polarized light. We furthermore studied the gain properties at 1064 nm using a pump centered at 808 nm for different levels of seed and pump. Under the same conditions, the 10 μm double tracks show the best gain properties compared to the bigger double tracks or the non guided gain in the same bulk material.

ACKNOWLEDGEMENTS

This study has been carried out with financial support from the French State, managed by the French National Research Agency (ANR) in the frame of "the Investments for the future" Programme IdEx Bordeaux – LAPHIA (ANR-10-IDEX-03-02).

REFERENCES
[1] K.M. Davis, K. Miura, N. Sugimoto, and K. Hirao, "Writing waveguides in glass with a femtosecond laser," Opt. Lett. vol. 21 (no. 21), pp. 1729-1731 (1996).

[2] A.H. Nejadmalayeri, P. Herman, J.Burghoff, M. Will, S. Nolte, and A. Tünnermann, "Insription of optical waveguides in crystalline silicon by mid-infrared femtosecond laser pulses," Opt. Lett. Vol. 30 (no. 9), pp. 964-966 (2005).

[3] J. Siebenmorgen, K. Petermann, G. Huber, K. Rademaker, S. Nolte, A. Tünnermann, "Femtosecond laser written stress-induced $Nd:Y_3Al_5O_{12}$ (Nd:YAG) channel waveguide laser," Appl; Phys. B 97, pp. 251-255 (2009).

[4] J. Siebenmorgen, T. Calmano, K. Petermann, and G. Huber, "Highly efficient Yb:YAG channel waveguide laser written with a femtosecond-laser," Opt. Express, vol. 18, no. 15, pp. 16035–16041 (2010).

[5] T. Calmano, J. Siebenmorgen, O. Hellmig, K. Petermann, and G. Huber, "Nd:YAG waveguide laser with 1.3 W output power, fabricated by direct femtosecond laser writing," Appl. Phys. B, vol. 100, no. 1, pp. 131–135 (2010).

[6] T. Calmano, A.-G. Paschke, S. Müller, C. Kränkel, and G. Huber, "Curved Yb:YAG waveguide lasers, fabricated by femtosecond laser inscription," Opt. Express, vol. 21, no. 21, pp. 25501–25508 (2013).

[7] G. A. Torchia, A. Ródenas, A. Benayas, E. Cantelar, L. Rosso, and D. Jaque, "High efficient laser action in femtosecond-written Nd:yttrium aluminium garnet ceramic waveguides", Appl. Phys. Lett., vol. 92, 111103 (2008).

[8] A. Ródenas, G. Zhou, D. Jaque, and M. Gu, "Direct laser writing of three-dimensional photonic structures in Nd:yttrium aluminium garnet ceramics", Appl. Phys. Lett., vol. 93, 151104 (2008).

[9] H. Liu, Y. Jia, J.R. Vazquez de Aldana, D. Jaque, and F. Chen, "Femtosecond laser inscribed cladding waveguides in Nd:YAG ceramics: Fabrication, fluorescence imaging and laser performance", Opt. Express, vol. 20, no. 17, pp. 18620-18628 (2012).

[10] T. Calmano, A. G. Paschke, J. Siebenmorgen, S. T. Fredrich-Thornton, H. Yagi, K. Petermann, and G. Huber, "Characterization of an Yb:YAG ceramic waveguide laser, fabricated by the direct femtosecond-laser writing technique," Appl. Phys. B, vol. 103, no. 1, pp. 1–4 (2011).

[11] A. Ródenas, A. Benayas, J.R. Macdonald, J. Zhang, D.Y.Tang, D. Jaque, and A.K. Kar, "Direct laser writing of near-IR step-index buried channel waveguides in rare earth doped YAG", Opt. Lett., vol. 36, pp. 3395-3397 (2011).

[12] Y.Ren, G. Brown, A. Rodenas, S. Beecher, F. Chen, and A.K. Kar, "Mid-infrared waveguide lasers in rare-earth-doped YAG", Opt. Lett., vol. 37, pp. 3339-3341 (2012).

[13] A. Ródenas, G. A. Torchia, G. Lifante, E. Cantelar, J. Lamela, F. Jaque, L. Roso, and D. Jaque, "Refractive index change mechanisms in femtosecond laser written ceramic Nd:YAG waveguides: Micro-spectroscopy experiments and beam propagation calculations," Appl. Phys. B, vol. 95, no. 1, pp. 85–96 (2009).

[14] F. Cheng an J.R. Vazquez, "Optical waveguides in crystalline dielectric materials produced by femtosecond-laser micromachining", Laser & Photonics Reviews, vol. 8, no. 2, pp. 251-275 (2014).

Additive Manufacturing and 3D Printing Technologies

COMPARISON OF DYNAMIC MASK- AND VECTOR-BASED CERAMIC STEREOLITHOGRAPHY

S. Baumgartner, M. Pfaffinger, B. Busetti and J. Stampfl
Institute of Materials Science and Technology, TU Wien, Vienna, Austria

ABSTRACT

Additive manufacturing (AM) has developed into a promising technology allowing the parallel production of several complex parts with high resolution. Especially AM of ceramics shows great potential in the field of medical and dental applications where personalized aesthetic restorations are demanded. Accuracy is therefore a must to meet the high standards of those fields. At TU Wien we developed a special lithography based manufacturing technology where layer-by-layer a photosensitive slurry with ceramic or glass-ceramic filler is cured. To get dense ceramics the so called green body is then debinded and sintered. Two printing systems were evaluated: The first system uses a Digital Light Processing (DLP) approach, where a digital mirror device projects visible light (460 nm) and triggers polymerization. A second system is based on a diode-laser with 405 nm wavelength. In this case a galvanoscanner is used for structuring. The second system enables a feature resolution down to 20 μm.

The aim of this study is to compare both of this technologies for multiple ceramics such as zirconium oxide, tricalcium phosphate and composites. In order to get the best accuracy, printing parameters such as laser speed, hatching style, exposure time and intensity are varied.

INTRODUCTION

During the past decades, industry has recognized additive manufacturing as valuable addition to conventional manufacturing technologies. The possibility of producing customized parts in very small to medium lot sizes puts AM technologies more than ever in the center of attention, by now also for direct end-application production additional to traditional rapid prototyping purposes[1]. Further, the layer-by-layer approach of these processes allows the production of geometries with a degree of complexity exceeding the possibilities of any conventional production technology available. Still, each additive manufacturing method has its advantages and drawbacks, and therefore often a tradeoff between precision, surface quality, material properties and economy has to be made. Depending on the processed material and the application area of the produced parts, further requirements have to be satisfied. While for biomedical applications the biocompatibility of the material is a crucial factor, in case of ceramics density is often a weak point of current additive manufacturing technologies (AMTs) available.

When it comes to produce ceramic parts with high dimensional accuracy and good to excellent mechanical properties, lithography-based AMTs are to favor[2]. Stereolithography ceramic manufacturing (SLCM) turned out to be the technology of choice if ceramic parts with high complexity, surface quality and density are desired. The whole production cycle, consisting of the printing process, a thermal debinding and a sintering step works well for a range of ceramic and glass-ceramic filled slurries and different batch sizes, creating dense ceramic parts with mechanical properties comparable to conventionally processed ones. Up to now this could be shown in various experiments for technical ceramics like zirconium oxide (ZrO_2), aluminium oxide (Al_2O_3)[3, 4], bioactive glasses[2], tricalcium phosphate (TCP)[5, 6] and glass ceramics.

In order to show the possibilities and restrictions of stereolithography, two different printing approaches were compared. While a prior system is based on digital light processing (DLP)[8], the recently developed one uses a diode-laser as light source to polymerize the slurry. The aim was to identify the differences between those two regarding resolution, printing parameters and conditions and final part quality.

EXPERIMENTAL
For the SLCM process highly filled photosensitive slurries are used. The particle size of the ceramic powders used for those slurries varies between 200 nm for ZrO_2 and 5 μm for β-tricalcium phosphate. The system consists of a variety of acrylate-based monomers, an organic solvent (either polyethylene glycol or polypropylene glycol), dispersing agents, a light absorber and a photoinitiator. Depending on the filler type some additives affecting the viscosity and processability might be added further. Also the solid loading ranges from 42% for ZrO_2 to 50% for β-tricalcium phosphate, which is at least necessary to get dense ceramic parts comparable to conventionally manufactured ones[8].

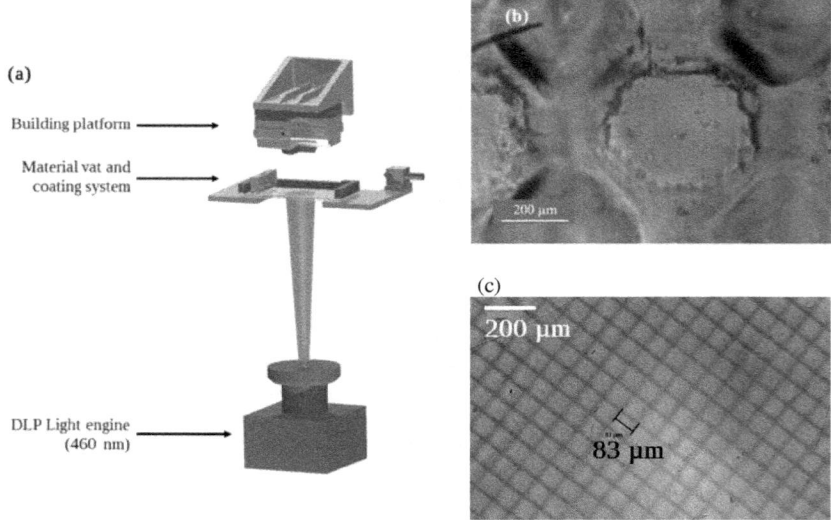

Figure 1 (a) DLP-based system with circular coating system and tiltable vat; (b) Circular structure printed on the DLP-based machine using unfilled photopolymer; the contours clearly show the pixels of the light engine; (c) Structure showing the pixel size of a DLP machine with 80 μm optics

Two stereolithography-based systems were used to solidify the slurries. The first system is based on the principal of digital light processing (DLP, Fig.1a) where visible blue light is projected onto the resin by a digital micromirror device (DMD). The number of mirrors determines the resolution, while each one represents one or more pixels of the projected image.

Fig. 1c shows the pixel of an 80 μm optics. For this study an optics with 25 μm was used. Light emitting diodes with a wavelength of 460 nm trigger polymerization of the slurry between the vat and building platform. In case of the DLP-based system, resolution in lateral layer depends on the DMD-chip and optical system used. Larger optics allow exposure of larger building areas, while reducing the overall resolution. In z-direction, however, the layer thickness and printing parameter settings are the main factors to determine the resolution. The limits of this dynamic mask approach are shown in Fig. 1b, displaying a circular structure with clearly visibly pixelated borders.

The second system uses a vector-scanning approach where a diode-laser with 405 nm wavelength is used to cure the slurry layer-by-layer (Fig.2a). In this case a galvanometer scanner positions the beam and does the structuring. For the laser-based approach, the resolution strongly depends on the laser used and its parameters. Similar to other laser-applications, the laser beam quality and diameter determine the final parts quality. While light intensity and exposure time are crucial factors to be identified for the DLP-system, laser speed, power and hatching style have to be varied in the laser-based machines. With the laser used for this study a feature resolution of 20 μm can be achieved which is shown in by a pattern with 20 μm hatching distance in Fig. 2c. With this approach even small circular structures with a smooth borderline can be printed (Fig. 2b).

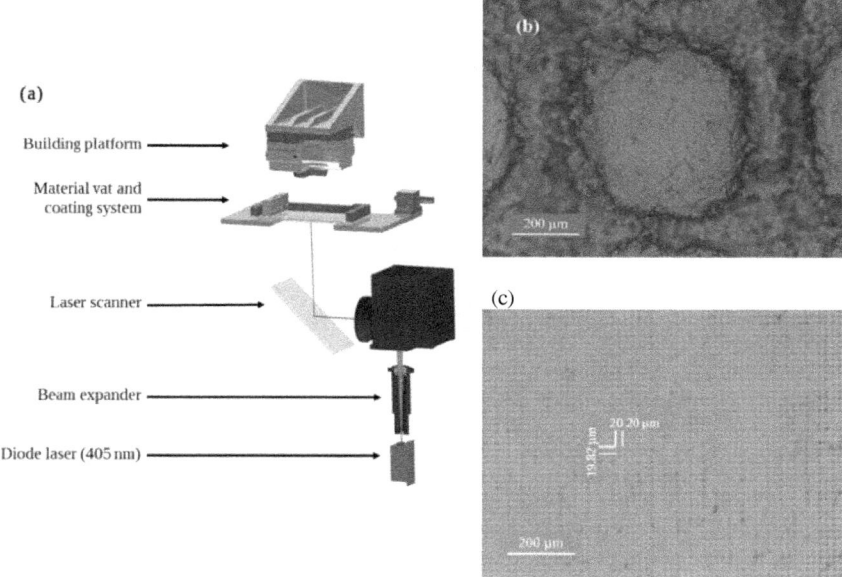

Figure 2 (a) Vector-based laser system with linear coater and tiltable vat, (b) Circular structure printed with laser-based machine using unfilled photopolymer; (c) Structure showing a hatching distance of 20 μm

In both systems a doctor blade (Fig. 3) ensures a total covering of material in the vat when the building platform moves up after each layer.[9] Figure 3 shows the setup of the recoating system used in this study. This special arrangement with two parallel doctor blades serves as reservoir for fresh slurry or resin during the printing process and allows to form a new material layer in both movement directions. During the movement of the coater it pushes excess material ahead of it, filling possible holes or empty spaces in the material layer. When the doctor blade reaches the other end of the vat the excess material is pushed back into the reservoir through the holes in the blade. This setup allows a good mixing of the slurry and an even deposition during the job. Furthermore, very thin coated layers of material in the vat reduce the cleaning effort, especially for delicate and porous parts, greatly. To reduce the detaching-forces to a minimum in the separation step, the vat is slightly tilted. This is one of the most time consuming steps, but necessary in order to avoid delamination or even detaching

(a)   (b)

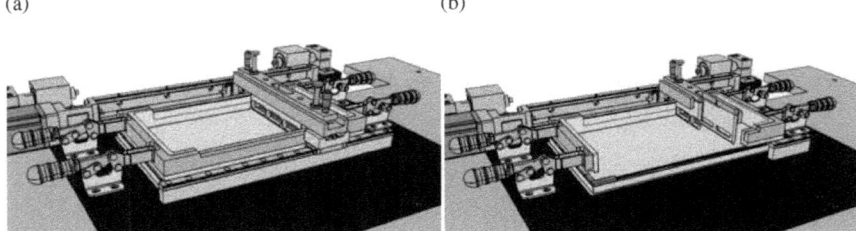

Figure 3 Recoating system developed and used for the stereolithography systems at TU Wien; (a) View of the whole system, showing the coater, material vat and clamping frame; (b) section view of the two parallel doctor blades

After the printing process a so-called green body is obtained which has to undergo a thermal post-processing in order to get dense ceramic parts. At lower temperatures as an initial step the solvent and later the whole organic content is removed. The polymeric binder decomposes during this debinding step. For each slurry composition and part geometry several TGA measurements are necessary to find the optimal temperature program. Unsuitable heating rates or insufficient dwelling times either result in the destruction of the part by cracks or incompletely debinded, grey structures caused by carbon residuals in the material. Perfect debinding leaves only the inorganic filler material which then undergoes a sintering procedure at up to 1600 °C, depending on the ceramic material. This sintering step usually causes an overall shrinkage of the parts of up to 25%, which has to be considered before the printing process, having again an influence on the final part's accuracy.

For this study different test structures and patterns were printed in order to compare the possible resolution of both systems. Further, the printing parameters (light intensity, exposure time, laser beam power, hatching style and speed) were varied. Influences on the thermal treatment process are discussed.

RESULTS AND DISCUSSION

To transfer the theoretical resolution of both systems to real parts, the exposure parameters and machine settings have to be adjusted for each material. This, however, is far more complicated for complex systems like filled slurries, than for unfilled resins. Glass or ceramic fillers highly influence the refractive index of the photosensitive system which is a crucial factor for absorbance and refraction behavior and therefor curing depth, overpolymerization and accuracy. To investigate the influences on filled systems a glass filled composite with monomers widely used in dental applications was used (see Table I).

Table I Composition of the composite material for DLP and laser machine

Component	DLP	Laser
	wt%	
BisGMA	40	40
TEGDMA	10	10
Photoinitiator	0,2	0,2
Light Absorber	0,05	0,01
Glass filler (Schott®)	49.75	49.79

DLP-based System

Most important for the outcome is the light dose (light intensity x exposure time), which was determined for each composition. Maximum available intensity was 85 mW with a 25 µm optics. For the printed test structures the light dose was set so as to get a curing depth of three times the layer thickness (here 75 µm). This is sufficient to avoid detaching of the layers while averting overpolymerization in z-direction. Best results were achieved at 20 mW intensity and 4s exposure time.

Laser-based System

For printing on the laser-based system the absorber content of the slurries had to be adapted. Higher intensities demand a higher light absorber concentrations. Having much more tunable influence factors, the laser system requires a more detailed investigation of the curing of a layer. Due to high intensities the exposure time to reach the necessary light dose has to be very short, leading to high scanning speed. Best results were achieved with a scanning speed of 10000 mm/s at a laser power of 64 mW. Lower speed resulted in overpolymerization in lateral direction. In contrast to the DLP-based systems where the time for each layer is independent of the size of the projected image, the time needed for a laser-cured layer varies strongly with size and shape of the printed structure. Every single line has to be scanned at least once, depending on the hatching strategy used. Besides the decision to hatch in one (X- or Y-hatch) or both (XY-hatch) directions, the combination of different delay settings result in a huge amount of different hatching strategies. Fig. 4 shows two ones used in this study with the best results regarding surface quality.

(a)

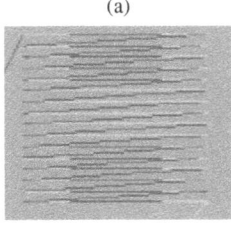

When one line is written, the laser jumps back to the other side. Each line is therefore scanned from the same direction. This scanning method results in very high surface quality but increases the time needed for one layer.

(b)

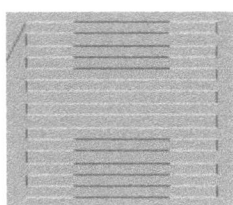

The laser does not jump back to the starting point, but the lines are written form one and then form the other direction. This safes time compared to other hatching strategies like the one showed above, but results in slightly worse surface quality.

Figure 4 Two different hatching strategies used in this study

To get an even surface the decision fell upon a double hatching in X- and Y-direction. The hatching distance was reduced to 15μm, resulting in a continuous surface. Bigger distances lead to unpolymerized or patchy areas.

Figure 5 and 6 show a test structure printed as well on the DLP- as the laser-based machine. It was designed as to show the possibility to print delicate structures of different sizes as well as the accuracy of the system.

(a)   (b)   (c)

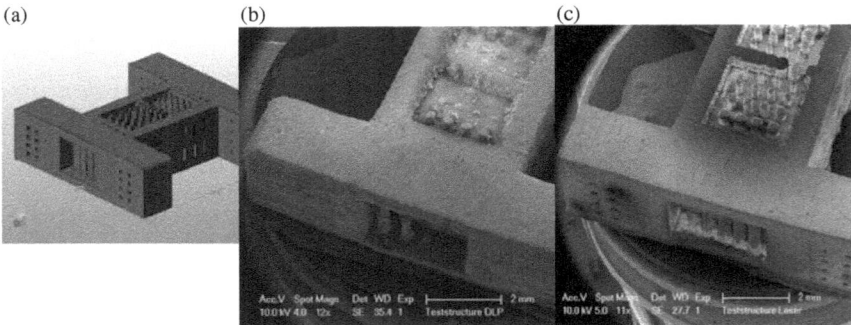

Figure 5 (a) CAD-drawing of the test structure which was printed with (b) DLP and (c) laser.

The first thing to realize in the DLP part (Fig. 5b and 6a) is the missing structures with small diameters and the closed pores on the side. This might be explained on the one hand by the resolution of the system, making it difficult to structure details with dimensions about the pixel size (here 25 μm). On the other hand, machine settings like tilting speed of the vat and therefore detaching force are not appropriate for this material and lead to rupture of the small pillars. The laser-printed part (Fig. 5c and 6b), however, shows every detail of the test structure. The round pillar structures on the top are slightly overpolymerized which can be lead back to the delay settings of the laser-scanning system, since it could not be changed by varying the laser power or scanning speed.

To remove unpolymerized residues from the parts, they were cleaned with Ethanol for 2 minutes. In the case of the composite used for this test structure, the solvent attacked the surface significantly, visible in Fig. 5c. A suitable solvent has yet to be found.

(a)   (b)

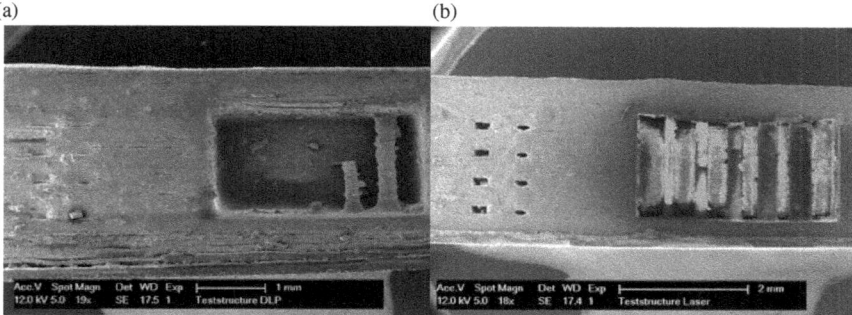

Figure 6 Side part of the test structure showing cavities and pillar structures of different diameters (50 – 250 μm) printed with (a) DLP and (b) laser

With a width of 18 μm of a single laser-written line the hatching distance of 15 μm might still be too high, although the surface seems smooth enough. Therefore, also the hatching distance has to be further optimized in order to get an even surface.

Mechanical Properties

The high resolution in combination with proper post treatment results in parts with excellent surface quality. This allows the fabrication of not only dense, but also smooth ceramic parts which is essential when it comes to mechanical properties. The lesser surface damages, the higher the bending strength values which can be achieved. Fig. 7 shows sintered ceramic parts of ZrO_2 and TCP.

(a)　　　　　　　　　　　　　　(b)

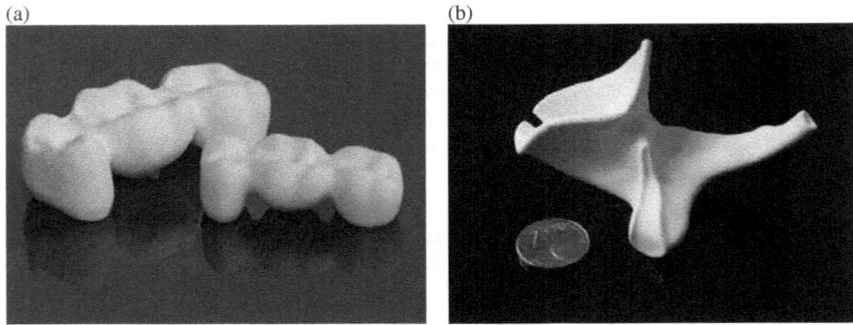

Figure 7 (a) Molar brigde printed with DLP out of ZrO_2, showing the part after debinding (left and after sintering (right); (b) Bone structure printed with DLP out of TCP after sintering

Defect free sintered ceramic parts lead to outstanding mechanical properties (Tab. II).

Table II Material properties of ZrO2 and TCP

	Material	
	ZrO_2	TCP
Measured biaxial bending strength [MPa]	1098	32
Density [g/cm³]	5.9	3.14
Relative Density [%]	99.92	88
Solid loading – green parts [vol%]	42	50

CONCLUSION

　　As could be shown so far, the already good results of the older DLP-based system could be further improved with introducing the laser-approach into the stereolithography system. Although the duration of printing a single layer is extended due to the exposure strategy, the overall improvement cannot be denied. The small laser beam diameter allows the writing of delicate structures, which were impossible up to now. Developing new hatching strategies and parameter combinations may further improve the parts' quality and speed up the printing process.

ACKNOWLEDGEMENTS

　　We gratefully acknowledge the financial support by the Christian Doppler Laboratory "Photopolymers in Restorative and Digital Dentistry".

REFERENCES
[1]　　Spierings A, Levy G (2009). Rapid Manufacturing - auch mit SLM. RTejournal - Forum für Rapid Technologie, Vol. 6.

[2]   Gmeiner, R., Mitteramskogler, G. and Stampfl, J. 'Stereolithographic Ceramic Manufacturing of High Strength Bioactive Glass', Int. J. Appl. Ceram. Technol., 12 [1] 38–45 (2015)

[3]   R. Felzmann, et al., Adv. Eng. Mater. 14 [12] 1052–1058 (2012).

[4]   S. Gruber, "Lithography-Based Additive Manufacturing of Alumina Parts"; PhD Thesis, Vienna University of Technology, Vienna, Austria, 2013.

[5]   R. Felzmann, S. Gruber, G. Mitteramskogler, M. Pastrama, A. R. Boccaccini, and J. Stampfl. "Lithography-based Additive Manufacturing of Customized Bioceramic Parts for Medical Applications", in Biomed. Eng. ACTAPRESS, Innsbruck, Austria, 2013.

[6]   G. Mitteramskogler, „Generative Fertigung von Bauteilen aus TCP"; Master Thesis, Vienna University of Technology, Vienna, Austria, 2011.

[7]   Liska R, Patzer J, Stampfl J, Wachter W, Appert C, Technische Universität Wien. „Device and method for processing light-polymerizable material for building up an object in layers." WO 2010/045951 A1; 2010.

[8]   Stampfl J, Liu HC, Nam SW, Sakamoto K, Tsuru H, Kang S, et al. „Rapid prototyping and manufacturing by gelcasting of metallic and ceramic slurries." Mater Sci Eng A 2002; 334:187–92.1.

[9]   Gruber, S., Stampfl, J., Ebert, J., "Device For Processing Photo-Polymerizable Material For Layer-By-Layer Generation Of A Shaped Body", WO2015075094 A1, 2015 May 28

ADDITIVE MANUFACTURING (3D PRINTING) OF CERAMICS: MICROSTRUCTURE, PROPERTIES, AND PRODUCT EXAMPLES

P. Karandikar, M. Watkins, A. McCormick, B. Givens, and M. Aghajanian
II-VI M Cubed
1 Tralee Industrial park
Newark, DE 19711

ABSTRACT

Additive manufacturing offers significant advantages (over conventional processing) including enabling design freedom, complex shape capability (e.g. cooling channels), near-net shape capability, the resultant ability to eliminate joining and reduce part count, elimination of some machining, reduced raw material consumption, reduced tooling cost, reduced prototyping and production time, and overall lower cost. However, several challenges have to be overcome to make additive manufacturing a viable production process including achieving the same composition, material properties (static, dynamic, mechanical, physical, chemical, electrical, etc.), isotropy of properties, feature definition, and overall production cost. Also, for some simple shapes (e.g. blocks, ingots) and extremely high volume (100s of thousands/month) small parts, other processes can be more cost effective. In this work, microstructure and properties of additively manufactured (3D printed) reaction bonded (RB) ceramics are presented. The ability to systematically vary process parameters to produce RB ceramics of varying compositions and properties is demonstrated. The properties and microstructures of additively manufactured RB ceramics are compared with the properties and microstructure of the RB ceramics made by conventional processing. Finally, the ability to produce complex-shaped components is demonstrated with some examples.

INTRODUCTION

Conventional component manufacturing involves making a billet (casting/forging/HIPing etc.) and CNC machining to shape. Parts could also be cast or HIPed near net shape and CNC machined to final tolerances. All these processes require removing unwanted material and thus are subtractive. Also, when complex shaped parts have to be made, e.g. with cooling channels and significantly varying cross sections; or systems have to be built with multiple parts; joining processes such as welding, brazing, soldering, preform bonding, etc. have to be employed, increasing manufacturing cost.

Additive manufacturing[1-18], on the other hand, is a process of joining materials to make objects directly from the 3D solid model data, usually layer-by-layer (ASTM Standard F2792 Committee F42). Benefits of additive manufacturing include the following:

- Eliminate costly machining
- Eliminate patterns/tooling (cost & time, iterations for shrinkage)
- Minimize/eliminate green machining (ceramics) – reduce raw materials wastage
- Eliminate preform bonding, welding, brazing, soldering – reduce part count
- Create flow-through components directly
- Reduce cycle time
- Improve uniformity, enable functional grading
- Design freedom – complexity, quick changes, unique new products

There are a variety of additive manufacturing processes that have been explored for more than 30 years[1-18]. These vary depending on the material being processed and have been traditionally given many different names such as stereo lithography (SLA), laminated object manufacturing (LOM), 3D printing, rapid prototyping, etc. More recently, the ASTM committee F42 has developed official categories of AM processes. These are described in brief below. Schematics of these processes are shown in Figure 1.

Vat Polymerization

This is one of the oldest AM processes and is also known as stereo lithography (SLA)[8-10]. A schematic of the process is shown in Figure 1a. In this process, a photo polymer is contained in a bath. The part is built layer-by-layer on a build bed by curing a thin layer of the photo polymer with a UV laser at locations where the part exists. After curing of the layer, the build bed moves down by a distance equivalent to the layer thickness and the laser curing process is repeated for the freshly exposed photopolymer layer. Some examples of commercial machines for this process are 3D Systems SLA Viper, Materialise Mammoth, and Envisiontec. This process is extensively used for making plastic patterns for making metal and ceramic casting tooling. Based on the layer thickness, some post processing of the parts may be needed to get a very smooth surface finish.

Material Jetting

A schematic of the material jetting process is shown in Figure 1b. In this process, droplets of multiple materials can be jetted onto the build bed. For example, an ink containing fillers can be deposited via a print head. In addition, a liquid polymer can be jetted using nozzles. For each layer, material is selectively jetted only where the part exists. One of the materials can be used to define the part and the other can be used to deposit a support/containment structure. A lamp can provide the light to cure the deposited polymer. After curing of one layer, the build bed moves down by a distance equivalent to the layer thickness and the material jetting and curing processes are repeated to build the next layer. Some examples of commercial machines for this process are Stratasys Object Connex, 3D Systems ProJet, Optomec 3D, and nScrypt 3Dn. This process can be used, in combination with the material extrusion process, for printing electronic components and systems. The support layer could be wax which can be melted or burnt away. Alternatively, the support layer could a dielectric containing conducting lines.

Powder Bed Fusion

A schematic of the powder bed fusion process[11, 12] is shown in Figure 1c. In this process, a small amount of powder is dispensed from a hopper in front of a roller. The roller spreads the powder on the build bed in the form of a thin layer (e.g. 100 μm). Next, energy (e.g. laser or electron beam) is applied selectively where the part exists to melt and sinter the powder at that location. The powder where energy is not applied, serves as the support structure. After the energy is applied to one layer, the build bed moves down by a height equivalent to the layer thickness and the energy application process is repeated to build the next layer. Since, very high temperatures are needed for melting metal powders, an inert atmosphere or vacuum are needed to prevent oxidation/reaction of the powder. This adds significantly to the cost of the equipment. After the build is complete, the un-bound powder is removed by vacuuming or blowing away leaving behind a part. Both polymer and metal components have been successfully produced by this method. The process is also called selective laser sintering (SLS) or direct metal laser sintering (DMLS). The as processed metallic components may need hot isostatic pressing (HIPing) to achieve properties equivalent to wrought products[3]. Also, attention has to be paid to the directionality (anisotropy) of microstructure and properties. Some examples of commercial machines for this process are 3D Systems SPro (laser), EOSINT M280 (laser), and Arcam A2 (e-beam). Some of the examples of components produced by this method include the fuel nozzle and T25 sensor by the General Electric Company[13, 14].

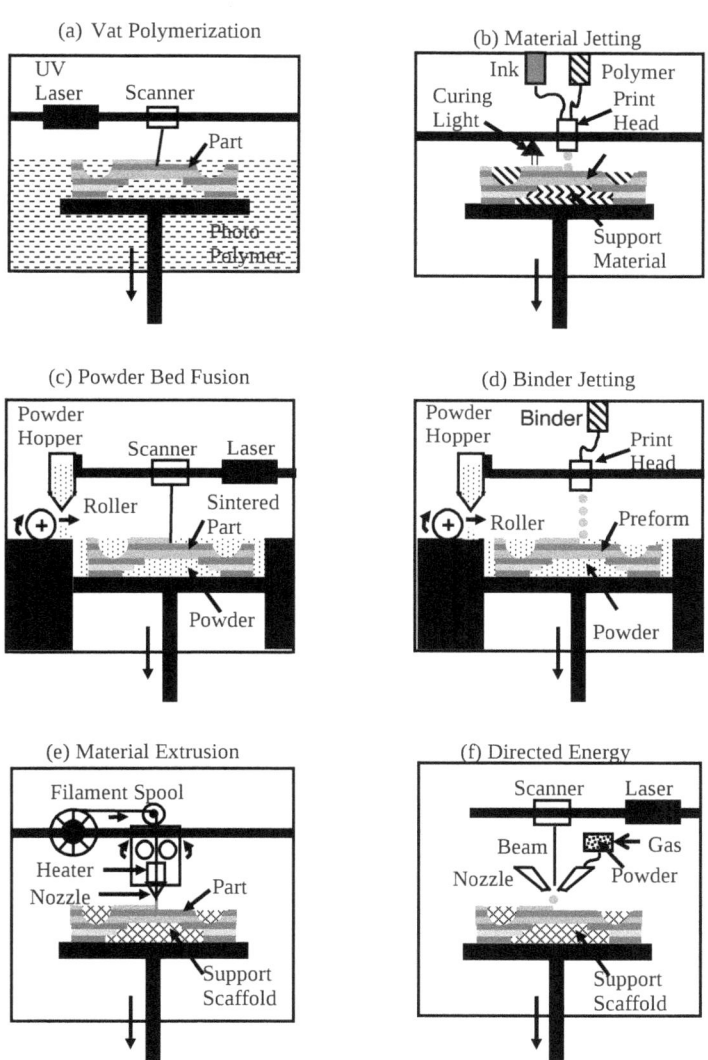

Figure 1. Various additive manufacturing processes per the ASTM classification [1-3].

Using AM the part count of the fuel nozzle assembly was reduced from 20 to 1, and its weight was reduced by 25%. Each CFM LEAP engine will have 19 such nozzles, and thousands of such engines are expected to be produced over the next 5 years. Thus, the fuel nozzles are going to be in high volume production by additive manufacturing by 2018 [13, 14].

Binder Jetting

A schematic of the binder jetting process [15] is shown in Figure 1d. In this process, a small amount of powder is dispensed from a hopper in front of a roller. The roller spreads the powder on the build bed in the form of a thin layer (e.g. 100 µm). Next, droplets of a binder material are deposited selectively where the part exists via a print head, much like a commercial paper ink-jet printer. The binder reacts with the powder and binds it locally. The powder where there is no binder deposited, serves as the support structure. After the binder is jetted on one layer, the build bed moves down by a height equivalent to the layer thickness and the binder jetting process is repeated to build the next layer. The build bed can be warmed up after the part build is complete to further strengthen/cure the binder. After this, the un-bound powder is removed by vacuuming or blowing away leaving behind a powder preform. This preform can then be infiltrated with other materials (polymers, metals etc.) to fill all the porosity or it can be sintered to full density. This is the process that is commonly known as 3D printing. Some examples of commercial machines for this process are 3D Systems ProJet 860 Pro (formerly Z-Corp), ExOne S-print, and Voxeljet VX1000. Some of these machines are extensively used for creating sand molds for casting metals.

Material Extrusion

A schematic of the material extrusion process is shown in Figure 1e. In this process, a polymeric filament is pulled through a nozzle, locally heated, and deposited selectively [16]. Typically, thermoplastic filaments are used as they can be softened on reheating. This process is predominantly used for polymeric materials such as ABS, polycarbonate, and Ultem. Most of the table top, low cost (as low as $300), do-it-yourself (DIY) printers use this methodology. Example printers include Stratasys Fortus, Stratasys Dimensions, Bukito, Makerbot, RepRap, etc. In this process, a support structure may have to be deposited for certain geometric features such as overhangs.

Directed Energy

A schematic of the directed energy processes is shown in Figure 1f. In this process, focused thermal energy is used to fuse or melt a material as it is being deposited. The material could be supplied in the form of a powder or wire into a laser or electron beam, melted and then deposited selectively. Examples of equipment for this process include Optomec LENS 850R, and Sciaky EBAM.

Sheet Lamination (Laminated Object Manufacturing)

This is one of the oldest AM processes. In this process, individual sheets of a material (e.g. paper, polymer, or metal) are cut selectively or cured selectively [17]. Layers are subsequently stacked to build the part. In the cases where the sheets are selectively cured, the uncured portion is removed by means such as grit blasting leaving behind a part. Examples of equipment for this process are LOM2030 (Helisys – Cubic Technologies) and MCOR ARKE. This process has been extended to make C_f/polymer composites [18].

LIMITATIONS OF ADDITIVE MANUFACTURING

While part manufacturing based on AM shows significant promise and advantages, there, however, are some limitations to these processes. These include:

- AM may not be cost competitive for making large chunks of materials.
- AM may not be cost competitive for extremely high volume: e.g. injection molding could still be more cost effective for small parts to be made in millions/day volumes
- Fugitive support structures have to be used in certain AM processes
- Materials available for AM processes are limited and expensive (captive supply)
- There is potential for property anisotropy in parts produced by some AM processes.
- Significant investment in AM equipment is needed to displace installed conventional production equipment. Thus, significant ROI (return-on-investment) is needed.

While significant work has been done to develop AM process for polymers and metals, and significant publications exist in this regard, very limited work has been reported in terms of AM of engineering ceramics[5-7], especially with respect to demonstrating shape, size, complexity capability and application to product development. This is the focus of this work.

CONVENTIONAL CERAMIC COMPONENT MANUFACTURING BY REACTION BONDING (RB)

Reaction bonding is a very versatile process for making components out of materials such as SiC, B_4C, diamond, TiB_2, carbon fibers, and SiC fibers. M Cubed currently supplies components made by reaction bonding to a variety of markets[19-25]. In the conventional reaction bonding process (silicon-based matrices), good wetting and highly exothermic reaction between liquid silicon and carbon are utilized to achieve pressure-less infiltration of a powder preform. This process has been given many names such as reaction-bonding, reaction-sintering, self-bonding, and melt infiltration. A schematic of the reaction bonding process[22] is shown in Figure 2. The steps in the process are as follows: (1) Mixing of ceramic powder, reinforcements (e.g. carbon fibers, SiC fibers, carbon nanotubes-CNT), and a binder to make a slurry, (2) Shaping the slurry by various techniques such as casting, injection molding, pressing etc., (3) Drying and carbonizing of binder, (4) Green CNC machining, (5) Infiltration (reaction bonding) with molten Si (or alloy) above 1410°C in an inert/vacuum atmosphere, and (6) Solidification and cooling. During the infiltration step, carbon in the preform reacts with molten Si forming SiC around the original ceramic particles, bonding them together – hence the term reaction bonding.

Due to the good wetting and high capillary forces this process is very robust and large complex shaped components can be manufactured (can be as big as 2.5 m). Also, the dimensional change from green to finished state is < 0.5%, so most machining can be done in the green state. RB SiC material is EDMable which opens it up to machining unique features. There are also patented brazing (US patent 7,270,885) and bonding (US patent 6,863,759) technologies for making flow-through components (cooled mirror). The process is amenable to significant tailorability via control of particle size, Si-alloying, residual Si etc. Using this technology, M Cubed produces directly polishable SiC[25] (OptiMum®), extremely flat SiC (MESA®), very high thermal conductivity SiC/diamond[24] (TherMadite® , 540 W/mK, CTE = 1.5 ppm/K)); high ballistic resistance B_4C/diamond[23] (DyMonite®); and higher toughness SiC with carbon fiber (C_f/Si/SiC) and carbon nanotube- CNT (CNT/Si/SiC)[20]. Figure 3 shows examples of components made by this process.

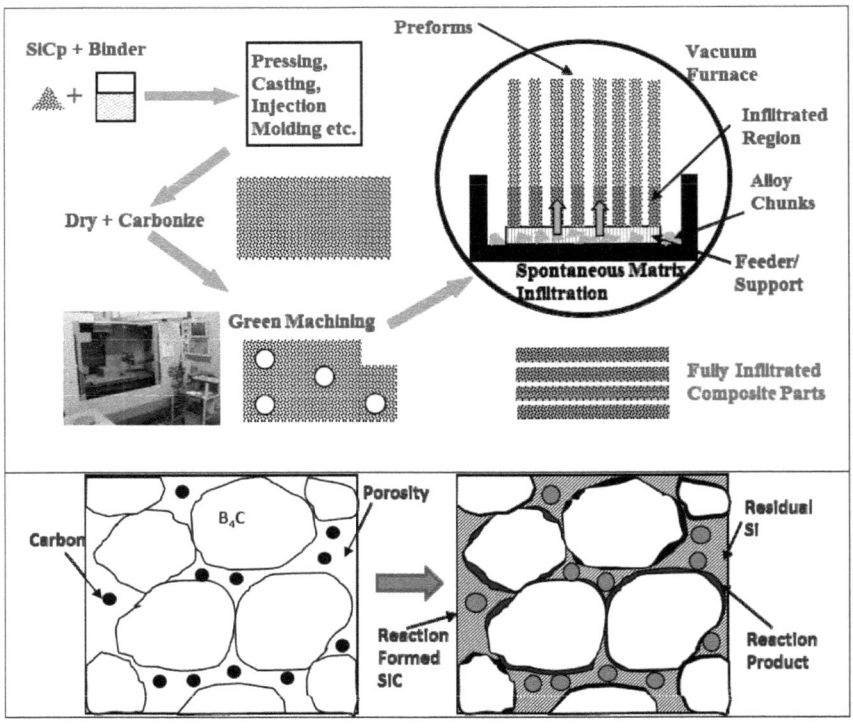

Figure 2. A schematic of conventional RB SiC component fabrication at M Cubed[22]

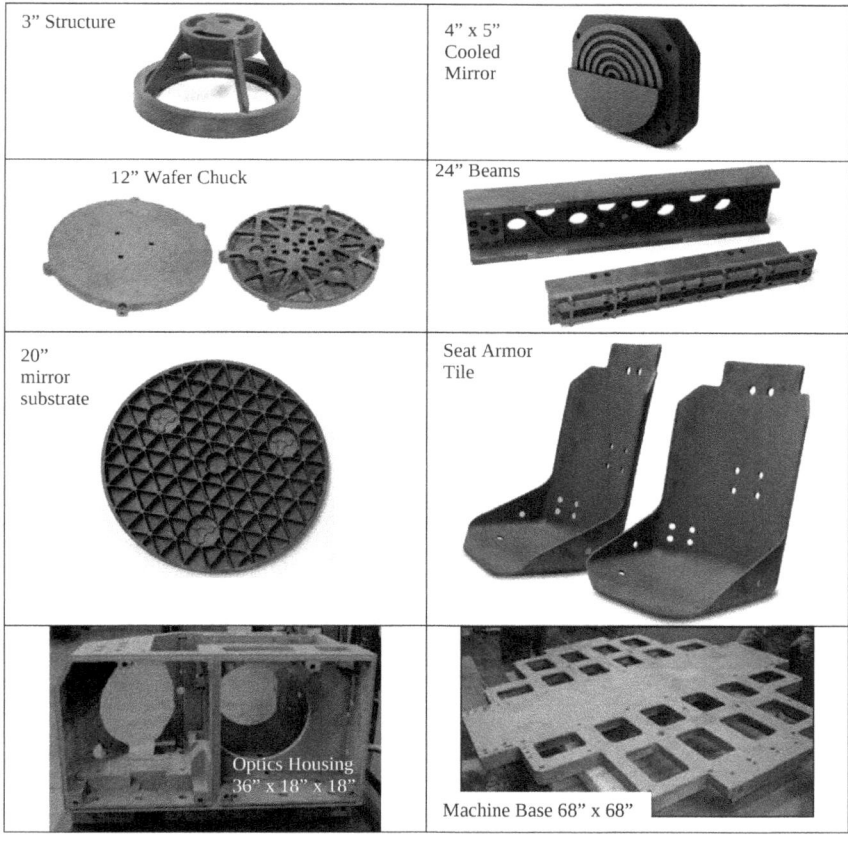

3" Structure		4" x 5" Cooled Mirror	
12" Wafer Chuck		24" Beams	
20" mirror substrate		Seat Armor Tile	
Optics Housing 36" x 18" x 18"		Machine Base 68" x 68"	

Figure 3. Examples of components made by conventional reaction bonding at M Cubed.

3D PRINTED RB SiC FABRICATION AND CHARACTERIZATION

The goal of this study is to eventually be able to fabricate preforms for all these components by additive manufacturing. However, there is a long validation process to achieve this goal. The steps in this process include: (1) Demonstrate that equivalent material properties can be produced, (2) Demonstrate the capability to produce shapes, (3) Demonstrate the ability to achieve component sizes, (4) Demonstrate the ability to achieve feature fidelity competitive with CNC green machining, and (5) Demonstrate that all of the above can be done at a lower overall cost (including raw materials, labor, machining costs, turn-around time, and quality).

This study is thus just the beginning of this process, addressing a few of the above mentioned steps to various degrees. Specifically, for this study, 4" x 4" x 0.25" SiC preforms were made by a binder jetting (3D printing process). These preforms were then subjected to binder removal and reaction bonding. The reaction bonded plates were subjected to microstructural, physical, and mechanical characterization. The densities of the test samples were determined using the Archimedes principle per ASTM C373. For microstructural observations, specimens were sectioned and polished and observed by optical microscopy. Specimen elastic moduli were measured using the ultrasonic pulse-echo technique (ASTM E494-05). The flexural strengths were determined using a four-point bend testing apparatus per ASTM C1161. Fracture toughness measurements were made by the Chevron notch method (ASTM C1421). Strength and toughness measurements were made on at least 10 specimens of each type and average values were reported.

MICROSTRUCTURE AND MECHANICAL PROPERTIES

Microstructures of RB SiC of various SiC volume fractions with monomodal SiC particles made by 3D printing are shown in Figure 4. Microstructures of RB SiC of various SiC volume fractions with bimodal SiC particles made by 3D printing are shown in Figure 5. In both figures 4 and 5, microstructure of conventionally made RB SiC with bimodal SiC particles is also included for comparison. As can be seen, microstructures similar to that of conventionally made material are achieved by 3D printing.

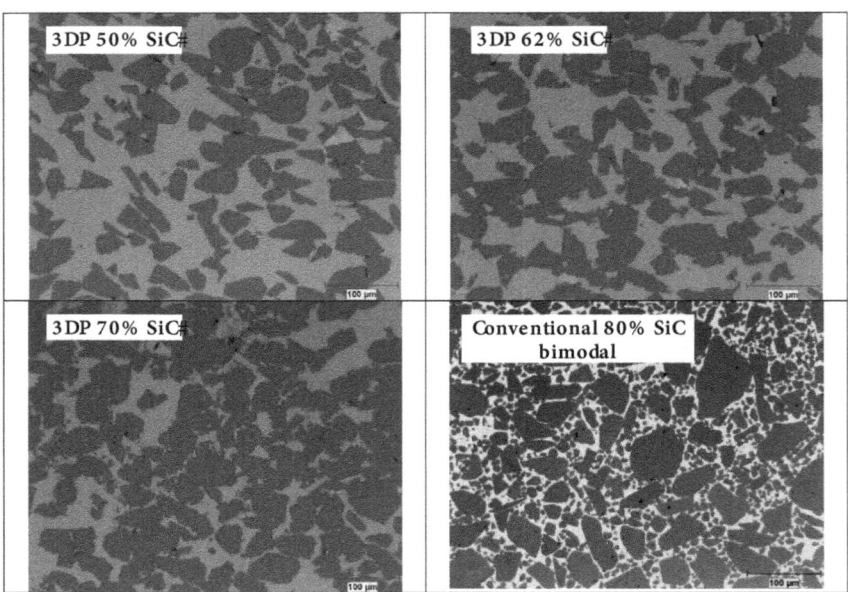

Figure 4.   Microstructures of RB SiCs made by 3D printing using monomodal SiC particles and comparison with microstructure of RB SiC made by the conventional approach (bimodal SiC particles). Darker particles are SiC, lighter phase is Si.

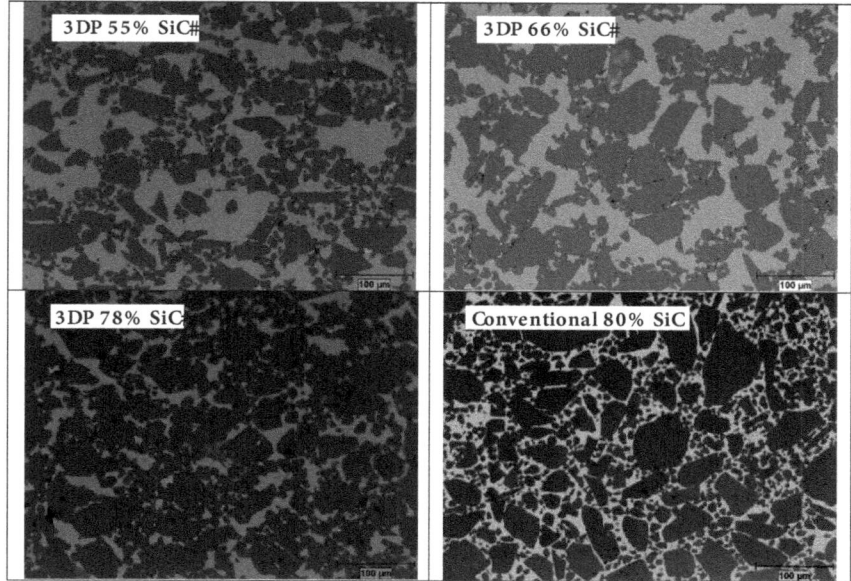

Figure 5. Microstructures of RB SiCs made by 3D printing using bimodal SiC particles and comparison with microstructure of RB SiC made by the conventional approach (bimodal SiC particles). Darker particles are SiC, lighter phase is Si.

Densities and elastic moduli of RB SiC of various SiC volume fractions with monomodal as well as bimodal SiC particles made by 3D printing are shown in Figure 6. As can be seen, density and elastic modulus almost as high as those of the conventionally made material were achieved by 3D printing with the material with bimodal SiC particles.

Flexural strength and fracture toughness of RB SiC of various SiC volume fractions with monomodal and bimodal SiC particles made by 3D printing are shown in Figure 7. The flexural strength of the RB SiC made with bimodal SiC particles made by 3D printing is comparable to that of the conventionally made bimodal material. The monomodal 3D printed material with larger grain size shows lower strength. In RB materials, flexural strength is known to be lower for materials with coarser particle size[21]. The fracture toughnesses of 3D Printed material with both monomodal and bimodal particles are slightly lower than that of the conventionally made bimodal material.

In summary the density, elastic modulus, and flexural strength very similar to those of conventional material have been achieved by 3D printing, while the fracture toughness of 3D printed material is slightly lower.

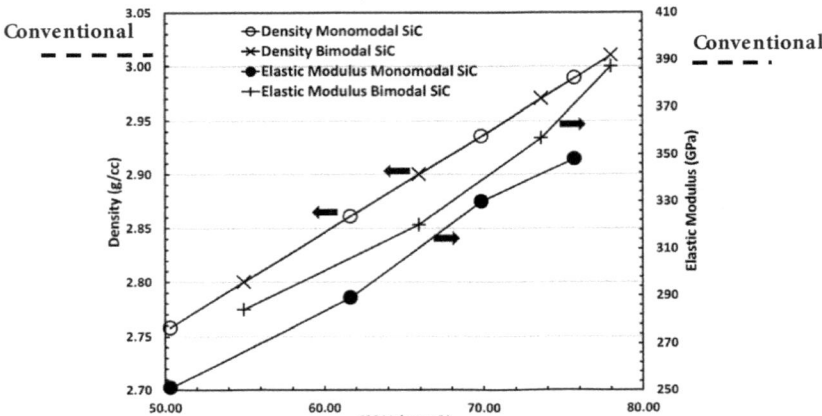

Figure 6. Density and elastic modulus of 3d Printed RB SiC (monomodal and bimodal SiC) as a function of SiC content.

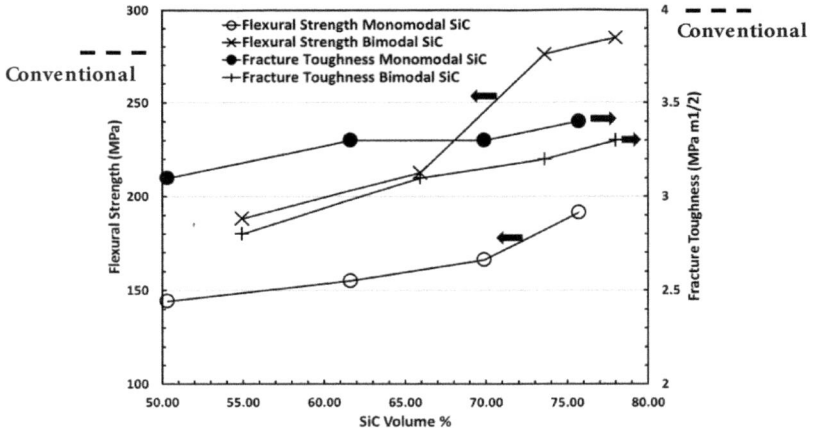

Figure 7. Flexural strength and fracture toughness of 3D printed RB SiC (monomodal and bimodal SiC) as a function of SiC content.

PRODUCT EXAMPLES

For evaluating and demonstrating the process capability, components of a variety of shapes and complexities were printed with monomodal and bimodal SiC particles and then reaction bonded. Figure 8 shows as printed and reaction bonded chain links, bolts, nuts, universal joints, bearing housings, thick and thin walled lightweighted M Cubed logos, and flow-through components.

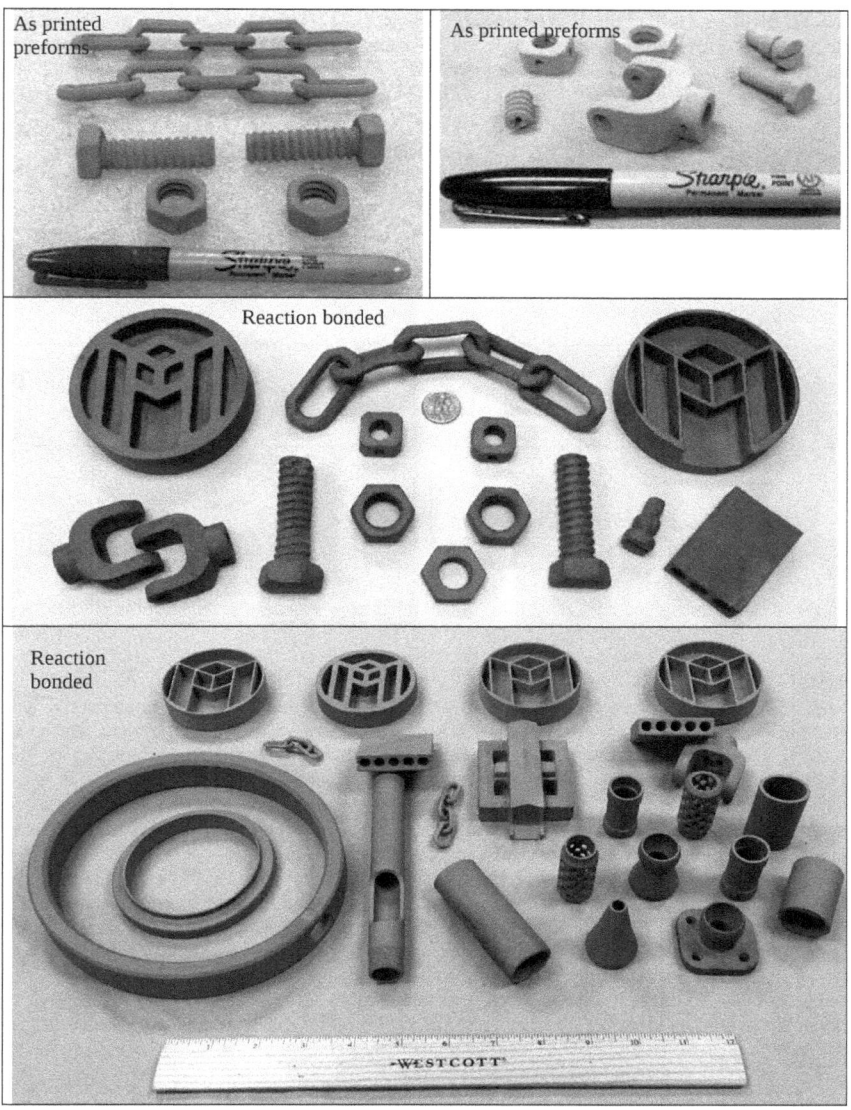

Figure 8. Example of products of various shapes and complexities fabricated by 3D Printing

Reasonable feature definition was obtained on most components. To evaluate and demonstrate the capability to make larger components, 10" (250 mm) diameter lightweighted flat and spherical mirror components were printed (Figures 9 and 10) using the coarse monomodal SiC particles and were reaction bonded. Again, reasonable feature definition was achieved. The mirror face of the flat substrate was also ground to assess grinding ability.

Figure 9. Lightweighted 10" (250 mm) flat mirror substrate fabricated by 3D Printing - Left: As printed preform, Right: Reaction bonded back and front (ground).

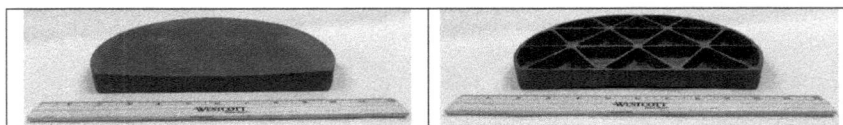

Figure 10. Lightweighted 10" (250 mm) spherical mirror substrate fabricated by 3D Printing and reaction bonding. Curvature of the front face is apparent in the photo on the left.

SUMMARY

A process was developed to 3D Print SiC ceramic preforms using monomodal and bimodal SiC particles. As expected, higher ceramic loading was achieved with bimodal particles. Density, elastic modulus, and flexural strength comparable the respective properties of the conventionally made material were achieved using 3D printing with bimodal SiC particles. Fracture toughness of 3D printed material was slightly lower. A variety of components were successfully printed, debindered, and infiltrated (reaction bonded) to demonstrate process capability. Components with many different shapes and complexities were printed such as: threaded bolts, nuts, universal joints, bearing cages, tubes, cones, threaded tubes, nozzles, rings, flow through components, novel design armor ceramic tiles, finned thermal management components, and up to 12" diameter thin ribbed components (lightweighted mirrors and chucks).

Work continues to further optimize the process to achieve even higher ceramic loading and even better feature definition. In addition, head-to-head manufacturing cost comparison is being conducted with the conventional processes. The 3D printing process developed is applicable to making preforms of other ceramics, preforms for other ceramic processes or metal fabrication processes (e.g. sintering), and for metal matrix composites (MMCs).

REFERENCES

[1] Yan X. and Gu, P., "A review of rapid prototyping technologies and systems," Computer Aided Design, Vol. 28[4] (1996) 307-318.

[2] Stansbury, J. W. and Idacavage, M. J., "3D printing with polymers: challenges among expanding options and opportunities," Dental Materials 32 (2016) 54-64.

[3] Frazier, W. E., "Metal additive manufacturing: a review," Journal of Materials Engineering and Performance, Volume 23(6) (2014) 1917-1928.

[4] Herbert, R. J., "Viewpoint: metallurgical aspects of powder bed metal additive manufacturing," J Mater Sci 51 (2016) 1165-1175.

[5] Zocca, A., Colombo, P., Gomes, C. M., and Gunster, J., "Additive manufacturing of ceramics: Issues, potentialities, and opportunities," J. Am. Ceram. Soc. 98 [7] (2015) 1983-2001.

[6] Moon, J., Grau, J. E., Knezevic, V., Cima, M., and Sachs, E. M., "Ink-jet printing of binders for ceramic components," J. Am. Ceram. Soc. 85 [4] (2002) 755-762.

[7] Lewis, J., Smay, J. E., Stuecker, J., and Cesarano, J., "Direct ink-jet writing of three dimensional ceramic structures," J. Am. Ceram. Soc. 89 [2] (2006) 3599-3609.

[8] Baese, G., "Photographic process for the reproduction of plastic objects," US Patent 774,549 (1902).

[9] Swainson, W., "Method, medium, and apparatus for producing there-dimensional figure product," US Patent 4,041,476. (1977)

[10] Herbert, A., "Solid object generation," Journal of Applied Photographic Engineering Vol. 8 [4] (1982).

[11] Ciraud, L., "A method and apparatus for making any belongings from any fusible material," German patent application DE 19722263777A1 (1973).

[12] Deckard, C., "Method and apparatus for producing parts by selective laser sintering," US Patent 4863538 (1989).

[13] http://www.geglobalresearch.com/innovation/3d-printing-creates-new-parts-aircraft-engines

[14] http://www.gereports.com/post/116402870270/the-faa-cleared-the-first-3d-printed-part-to-fly/

[15] Sachs, E., Haggerty J., Cima, M., Williams, P., "Three dimensional printing techniques," US Patent 5,204,055 (1993).

[16] Crump, S., "Approaches and method for creating 3D objects," US Patent 5,121,379 (1992)

[17] Zang, E., "Vitavue relief model technique," US Patent 3,137,080 (1964)

[18] http://www.impossible-objects.com

[19] Karandikar, P., Aghajanian, M., and Morgan, B., "Complex, net-shape ceramic composite components for structural, lithography, mirror and armor applications," Ceramic Engineering Science Proceedings (CESP) Vol. 24, [4] (2003) 561-566.

[20] Karandikar, P., Evans, G., and Aghajanian, M., "Carbon nanotube and carbon fiber reinforced high toughness reaction bonded composites," CESP Vol. 28 [6] (2007), 53-64.

[21] Salamone, S., Karandikar, P., Marshall, A., Marchant, D., and Sennett, M., "Effects of Si:SiC ratio and grain size on the properties of RBSC," CESP Vol. 28 [2] (2007) 101-109.

[22] Karandikar, P., Wong, S., Evans, G., and Aghajanian, M., "Microstructural development and phase changes in reaction bonded B_4C," CESP Vol. 31 [5] (2010) 251-259.

[23] Karandikar, P. and Wong, S., "Development of reaction bonded B_4C-diamond composites," CESP Vol. 33 [5] (2012) 51-59.

[24] Salamone, S., Neill, R., Aghajanian, M., "Si/SiC and diamond composites: microstructure-mechanical properties correlation," CESP Vol. 31 [2] (2010) 97-106.

[25] Aghajanian, M., Emmons, C., Rummel, S., Barber, P., Robb, C., and Hibbard, D., "Effect of grain size on microstructure, properties and surface roughness of reaction bonded SiC ceramics," Proc. of SPIE Vol. 8837 88370J-1 (2013).

Geopolymers, Chemically Bonded Ceramics, Eco-Friendly, and Sustainable Materials

IMPACT OF VARIOUS ALUMINOSILICATE COMPOUNDS IN GEOPOLYMER FOAM FORMATION TO A SI/M=0.7 OF SILICATE SOLUTION

M. Arnoult[1], M. Perronnet[2], A. Autef[2], G. Gasgnier[2], S. Rossignol[1*]
[1] Science des Procédés Céramiques et de Traitement de Surface, Centre Européen de la Céramique, 12 rue Atlantis, 87068 LIMOGES CEDEX
[2] Imerys Ceramics, Imerys Ceramic Centre, 8 rue Soyouz 87068 Limoges, France
* Corresponding author: sylvie.rossignol@unilim.fr, tel.: 33 5 87 50 25 64

ABSTRACT
Mineral foams are expected to be used in many technological applications. Due to their low thermal conductivity, good heat resistance and acoustic properties, this type of material is suitable for insulation applications. This work focuses on the impact of reactants used for geopolymer foams synthesis. The results provide evidence of relationship between the chemical composition, the foam morphologies and the thermal conductivity. Reactants (alkaline solution and metakaolin), are the key parameters to understand the foam formation and the working properties. The alkaline solution impacts the foam morphologies. The solution's water content impacts the volume expansion of the foam. Moreover, the type of the cation influences the pore size distribution. Finally, relationship were demonstrated between the thermal conductivity, the [Si] molar concentration and the ratio M/Al (M=alkali cation Na, K). The optimal thermal conductivity was obtained for foams based on sodium solution with a [Si]=3.52 Mol.l^{-1} and a M/Al = 3.92. Thus, this work shows that chemical composition controls the working properties of geopolymer foams.

INTRODUCTION

Geopolymers are amorphous three dimensional aluminosilicate binder materials, which were first introduced to the inorganic cementitious world by Davidovits in 1978[1]. Geopolymers are synthetized at room temperature by alkaline activation of alumino-silicates[2]. Since geopolymers are intrinsically nanoporous materials, it could be used to develop porous materials covering pore size ranging from nanometer to millimeters[3]. Those materials have gathered increasing interest because of their synthesis method, high working performances, wide range of applications and low environment impact. Geopolymer foam can be obtained by in situ formation at low temperature by adding a foaming agent to a geopolymer reactant mixture[4]. Among foaming agents, it can be cited hydrogen peroxide, silica fume, silicon carbide, aluminium powder[5,6,7,8]. As example, silica fume used as foaming agent, containing small amount of free silicon, under basic conditions, free silicon is oxidized by water, induces the formation of hydrogen (1)

$$4H_2O + Si^0 \rightarrow 2H_2 + Si(OH)_4 \qquad (1)$$

Regardless of the used foaming agent to synthetize foams, the chemical composition controls the pore distribution [9] and, therefore, defines the application. Geopolymer foams could be applied for many applications because of their high gas permeability, high temperature resistance, thermal shock resistance and low toxicity thermal insulators[10,11].

In the last few years, numerous studies have focused on the impact of reactants used for geopolymer synthesis. Alkaline solution and aluminosilicate source strongly affect the geopolymer properties. Concerning the alkaline silicate solution, there are various parameters that could influence the polycondensation reaction such as the alkali cation nature, the water content, the Si/M ratio where (M=Na or K) or the siliceous species present in the solution. Recently, Vidal et al.[12] investigated the role of Si/M of the solution on geopolymerization reaction. It was demonstrated that the most reactive solution is obtained with a Si/M molar ratio of 0.7. The nature of alkali cation has an effect on the formation of the rings, determined by Raman spectroscopy, and thus affects the geopolymerization reaction[13]. In addition, the role of aluminosilicate reactants has been extensively studied. Metakaolin is the most commonly used raw material because of its high reactivity and purity[14]. Gharzouni et al.[15] highlighted that the metakaolin reactivity is linked to the amorphous phase and the wettability of the material. A previous work[16] evidenced a correlation between the metakaolin features and the reactivity of the geopolymer mixture. The metakaolin reactivity is responsible of either the formation of one or several networks. A notably reactive metakaolin dissolves quickly and allows the formation of a perfect geopolymer network. A highly reactive alkaline solution favors the geopolymer network towards other possible formed networks.

This work aims to understand the impact of the influence of alkaline silicate solution and aluminosilicate source on geopolymer foam properties. This study focuses on the chemical features such as pore size, volume expansion and working properties such as thermal conductivity of geopolymers. A correlation was established in function of the alkaline solutions and the different metakaolins.

EXPERIMENTAL
Sample preparation and nomenclature
    Two commercial sodium and potassium silicate solutions (supplied by Wöllner, France) (named S_{Na}, S_K) (Table I) were used with different metakaolins (M_K) provided by Imerys (France) to synthesize the geopolymer foam samples. The different metakaolins used for this study are classified in Table II. A silica fume (0.38 wt.% of free silicon, 0.22 wt.% of free carbon, 98 wt.% of silica) supplied by Ferropem (France) was also used as a pore foaming agent in each mixture. The experimental protocol of geopolymer foam synthesis is described in Figure 1. The NaOH and KOH hydroxide pellets were mixed with an automatic stirring device in the alkaline solutions and then allowed to cool to room temperature. Then, the aluminosilicate material M_K was added. The reactive mixture was mixed with silica fumed and then placed in a closed sealable polystyrene mold at 70 °C for 24 hours in order to complete the polycondensation reaction. For example, foams are synthetized with 10 g of metakaolin, 17 g of silica fume and 29.09 g, 28.46 g, 28.80 g and 31.3 g for S_{KK}, S_{KNa}, S_{NaNa} and S_{NaK} respectively.

Table I. Physical and chemical features of silicate solutions before and after the addition of hydroxide pellets

Solutions	Si/M	Si/H$_2$O
SiO$_2$ / Na$_2$O (S_{Na})	1.69	0.14
SiO$_2$ / K$_2$O (S_K)	1.71	0.07
S_{Na} + NaOH (S_{NaNa})	0.72	0.14
S_K + KOH (S_{KK})	0.70	0.07
S_K + NaOH (S_{KNa})	0.70	0.07

    The nomenclature used is $S_{XM}M_K$, where S_X is the type of solution with x the alkali cation present in the starting silicate solution (S_{Na} or S_K); M is the type of hydroxide pellet (Na: NaOH or K: KOH); M_K is the type of metakaolin used in the reactive mixture. For example, the $S_{KK}M_1$ product is a geopolymer foam made from the dissolution of potassium hydroxide pellets in the potassium silicate solution S_K with the addition of metakaolin M_1 and silica fume.

Sample characterization
    After 24 hours of cure, the foam volume was measured. The volume expansion E_V (2), ie. the ratio of the foam volume after consolidation to the initial volume of liquid introduced V_0, was evaluated (cm^3).

$$E_v = \frac{V}{V_0} \tag{2}$$

The bulk density ρ of the foam is given by the mass of a cylinder of foam divided by its apparent volume. The density of the dense matrix ρ_0 is the theoretical density of solid material. It is determined using an Accupyc 1330 helium pycnometer (Micrometrics). The sample was crushed and passed through a 63μm sieve before measurement. Then an estimation of the pore volume fraction Xp can be determined with the relation (3):

$$X_P = 100 * \left(1 - \frac{\rho}{\rho_0}\right) \tag{3}$$

The pore size distribution of a sample cut at a given height (50%) is evaluated from pictures using the Image J software. The pores are assumed spherical and the mean pore diameter Γ_V is calculated for each cut using the equation (4)[17].

$$\Gamma_V = \frac{\sum_{i=0}^{n} n_i d_i^{\,4}}{\sum_{i=0}^{n} n_i d_i^{\,3}} \tag{4}$$

Where d_i=pore diameter for class i, n_i/n = number of pore inside the class i/total number of pores.

The thermal conductivity was measured using a hot-disk thermal analyser (TPS 1500, Hot disk) at room temperature. The measuring time, the radius of the disk sensor and the output powder used were 80 seconds, 6.394 mm and the 0.052 W respectively. To achieve a relatively precise result, measures were made on two different samples, three times.

Table II. Physical and chemical features of metakaolins (Si/Al data are extratected from X-ray fluorescence spectrometry)

Code	Si/Al	Amorphous phase (%)	Wettability (μL/g) ± 2	Mineral phases	
M_1	1.20	87	1250		
M_2	1.34	82	1065		
M_3	1.51	77	880		
M_4	1.48	72	805		Hematite; Rutile
M_5	1.52	70	807	Quartz; Anatase; Muscovite; Kaolinite; Metakaolinite	Hematite Rutile
M_6	1.54	78	731		Hematite; Rutile
M_7	1.53	67	705		Calcite; Hematite
M_8	1.53	58	734		Hematite; Rutile
M_9	1.44	64	530		Calcite; Hematite;
M_{10}	1.46	49	570		Calcite; Hematite; Rutile
M_{11}	1.41	53	602		Calcite; Hematite
M_{12}	1.38	67	595		Calcite; Hematite

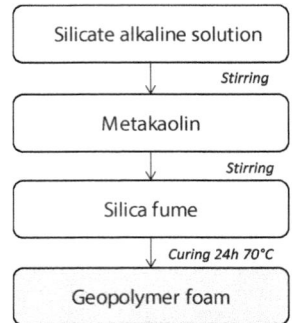

Figure 1. Synthesis protocol of geopolymer silica fume foam samples

RESULTS

Foam formation

In order to understand the foam formation, the volume expansion of the synthetized foams was observed. Figure 2 presents the evolution of the foam volume expansion as a function of the ratio $\frac{n(Si)}{n(Al)+n(M)}$. This ratio represents the geopolymer network formation similarly to vitreous materials[18]. Two types of samples are notable depending on the type of solution (cation used). At the exception of metakaolins M_1 and M_2, the synthesized foams from potassium silicate solution (S_{KK} and S_{KNa}) describe a first tendency and the other one is defined by the S_{NaNa} foams. The foams synthesized from the S_{KK} solution have a constant volume expansion regardless of the used aluminosilicate source. The substitution of KOH by NaOH in the potassium-based solution does not induce major modifications except for the two aluminosilicate sources M_1 and M_2. Nevertheless, the use of the alkaline sodium solutions S_{NaNa} or S_{NaK} lead to an increase in volume expansion values of the foam. The particular behavior of M_1 can also be notable whereas the M_2 based sample is located with the other aluminosilicate sources. These data reveal that the alkaline solution has a strong impact on foam volume expansion. The difference in behavior between the two solutions can be explained by their water content (Table I). Indeed, in the presence of a solution having high water content, the siliceous species are more polymerized and therefore the reactivity degree of the solution is lower limiting the oligomers formation[19]. Therefore, polycondensation reactions are modified and disrupt the H_2 production produced by water reduction [9]. Consequently, two behaviors can be distinguished; either the polycondensation reaction is favored at the expense of the gas formation thus decreasing the volume expansion of the foam or the inverse.

In the case of the $S_{KK}M_1$ sample, the volume expansion value obtained is 2.46. When the metakaolin changes to M_{10} ($S_{KK}M_{10}$), the volume expansion increases up to 3.11. Indeed, the metakaolins M_1 and M_2 react differently than the others since the oligomers are formed rapidly. Thus, the polycondensation reactions occur very quickly creating a solid network[20] limiting the volume expansion by trapping the gas. For $S_{KK}M_{11}$ foam, the low reactive solution associated with an aluminosilicate source containing a high amount of impurities (Table II) leads to a modification of the geopolymerization reactions. Indeed, it causes an incomplete dissolution and a slow oligomers formation[21] favoring the volume expansion of the foam. Moreover, the use of mixed cation solution characterized by a higher reactivity[22], with metakaolin containing impurities, leads to trap the produced gas. Thus, the volume expansion of the foam is promoted. Therefore, the H_2 production can be slowed, either by the kinetics or the thermodynamics of the reaction medium.

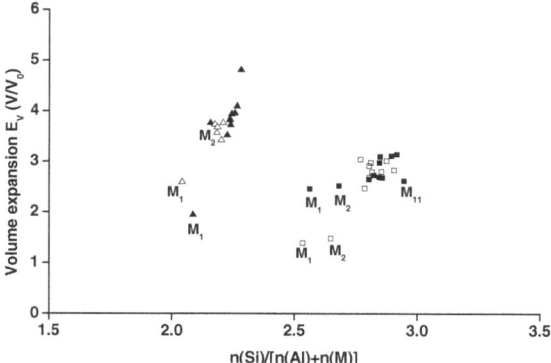

Figure 2. Evolution of the volume expansion value versus the chemical composition $(n_{Si}/(n_{Al}+n_M))$ of the different foams based on ■ S_{KK}, □ S_{KNa}, ▲ S_{NaNa} and △ S_{NaK} solutions

Foam morphologies

To assess the effect of aluminosilicate source and alkaline solutions on foam morphologies, the section of different synthesized foams were analyzed with Image J software, (Table III). For a fixed aluminosilicate source, differences in morphology are noticeable depending on the solution (a). In the case of M_{10}, the use of S_{KK} leads to the formation of small-sized pore, while with the S_{NaNa} solution the pore size of the foam increases. Foams synthetized with mixed solution (S_{KNa} and S_{NaNa}) present a smaller pore size distribution. For a given solution S_{NaNa}, different pore morphologies can be observed depending on the used metakaolin (b). The foam synthesized with the metakaolin M_1 shows a homogeneous distribution of small pores. Nevertheless, the use of M_6 metakaolin leads to the formation of larger pores with a heterogeneous distribution. The alkaline solutions (type of cation) as well as the aluminosilicate source modify the foams morphology. These data must be confirmed with the mean pore diameter of the pores.

Table III. Morphologies of foams (Ø = 5 cm) synthetized with different (a) metakaolins and (b) alkaline solutions

(a)

Alkaline solution	S_{KK}	S_{KNa}	S_{NaNa}	S_{NaK}
M10				

(b)

Metakaolin	M_1	M_4	M_6	M_8	M_{11}	M_{12}
S_{NaNa}						

In the interest of understanding the influence of the chemical composition on foam porous structure, the pores sizes were measured using the image software J. Figure 3 a,b shows pore size distribution of geopolymer foams samples synthetized with different alkaline solutions and two metakaolins. Whatever the used metakaolin, foams synthetized with S_{KK} solution present in each case a larger pore size distribution with at least two contributions. The addition of NaOH into the solution (S_{KNa}) leads to modification favoring a homogeneous distribution. This fact can be correlated with the size of the alkali cation of the silicate solution, modifying then the mixture viscosity and consequently the reaction rate[23]. In the presence of S_{NaNa} solution, the size distribution is narrow due to the mobility of sodium. Nevertheless, the substitution of NaOH by KOH in sodium solution leads to the increase of the pore size distribution. In presence of metakaolin with impurities, this phenomenon is highlighted due to the low amount of species in the mixture. Indeed, impurities (Table II) disturb the metakaolin dissolution. Consequently, the kinetic of the reaction is very low allowing the formation of several networks inducing a large pore distribution whatever the alkaline solution.

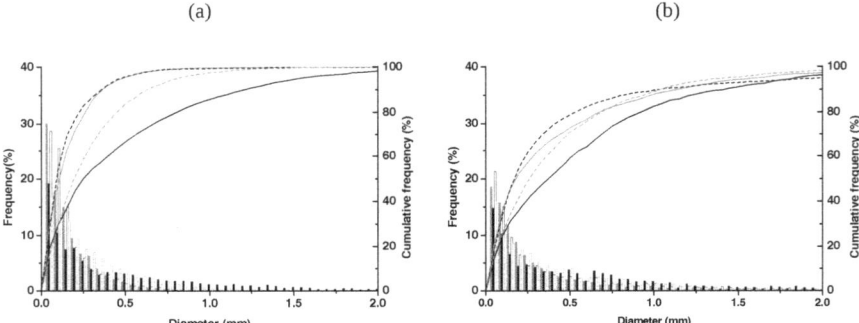

(a)   (b)

Figure 3. Pores size distribution of geopolymer foams (a) S_{M1} and (b) S_{M11} (■ S_{KK}, □ S_{KNa}, ■ S_{NaNa} and S_{NaK})

Foam working properties

To correlate the thermal properties and the porosity rate, the evolution of the thermal conductivity was plotted as a function of the porosity rate (Xp). The thermal conductivity decreases when the porosity rate increases. The $S_{KK}M_{11}$ sample displays a lower thermal conductivity while the $S_{KNa}M_1$ and the $S_{NaNa}M_1$ foams exhibit the highest thermal conductivity due to the use of metakaolin with lower amount of impurities (Figure 4). Globally, foams synthesized with "quasi pure" metakaolins show a higher thermal conductivity value and lower porosity rates than foam based on low reactive metakaolin. In fact, the rapid oligomer formation, which is linked to the metakaolin, can trap the system earlier and blocks the H_2 production. Thus, the porosity rate of those foams is smaller and the thermal conductivity is higher. Regardless of the aluminosilicate source, the same behavior is observed with the three different solutions. In fact, the global behavior of the thermal conductivity is controlled by the pore volume fraction. Indeed, the most porous sample exhibits the lowest thermal conductivity. The porous phase permits to decrease the thermal conductivity as described by the literature[24, 25, 26]. Analytic models, such as Maxwell Eucken, Landauer, Hashin-shtrikman, show the decrease of thermal conductivity value as a function of porosity rates.

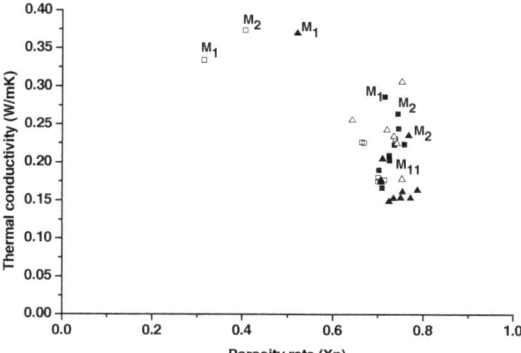

Figure 4. Evolution of the thermal conductivity value versus the porosity rate Xp for samples based on ■ S_{KK}, □ S_{KNa}, ▲ S_{NaNa} and △ S_{NaK} solutions

The alkaline solutions as well as the aluminosilicate sources impact the reaction medium. In order to correlate the thermal properties and the chemical composition of the synthetized foams, the evolution of the thermal conductivity value was plotted as a function of silicon concentration [Si] and the M/Al ratio in Figure 5. The same tendency is noted for the four solutions, the thermal conductivity decreases when the Si molar concentration decreases and when the M/Al ratio increases. Indeed, for the samples based on S_{KK} and S_{KNa} solutions, the lowest thermal conductivity value is reached for [Si] = 5.63 M and M/Al= 1.13, while for those based on S_{NaNa} and S_{NaK} solutions, the values are 3.52 M and 3.92 respectively. This difference is once again due to the higher content of water in the S_{KK} solution than S_{NaNa} leading to different natures of oligomers. The decrease of the thermal conductivity value with the silicon content must be correlated to the different values of specific heat. Different works showed that the addition of silicon promotes an increase of the thermal conductivity[27] as it is observed in the case of these foams. Nevertheless, the aluminum and alkali content must be considered too. Lower thermal conductivity for higher ratio M/Al value highlights that a low level of aluminum is needed. Indeed, aluminum increases the heat capacity and therefore the thermal conductivity[28].The difference between the two solutions can be correlated with the specific heat values. The cation size has an impact on the heat capacity since, the heat capacity increases with the size of the cation[29].

This behavior demonstrates that the chemical composition governs the thermal conductivity. It is also very important to determine the thermal conductivity value of the skeleton. This work is in progress. These data highlight also the importance of the porosity which is directly linked to the mixture reactivity.

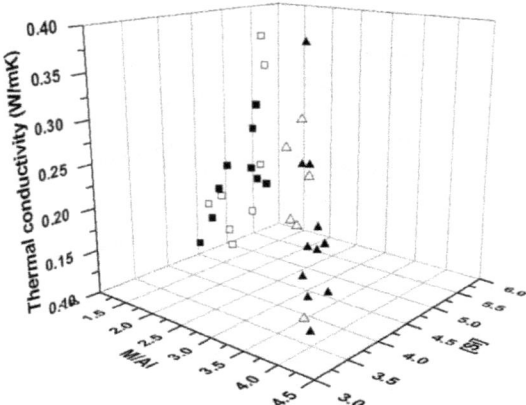

Figure 5. Thermal conductivity variation versus the chemical composition of the synthetized foams based on ■ S_{KK}, □ S_{KNa}, ▲ S_{NaNa} and △ S_{NaK} solutions

CONCLUSION

To understand the impact of reactants on geopolymer foam properties, foams synthetized with different alkaline solutions and aluminosilicate sources were characterized.

(i) Reactants, alkaline solution and aluminosilicate source, impact the volume expansion of the foams. The water content of the solution determines the solution reactivity and consequently the volume expansion. In the case of less pure metakaolins the alkaline solution governs.

(ii) The alkaline solution (water content, alkali cation) influences the foam pore size. The addition of a different cation permits to control the pore size. Mixed solutions (S_{KNa} and S_{NaK}) do not have the same behavior. Nevertheless, a "quasi-pure" metakaolin counterbalances the silicate solution ability.

(iii) The chemical composition, namely the silicon and aluminum molar concentrations, of the foams impact the thermal conductivity properties.

In summary, both are the key parameters to understand the foam formation and the working properties. In perspective, other properties such as mechanical and acoustic properties will be evaluated.

REFERENCES

[1] J. Davidovits, Geopolymer Chemistry and Application, 2nd ed., Saint-Quentin, France: Institut Géopolymère, 2008

[2] V. Glukhovsky, E. Starchenskaya, P.Krivenko, P Studies on formation of silicates in mixes of clays, quartz sand and sodium carbonate. Ukrainian Chemical Journal, 35 [4] (1969), 65–68

[3] E. Landi, V. Medri, E. Papa, J. Dedecek, P. Klein, P. Benito, A. Vaccari, Alkali-bonded ceramics with hierarchical tailored porosity, Applied Clay Science, 73 (2013) 56-64

[4] E. Prud'homme, P. Michaud, E. Joussein, C. Peyratout, A. Smith, S. Rossignol, In situ inorganic foams prepared from various clays at low temperature, Applied clay science, 51 (2011) 15-22

[5] G. Masi, W.D.A. Rickard, M.C. Bignozzi, A.V. Riessen, The influence of short fibres and foaming agents on thermal behavior of geopolymer composites, advances in Science and Technology, 92 (2014) 56-61

[6] E. Prud'homme, P. Michaud, E. Joussein, M-J. Clacens, S. Rossignol, Role of alkaline cations and water content on geomaterial foams: Monitoring during formation, Journal of Non-Crystalline Solids, 354 (2011) 1270-1278

[7] V. Med ri, E. Papa, E. Landi, Behavior of alkali bonded silicon carbide foams in modified synthetic body fluid, materials letters, 106 (2013) 377-380

[8] G. Masi, W.D.A. Rickard, L. Vickers, M.C. Bignozzi, A.V. Riessen, A comparison between different foaming methods for the synthesis of light weight geopolymers, Ceramics International, 40 (2014) 13891-13902

[9] R. Beloborodov, M.Pervukhina, V. Luzin, C. DellePiane, M. Clenelle, S. Setayesh and M Lebedev, . Peyratout, A. Smith, S. Arriii-Clacens, J-M S. Rossignol, Compaction of quartz-kaolinite mixtures: The influence of the pore fluid composition on the development of their microstructure and elastic anisotropy, Marine and Petroleum Geology 78 (2016) 426-438.

[10] E. Prud'Homme, P. Michaud, E. Joussein, C. Peyratout, A. Smith, S. Arriii-Clacens, J-M S. Rossignol, Silica fume as porogent agent in geo-materials at low temperature, Journal of European Ceramic Society, 30 (2010) 1641-1648

[11] M. Strozi Cillaa, P. Colombob, M. Raymundo Morellia, Geopolymer foams by gelcasting, Ceramics International, 40 (2014) 5723–5730

[12] L. Vidal, E. Joussein, M. Colas, J. Cornette, J. Sanz, I. Sobrados, J-L. Gelet, J. Absi, S. Rossignol, Controlling the reactivity of silicate solutions : A FTIR, Raman and NMR study, Colloids ans Surfaces A; Physicochem, Eng Aspects, 503 (2016) 101-109

[13] V. Benavent, P. Steins, I. Sobrados, J. Sanz, D. Lambertin, F. Frizon, S. Rossignol, A. Poulesquen, Impact of aluminum on the structure of geopolymers from the early stages to consolidated material, Cement and Concrete Research, 90 (2016) 27-35

[14] Y.J. Zhang, S.Li, Y.C, Wang, D.L. Xu, Microstructural and strength evolutions of geopolymer composite reinforced by resin exposed to elevated temperature, Journal non cryst solids, 358 (2012) 620-624

[15] A. Gharzouni, I. Sobrados, E. Joussein, S. Baklouti, S. Rossignol, Predictive tools to control the structure and the properties of mletakaolin based geopolymer materials, Colloids and Surfaces A: Physicochemical and Engineering Aspects, 511 (2016) 212-221

[16] A. Autef, E. Joussein, A. poulesquen, G. Gasgnier, S. Pronier, I. Sobrados, Role of metakaolin purities on potassium geopolymer formulation: the existence of several network, J. colloid Interface SCi, 408 (2013) 43-53]

[17] L. Courthèoux, F. Popa, E. Gautron, S. Rossignol, C. Kappenstein, Platinium supported on doped alumina catalysts for propulsion application. Xerogel versus aerogel, Journal of Non Crystallin solids 350 (2013) 113

[18] A. Autef, E. Joussein, G. Gasgnier , S. Rossignol, Feasibility of aluminosilicate compounds from various raw materials: Chemical reactivity and mechanical properties, Powder Technology 301 (2016) 169–178

[19] L. Vidal, E. Joussein, J. Absi, S. Rossignol, Coating of unreactive and reactive surfaces by aluminosilicate binder, Submitted to Ceramics international

[20] E. Prud'homme, A. Autef, N. Essaidi, P. Michaud, B. Samet, E. Joussein, S. Rossignol, Defining existence domains in geopolymers through their physicochemical properties, Applied Clay Science, 73 (2013)26-34

[21] A. Gharzouni, Contrôle de l'attaque des sources aluminosilicates par la comprehension des solutions alcalines, Université de Limoges et Université de Sfax (2016)

[22] L. Vidal, E. Joussein, M. Colas, J. Cornette, J. Absi, S. Rossignol, Identification of chains and rings in alkaline silicate solution by Raman spectroscopy: Effect of alkali cation, Submitted to Spectrochimica Acta Part A: Molecular and Biomolecular Spectroscopy

[23] P. Steins, A. Poulesquen, F. Frizon, O. Diat, J. Jestin, J. Causse, D. Lambertin, S. Rossignol, Effect of aging and alkali activator on the porous structure of a geopolymer, Journal of Applied Crystallography, 47 (2014) 316-324

[24] J.C. Maxwell, A treatise on electricity and magnetism, Clarendon Press Oxford, 14 (1892)

[25] R. Landauer, The electrical resistance of binary metallic mixtures, Journal of Applied Physics 23 (1952) 779-784

[26] Z. Hashin, S. Shtrikman, A variational approach to the theory of the effective magnetic permeability of multiphase materials. Journal of applied Physics, 33 (1962) 3125-3131

[27] W. Chen, C. Zou, X. Li, L. Li, Experimental investigation of SiC nanofluids for solar distillation system: Stability, optical properties and thermal conductivity with saline water based fluid, International Journal of Heat and Mass transfer, 107 (2017) 264-270

[28] Z. Wang, H. Wang, X. Li, D. Wang, Q. Zhang, G. Chen, Z. Ren, Aluminium and silicon based phase change materials for high capacity thermal energy storage, Applied Thermal Engineering, 89 (2015) 204-208

[29] P. Richet, A. Nidaira, D. R. Neuville, T. Atake, Influence of cation size on the low-temperature heat capacity of alkaline earth metasilicate glasses, American Mineralogist, 94 (2009) 1591-1595

GEOPOLYMERS BASED ON NATURAL AND SYNTHETIC METAKAOLIN
A CRITICAL REVIEW

Joseph Davidovits
Geopolymer Institute, 02100 Saint-Quentin, France.

ABSTRACT
Much of the original research into geopolymers was conducted on calcined kaolinitic clay precursors known as metakaolins. The chemical formula for kaolinite is $Si_2O_5Al_2(OH)_4$. From a geopolymer standpoint we may write $\equiv$Si-O-Al-$(OH)_2$ with the covalent aluminum hydroxyl - Al-$(OH)_2$ side groups of the poly(siloxo) hexagonal macromolecule $[Si_2O_5]_n$. Metakaolin results from the dehydroxylation of the -OH groups in kaolinite. The reactive molecule is an alumino-silicate oxide $Si_2O_5Al_2O_2$, coined MK-750 in order to pinpoint the calcination temperature. It comprises two Al-oxide types: $\equiv$Si-O-Al=O (alumoxyl), Al in 5-fold coordination together with $\equiv$Si-O-Al(O_2), Al in 4-fold coordination. The kiln technology determines the desired chemical reactivity. The geopolymer chemistry was invented 40 years ago with natural metakaolins. The review discusses the correlation between reactivity, calcination methods, MAS-NMR spectroscopy, reaction mechanism and applications, for natural and synthetic metakaolins. KEYWORDS: metakaolin, chemical microstructure, reaction mechanism, geopolymer.

INTRODUCTION
Much of the original research into geopolymers was conducted on calcined kaolinitic clay precursors known under the generic term of metakaolin. Although metakaolin reacts in alkaline as well as in acidic medium, the present issue focuses exclusively on the alkaline route.

Forty years ago, in October 1975, in our CORDI laboratory (later Cordi-Géopolymère) in Saint-Quentin, France, we were testing a new French metakaolin brand named Argical®. It was manufactured with an advanced technology in a flash calciner instead of being roasted in a rotary kiln or a vertical multiple-hearth oven. We discovered that this metakaolin was reacting very well with soluble alkali silicates. We recognized the potential of this discovery and presented an "Enveloppe Soleau" for registration at the French Patent Office. The hand-written text was filed on 29/12/1975, number 70528, at, INPI, Paris. It describes how metakaolin was reacting very well with soluble alkali silicates. It is reproduced with the English translation from French in Chapter 1 of the reference book by J. Davidovits, *Geopolymer Chemistry and Applications,* since the 3rd edition, 2011 onwards. The link to the free download PDF file of this Chapter 1 is located at <http://www.geopolymer.org/shop/product/geopolymer-chemistry-applications/>. The patent was filed with the self-evident title "Mineral polymer" (Davidovits, 1979)[1].

In 1983, at the Central laboratory of the American cement manufacturer Lone Star Industries, Houston, USA, we started the development of advanced cementicious materials. This research yielded the discovery of the first metakaolin-based geopolymer cement (the cement Pyrament®). However, we had to test at least 10 different metakaolin brands in order to find the right product, which would react as a geopolymeric precursor, in alkaline medium. Indeed, at that time, the bulk of the various metakaolins was used essentially as fillers in the paper making and plastic industry. Its specific chemical reactivity towards alkalis remained confined in the production of very special products, namely synthetic zeolites, especially the type Zeolite A. In addition, it was striking to discover that the metakaolin sources for zeolite manufacture were according to Breck (1974)[2] of two types, one calcined at 550°C (low temperature metakaolin) and the second at 925°C (high temperature metakaolin). Both metakaolins reacted weakly compared to the metakaolin we had been working with in France. We recognized that we had had luck when starting the geopolymer research, in Saint-Quentin. We had tested the right source of metakaolin, from the beginning. Furthermore, we became aware of one major parameter in geopolymer science, namely the calcining temperature of the geological kaolinitic clays.

The chemical formula for kaolinite is $Si_2O_5Al_2(OH)_4$. From a geopolymer standpoint we may write $\equiv$Si-O-Al-$(OH)_2$ with the covalent aluminum hydroxyl - Al-$(OH)_2$ side groups of the poly(siloxo) hexagonal macromolecule $[Si_2O_5]_n$. This new structural approach has profound consequences with regard to a better understanding of geopolymerization mechanisms. In particular, metakaolin results from the dehydroxylation of the OH groups in kaolinite according to the reaction:

$$Si_2O_5Al_2(OH)_4 \rightarrow Si_2O_5Al_2O_2 + 2H_2O$$

The reactive molecule is an alumino-silicate oxide $Si_2O_5Al_2O_2$, or $\equiv$ Si-O-Al=O. This suggests strong chemical reactivity for this aluminum oxide, as opposed to the traditional way of writing $2SiO_2 \cdot Al_2O_3$. Metakaolin is not alumina! See for details in Chapter 8 of the book *Geopolymer Chemistry and Applications*. We coined it KANDOXI, acronym for KAolinite, Nacrite, Dickite OXIde. Other researchers who continued to use the general term metakaolin did not adopt this label. Therefore, after several discussions, we decided to change the terminology in introducing MK-750, meaning metakaolinite calcined at 750°C.

However, in addition to temperature control, it is the kiln technology, which determines the feasibility and production of the alumino-silicate oxide MK-750. In calcination carried out in a rigid vertical multiple-hearth calciner, a sufficiently low water vapor pressure is maintained during the entire roasting process, providing the desired chemical reactivity (Al in 5-fold coordination). Same for products manufactured in a flash calciner. This is not the case for metakaolins obtained in a rotary kiln, commercialized as Portland cement additives. Unfortunately, this later product is more and more used in geopolymer research because it is easily available. This raises new concerns in terms of reactivity and reproducibility of the results obtained with this raw material essentially tailored for Portland cement applications, not for geopolymer technologies.

The present special issue, acknowledging the abundance of publications on metakaolin-based geopolymers, tries to collect the best of the scientific production for the understanding of geopolymer resins, binders and cement properties. Our review will also pinpoint those papers where the authors overlooked some important parameters.

The geopolymer chemistry was invented 40 years ago because we had the luck to get the right geological raw material and the appropriate calcination process. Today, new methods exist, based on the synthesis of alumino-silicates. In short, we have natural metakaolin MK-750 and artificial, or synthetic metakaolin SMK-750. Figure 1 below displays their microstructures. Therefore, this review is split into two subthemes, namely:
- Geopolymers based on natural metakaolins MK-750;
- Geopolymers based on synthetic metakaolins SMK-750;

GEOPOLYMERS BASED ON NATURAL METAKAOLIN MK-750

Our first contact with MK-750 in 1975 was very exciting. We had prepared a mixture of 1 kg powder with 0.3 kg NaOH solution 12 M and let it mature in a plastic bag for a while. After 1 hour, we were surprised by the high amount of water vapor and condensation seeping outside of the bag, with a temperature exceeding 100°C, and a polyethylene bag totally destroyed. We had discovered one of the major properties of MK-750, namely its powerful exothermicity and reactivity in alkaline medium. This characteristic was mentioned in all our earlier published papers[3,4,5]. It helped us to develop a standard method for quality control and selection of reactive raw materials. This technique was successfully used during our selection of the US metakaolin brands mentioned in the introduction. Chapter 8 of the reference book *Geopolymer Chemistry and Applications* provides several examples. Yet, the amount of research carried out and the number of published papers on this topic are surprisingly low. We only found one paper worth being cited in this review, published in 2009 by Yao et al.[6]. It confirms the various parameters involved in the development of this exothermicity, namely: curing temperature, alkalinity ratio $SiO_2:Na_2O$. The authors measured the exothermicity in relation with curing temperature from 20°C to 80°C and other parameters. However, like others, they explain the formation of intermediary ortho-sialate $(OH)_3$-Si-O-Al$(OH)_3$ by interaction of monomeric Si and Al, followed by polycondensation into what they call "small catenulate gels".

Figure 1: Microstructure of natural metakaolin MK and synthetic metakaolin SMK after Cui et al., 2008[32].

Another point to criticize in this work is the structural representation of MK-750. See in Figure 2 below, left, the wrong model. First, in the Al-O-Al layer, the Al^{3+} atom is represented being tetravalent (AlO_4), in the same way as the Si-O-Si network with its tetravalent Si^{4+} configuration. The oxygen atom is also trivalent O^{3-}, which is nonsense. This is a major error, also found in numerous scientific papers, namely the confusion between chemical valence and physical coordination. Their false structural representation is copy-pasted in numerous publications dealing with metakaolin-based geopolymer. In metakaolin, the Al atom is trivalent Al^{3+} but Al is tetracoordinated, Al(4), or pentacoordinated, Al(5) to oxygens. Actually, the reactive molecule comprises two Al-oxide types: ≡Si-O-Al=O (alumoxyl), Al in 5-fold coordination together with ≡Si-O-Al(O_2), Al in 4-fold coordination. See in Figure 2.

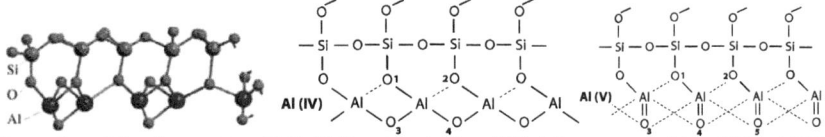

Figure 2: metakaolin structure $Si_2O_5Al_2O_2$ with trivalent Al^{3+}: left wrong model; middle, Al(IV), for aluminum oxide ≡Si-O-Al(O_2)- ; right, Al(V), for alumoxyl ≡ Si-O-Al=O (dotted line for coordination links, full line for chemical valence links). See Chapter 8 of the reference book *Geopolymer Chemistry and Applications*[5].

Reactivity, calcination methods and MAS-NMR spectroscopy:

The reactivity and exothermicity are closely related to the calcination method of the raw kaolinitic clay source. The standard parameters set at the Geopolymer Institute and in our industry partner's laboratories are described in Chapter 8 of the reference book *Geopolymer Chemistry and*

Applications. They involve the grinding of the clay, and calcining in an electric oven, in air, at 750°C for 3 hours; the heating time from 20°C up to 750°C is 1 h 30 min, the cooling time 1 h 30 min, totalizing 6 hours in the furnace. What we find in the literature is totally different, so that it is hard to compare any results deriving from these researches. The following non-exhaustive list is representative of what we came across during this critical review of publications; it gives the calcination temperature and the time (yet we do not know if it includes the heating and cooling times).

The list of papers selected for their data on calcination temperature or provenience, starts with Zibouche *et al.* (2009)[7]. Their choice of 800°C, 2 hours is not explained and the purpose of the paper is to demonstrate that any kaolinitic clays are suitable for geopolymerization. Zhang Yunshen et al, (2009)[8], use 700°C, 12 h, without any explanation, Rowles et al., (2009)[9], 750°C, 24 h, *Yao* et al., (2009)[6], 900°C, 6 h, Medri et al. (2011)[10], 750°C, 15 h. Others are simply using commercial brands: Zhang et al., (2009)[11], commercial MetaMax® from Engelhardt (now BASF); Favier et al., (2013, 2015)[12,13], commercial Argical® M1000 from Imerys;

Medri et al., (2010)[14], tested two metakaolins manufactured industrially by the company Imerys, France, with two different kiln technologies. One called M1000 is calcined in a rotary kiln and characterized by rounded massive aggregates of lamellar particles. The second, called M1200S, calcined in a flash kiln, is made up of fine lamellar particles with lower agglomeration. Granulometric distributions are broad and multimodal for both metakaolins. M1000 particles are coarser with a mean grain size d_{50}=6.5 microns, while M1200S has a d_{50}=1.7 microns. Scientists prefer the brand M1000 because of its low water demand. M1000 like the brand MetaMax and others are essentially commercialized as additive to Portland cement. The calcination cycle lasts 4 hours at 700–750°C in the rotary kiln (production 10 tones/h). On the opposite, the product M1200 is processed in a flash calciner with an air flow at 950°C during 1 second time (production 1 tone/h).

We mentioned in the introduction the beneficial influence of this flash calcined metakaolin in the discovery of geopolymer science. Some years later, in the 1980s onwards, the product was no longer manufactured and replaced by another brand, also very popular in geopolymer research, MetaStar 501 (same as Polestar 501) from ECC Int. UK, now Imerys. Highly refined kaolin clay is calcined in a vertically oriented multiple hearth furnace. The material is moved by mechanical rakes across each hearth and then drops to the next hearth below. Each hearth has separate temperature controls and, unlike with rotary kilns, the time that the material is inside the furnace and the temperature gradient that the material is exposed to is precisely controlled, ensuring the consistent production of high-purity highly reactive metakaolin. It is our reference material (coined Kandoxi and later MK-750) in our publications (see Chapter 8 of the reference book *Geopolymer Chemistry and Applications*).

In addition to exothermicity, the best investigation tool to determine the geopolymeric reactivity of metakaolin is the ^{27}Al MAS NMR spectroscopy. According to Medri *et al.*, (2010)[14], the simulation of the NMR spectra for the commercial brands showed that the relative concentration of the aluminum species Al(6), Al(5) and Al(4) is:
- M1000: 35% Al(6), 50% Al(5), 15% Al(4), with Al(5) + Al(4) = 65%.
- M1200S: 25% Al(6), 55% Al(5), 20% Al(4), with Al(5) + Al(4) = 75%.
In comparison, we had for
- MK-750 (MetaStar 501): 24% Al(6), 49% Al(5), 27% Al(4) with Al(5) + Al(4) = 76%.
The geopolymeric reactivity of M1200S is equivalent to the one of MK-750, and far better than M1000.

The introduction of MAS-NMR spectroscopy in the study of silicates by Lipmaa et al. (1980)[15], Mäller et al. (1981)[16], transformed our view and our knowledge on silicate structures. The most important contribution for our study was made in 1985 by MacKenzie's team from New

Zealand. The first paper by Meinhold, MacKenzie and Brown (1985)[17], reads: "*Metakaolinite contains 11 to 12% of the original kaolinitic water, associated with the 8% of aluminum atoms which remain in undistorted sites. Approximately one half of these well-defined aluminum sites are octahedral, one quarter are tetrahedral, and the remainder are either tetrahedral, or some other regular site with chemical shift intermediate between tetrahedral and octahedral.*" This chemical shift intermediate will be later assigned to pentahedral aluminum sites Al(5). But, in their second paper, MacKenzie et al. (1985)[18], the authors could not claim to have discovered this new Al(5). It was against the view of the main stream of scientists at that time.

Three years later, the paper by Sanz *et al.* (1988)[19] confirmed the presence of Al(4) and Al(5), in addition to Al(6). They applied MAS-NMR spectra to follow the kaolinite-mullite transformation with temperature, from 20°C up to 1200°C. The ^{27}Al MAS-NMR spectrum for raw kaolinite (20°C) contains one resonance at 0 ppm characteristic of Al(6). When the sample is heated, the intensity of this line decreases up to 980°C, then increases again at 1055°C (transformation to mullite). In the medium temperatures above 400°C, that is during the dehydroxylation phases of kaolinite, two new lines appear at 55–60 ppm assigned to Al(4) and at 25–30 ppm assigned to Al(5). The intensity of the Al(5) line increases with temperature from 450°C to 850°C, then it decreases and disappears at 980°C. The maximum intensity is spread 700–850°C. The Al(4) line remains with the same intensity.

Nonetheless, studies on the calcination time at definite temperatures are seldom. None of the papers cited above are mentioning the reason why the calcination was run at a particular temperature and during a given time. For example, why the authors in Zibouche et al. (2009) calcined their kaolinitic clays at 800°C for 2 hours, and not at 750°C, 24 hours long like in Rowles et al., (2009) ? Or why Xiao Yao et al., (2009), fired at 900°C for 6 hours, when we already knew, thanks to the research published by Sanz et al., (1988) that the intensity line of the most reactive Al specie, Al(5), diminishes above 850°C. due to the metakaolinite-spinel-mullite transformation.

Reaction mechanism:

Another important issue in metakaolin-based geopolymerization relates to its reaction mechanism. At the beginning of geopolymer research (Davidovits, 1976)[20] and afterwards for at least 25 years, it was assumed that the geochemical syntheses occurred through hypothetical oligomers (dimer, trimer). Further polycondensation of these hypothetical building units provided the actual structures of the three-dimensional macromolecular edifice. Review papers published at the First Geopolymer Conference in 1988, and at the second, 11 years later, in 1999, could not present scientific details describing the actual reaction mechanism (see the page Science at the Geopolymer Institute Internet site).

The most important contribution to this issue is the paper by North and Swaddle (2000)[21]. Using ^{29}Si and ^{27}Al NMR spectroscopy, they suspected the presence of solute species with Si-O-Al sialate linkages in concentrated solutions. One major improvement in their research was that their study was carried out at low temperature, at 5°C and below. Indeed, it was discovered that the polymerization of oligo-sialates was taking place on a time scale of around 100 milliseconds, i.e. 100 to 1000 times faster than the polymerization of ortho-silicate. At room temperature, or higher, the reaction is so fast that it cannot be detected with conventional equipment. They chose KOH over NaOH used in their previous study because concentrated KOH alumino-silicate solutions resist gelation longer than their NaOH analogues. Due to the very weak signal of ^{29}Si, the NMR experiments had to be run up to 3 days long to get significant detailed spectra. They successfully detected five solute species, displayed in Figure 3 below, namely two linear molecules and three cycles:
- one ortho-sialate $(OH)_3$-Si-O-Al$(OH)_3$ for Si:Al=1;

- one linear ortho(sialate-siloxo) $(OH)_3$-Si-O-Si$(OH)_2$-O-Al$(OH)_3$, one cycle ortho(sialate-siloxo), for Si:Al=2;
- two cycles ortho(sialate-disiloxo), for Si:Al=3.

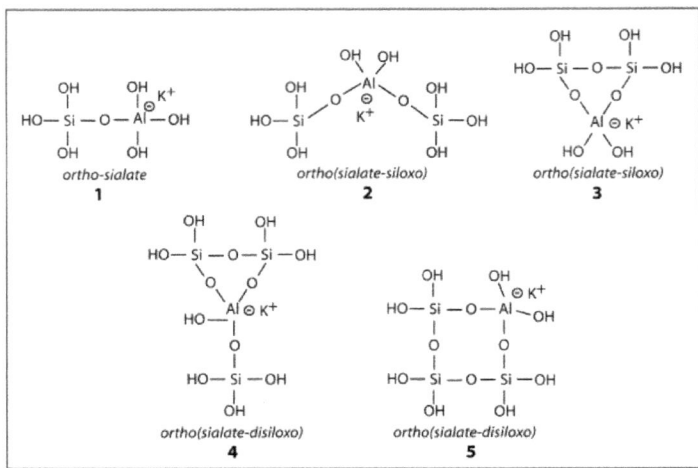

Figure 3: Five different orthosialate molecules soluble in KOH concentrated medium, adapted from North and Swadle (2000)[21].

The hypothetical oligomers set forth in geopolymer synthesis were no longer virtual molecules. As a matter of fact, they exist in soluble forms and are stable in concentrated solutions at high pH. Swaddle's study confirmed the polymerization mechanisms tentatively reported earlier by Davidovits (1976) with linear oligo-sialate, oligo(sialate-disiloxo) and rings or cycles, as starting geopolymer building units. As a matter of fact, Swaddle's study is practically unknown by the main stream of geopolymer scientists. This lack of knowledge explains their incorrect interpretation of the actual chemical mechanisms.

Zhang Yunshen et al. (2009)[8] worked at 700°C for 12 h. Their research contradicts the ideas presented by the mainstream of cement scientists on the geopolymerization mechanism. They stressed out that most of the researchers describe the formation process of metakaolin-based geopolymers (in his case MK-based geopolymer cement) by means of a traditional hydration mechanism as for Portland cement or alkali slag cement, which is wrong. For these cement scientists, the geopolymerization reaction involves a 3-step process, namely dissolution–reorientation– polycondensation. However, the setting and hardening of MK-based geopolymer cement is so rapid that this 3-step process almost takes place at the same time. Therefore, it is impossible to isolate the three steps by experimental study, yielding no better understanding of the details of each step until now.

In fact, we know from Swaddle's work that the reaction in strong alkaline medium between the monomeric species is so fast that it becomes undetectable. The suggested dissolution mechanism of metakaolinite should be similar to the one set forth in the Duxson and Provis (2008)[22] paper for alumino-silicate glass. The breakdown of the alumino-silicate glass molecular structure in high pH leads up to a well-defined molecule, namely the ortho-sialate $(OH)_3$-Si-O-Al$(OH)_3$ before the supposed final release of monomeric Si and Al. However, we already know that the monomers immediately polymerize into this same ortho-sialate in solution, according to Swaddle. In other words, there is no isolated monomeric Al and Si. The breakdown of metakaolinite in strong pH, which is always the case in alkaline geopolymerization, does not follow the mechanism set forth for acidic and low-alkaline media with its preferential solubilization of monomeric $Al(OH)_4$ followed by $Si(OH)_4$.

Therefore, the first step of the reaction mechanism is the formation of the ortho-sialate $(OH)_3$-Si-O-Al$(OH)_3$ molecules and results from the breakdown of the Si-O-Si chain (see the chemical mechanism displayed at the page Science of Geopolymer Institute internet site <http://www.geopolymer.org/science/about-geopolymerization/> and in Chapters 7–8 of *Geopolymer Chemistry and Applications*).

Supporting information to this claim is provided by another important paper by Bauer and Berger (1998)[23]. They followed the entire reaction of kaolinite in KOH solutions and examined the dissolution rate of kaolinite in high pH KOH solutions (0.1 to 4 M KOH) at temperatures of 35°C and 80°C in a batch reactor. They found that the aqueous concentrations of Si and Al increased linearly with log (t) whatever the temperature and the KOH concentration was. Moreover, the amounts of Si and Al are identical with time, i.e. one Si is released together with one Al, simultaneously. This result is in favor of the presence in solution of the stable otho-sialate $(OH)_3$-Si-O-Al$(OH)_3$ molecule. Since the 1970s, in all papers, the concentrations of Al in solutions are analyzed colorimetrically with an UV-visible spectrophotometer, using the Catechol violet method. Then, separately, Si concentrations are measured with the Molybdate blue method. Applied to the oligo-sialate molecule, this method gives 1 Al for 1 Si in solution, even when there is no monomeric Si and Al but rather the ortho-sialate molecule $(OH)_3$-Si-O-Al$(OH)_3$.

Despite the strong evidence towards the formation of the oligo-sialate molecule, several authors continue to propagate a model system based on the interaction of monomeric Na-silicate and Na-aluminate. Like many others, they do not take into account the microscopical structure of the raw material metakaolin, per se (see Figure 1 above). The calcination of kaolinite does not destroy the lamellar structure of the tabular shaped metakaolinite pellets. Electron microscope studies by German scientist Eitel showed, as early as 1939, that kaolinite particles retain their hexagonal outline far above the dehydration temperature, up to 750°C. The alkaline attack starts on the outer faces of the metakaolinite particle. It continues, layer by layer, from the edges to the inside. This is a very important feature, which induces two different geopolymerization mechanisms.

We discovered the implications of this structural parameter by chance. Our PhD student M. Zoulgami (1997-2000) prepared a MK-750-based Na–poly(sialate-siloxo) geopolymer (with Si:Al=2) dedicated to biomaterial applications. This implied a drastic diminution of the pH from pH 10.5–11 to a neutral value in the range of 7.5, by heat treatment at 750°C. In the paper by Zoulgami et al. (2002)[24], the X-ray diffraction of the dried geopolymer showed two major crystalline phases: nepheline ($NaAlSiO_4$), Si:Al=1, and albite ($NaAlSi_3O_8$), Si:Al=3. These two phases are coexisting in the ratio 50/50 and the bulk reactional mix corresponds to the formula

$$Na_{4.5}Al_{4.5}Si_{9.04}O_{27.11}, \text{ i.e. } NaAlSi_2O_6 \text{ or Si:Al=2}$$

The EDS analyses performed on this geopolymer treated at 750°C show that the bulk chemical composition is equal to $NaAlSi_2O_6$. This means that both geopolymer phases, namely nepheline and albite detected by X-ray diffraction, are in solid solution, on the nano scale. Duxson et al. (2005)[37] also noticed for temperature above 500°C the formation of nepheline with Na and of a not well-defined component similar to albite. One geopolymerization generates the formation of a Si:Al=1 geopolymer, the second a Si:Al=3 geopolymer, with an overall value Si:Al=2.

As recognized by MacKenzie's team in the Zhang et al. (2009)[11] paper, MK-750-based geopolymer binders and cements are different from the simple alkali activation with NaOH, published by several authors. This NaOH simple alkali-activation of metakaolin generates products of the type Zeolite A. For the authors, a geopolymer cement or a geopolymer binder implies the reaction of MK-750 with soluble (Na, K) alkali silicates. The authors suggest that the various silicate units (Q_0, Q_1, Q_2, Q_3) present in the alkali-silicate solution have a templating function during geopolymerization.

Rowles and O'Connor (2009)[9], calcined metakaolin at 750°C for 24 h. The SEM imaging of the resulting geopolymer reveals the presence of a two-phase microstructure; the matrix phase being the fully formed geopolymer with Si:Al=3, while the grain phase is reminiscent of, but chemically dissimilar to, the MK precursor with Si:Al=1. The problem is that the authors did not recognize that these 2 phases are due to the initial layered arrangement of the tabular shaped metakaolin particulates shown in Figure 1 above. It is strange to discover that this unique structural property of metakaolinite is ignored or not taken into consideration by the majority of scientists who try to explain the geopolymerization of MK-750-based geopolymers.

The relatively recent papers by Favier et al. (2013, 2015)[12,13] are the most important papers so far, dealing with this important and evident structural parameter. Their elastic modulus study on MK-750-based geopolymer suspensions confirms the 2 step reaction mechanism mentioned above. At the very early stage of the reaction (fewer than 15 min after the beginning of mixing), aluminosilicate molecules with Si:Al ratio in the range of 2 to 3 are formed at the grain boundaries, while the rest of the solution is still mainly composed of silicate/siloxonate oligomers of type Q_0, Q_1, Q_2 and also Q_3. In the second phase, the geopolymer hardens with a second rapid increase of the elastic modulus. This fast increase indicates that a new chemistry is taking over, with Si:Al=1, which does not involve any siloxonate molecules, only NaOH. As the geopolymerization progresses, at the scale of dozens to hundreds of nanometers, different sialate species coexist as a distribution of solid solutions between Si:Al=1 to Si:Al=4, more specifically poly(sialate) of the nepheline type, Si:Al=1, together with poly(sialate-disiloxo) of the albite type, Si:Al=3, and others. In their second study, Favier et al. (2015)[13] provide strong evidence for a heterogeneous formation and a two-step mechanism. They focused on the solid/liquid heterogeneous suspension. Using static liquid-phase NMR they could specifically monitor the chemical speciation of the mobile species in the suspending liquid (ions, oligomers or small gel particles). It is also one of the rare research, which takes into account the exact molecular composition of the alkali-silicate solution, namely the simultaneous presence of highly depolymerized species, isolated ortho-silicate Q_0, as well as polymerized silicate oligomers, with silicon speciation Q_1, Q_2, Q_3. The paper by Zhang. et al. (2009)[11], stresses the same importance for the molecular structure of the various silicate units (Q_0, Q_1, Q_2, Q_3) present in the alkali-silicate solution. They suggest that they have a templating function during geopolymerization.

Accordingly, the geopolymerization of MK-750 with (Na,K)-silicates results from two reaction mechanisms, illustrated in Figure 4 below, taking place in the following order:
- Phase 1: outer faces/edges reaction involving Na+, K+, OH− and Q_0, Q_1, Q_2 and Q_3 siloxonates; resulting geopolymer with atomic ratio Si:Al between 1 and 4;
- Phase 2: inner particulate reaction with only Na+, K+, OH−; resulting geopolymer with atomic ratio Si:Al=1.

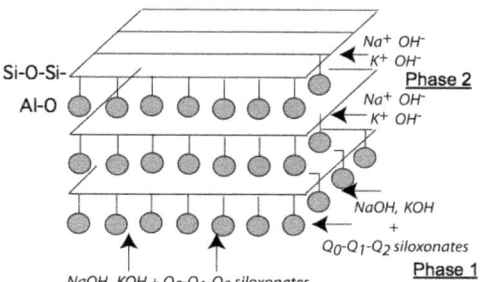

Figure 4: Schematic representation of the chemical attack on MK-750 layered particles with Phase 1 and Phase 2 (Davidovits, Chapter 8, Geopolymer Chemistry and Applications). The interval between each layer is 7.5 angstrom.

Applications

MK-750-based geopolymers have a variety of applications. They are new binders and resins for coatings and adhesives, fiber composites, waste encapsulation and new cement for concrete. The wide variety of applications includes: fire-resistant materials, decorative stone artefacts, thermal insulation, low-tech building materials, low energy ceramic tiles, refractory items, thermal shock refractories, bio-technologies (materials for medicinal applications), foundry industry, cements and concretes, composites for infrastructures repair and strengthening, high-tech composites for aircraft interiors and automobiles, high-tech resin systems, radioactive and toxic waste containment, arts and decoration, cultural heritage, archaeology and history of sciences.

The route to high-temperature ceramics via MK-750 geopolymerization was pioneered by W. Kriven and her team at the University of Illinois, USA. They converted fully condensed MK-750-based K–poly(sialate-siloxo) geopolymer into high valuable leucite ceramic $KAlSi_2O_6$. This demonstrated an alternative route in the production of these highly valuable refractory ceramics, without the necessity of expensive equipment, long processing times, and costly precursors (Kriven et al., 2003, 2006)[25,26]. One paper by Kriven et al. (2005)[27] was presented at the Geopolymer 2005 Conference. Later on, Dechang Jia and his team at Harbin Institute of Technology, China, also focused their research on this geopolymer route. When compared to hydrothermal methods, geopolymer technology is advantageous in low cost and short fabricating time; with the proper processing procedure, geopolymer can be directly converted into the final structural leucite, kalsilite or pollucite ceramic. See the video of the Keynote paper, *High-temperature geopolymer carbon fiber reinforced composites and their derived CMC*, by Dechang Jia at Geopolymer Camp 2011 at
http://www.geopolymer.org/conference/gpcamp/gpcamp-2011/.

He et al. (2010)[28] stressed how the geopolymer route is of great interest. They discovered that the addition of cesium stabilized the geopolymeric leucite. The cesium substitution for potassium was effective in stabilizing the cubic polymorph of leucite to room temperature. In the paper by He et al. (2011)[29], they used MK-750 containing a certain amount of silica (quartz). They obtained the transformation into leucite at a temperature as low as 800–900°C instead of 1050–1100°C with pure kaolin and the ceramic firing. Contrary to Kriven's paper , which produces fully reacted geopolymer, here, the method is not 100% satisfactory. Indeed, due to the existence of unreacted metakaolin, K-geopolymer was inhomogeneous at the molecular level. Thus a higher energy was required for the structural rearrangement during the change from amorphous to crystallized phase.

GEOPOLYMERS BASED ON SYNTHETIC METAKAOLIN SMK-750
The idea is to manufacture kaolinite-like products and to calcine them like for natural MK-750. Progress towards a simpler synthesis of kaolinite-like minerals, $Si_2O_5Al_2(OH)_4$, may be traced back to De Kimpe et al. (1981)[30]. They used a sol-gel method involving the simultaneous hydrolysis of TEOS tetraethylorthosilicate $Si(OC_2H_5)_4$ as Si source and aluminum isopropoxide $Al(OCH(CH_3)_2)_3$ for the Al source.

Different synthesis methods
A real breakthrough towards synthetic metakaolin SMK-750 occurred in 2005 with the research carried out by W. Kriven and her team at University of Illinois at Urbana Champaign, USA. Since 1995, W. Kriven used a patented technique known as the "steric entrapment method" to create a variety of ceramic oxides. Her team, Gordon et al. (2005)[31], explains how during the organic steric entrapment method, water solutions of aluminum nitrate nonahydrate $Al(NO_3)_3$ $9H_2O$ and sol-SiO_2, are added to a soluble polymer, typically polyvinyl alcohol PVA, dissolved into solution together and mixed. After mixing, the solute (water or alcohol) is removed by drying at 110°C and the resulting powder is calcined. During calcination, the anionic components of the salts are sublimed and the cation components oxidize. The resulting powder is chemically homogenous with a $Al_2O_3,1.3SiO_2$ composition and after calcination at 700-800°C, the SMK-750 synthetic metakaolin has a surface area of 166m²/g and a d_{50} particle size of 1.62 microns.
We have presently 3 published methods for synthetic metakaolin SMK-750, namely:
- Steric entrapment method, according to Kriven's team[31], involving aluminum nitrate nonahydrate $Al(NO_3)_3$ $9H_2O$, sol-SiO_2 and polyvinyl alcohol PVA.
- Sol-gel system according to Xuemin Cui and his team at Guangxi University, Nanning, China. Cui et al. (2008)[32], include aluminum nitrate nonahydrate $Al(NO_3)_3 \cdot 9H_2O$ and TEOS tetraethylorthosilicate $Si(OC_2H_5)_4$.
- Sol-gel procedure according to De Kimpe et al. (1981)[30], used by Chan and his team at National Taiwan University, Taipei, Taiwan. Tsai et al. (2010)[33] combine aluminum isopropoxide $Al(OCH(CH_3)_2)_3$ and TEOS tetraethylorthosilicate $Si(OC_2H_5)_4$.
Actually, the three methods are based on sol-gel, which yields SMK-750-types with high purity and homogeneous chemical composition. However, there are also differences between them. Kriven's and Cui's methods use aluminum nitrate nonahydrate as the Al source, but different Si sources are selected, i.e. sol-Silica or TEOS. Sol-silica is cheaper in price than TEOS and this could have an incidence on the choice of the technique. There are also differences in properties. The noticeable difference is the particle size. The SMK-750 prepared with Cui's method is nano-scaled, with an average particle diameter of about 0.25 microns, while the SMK-750 obtained according to Kriven's method is of micron scale, with an average particle diameter of about 1.6 microns or higher. This higher particle size is probably due to aggregation or densification. The authors noticed that "*.. it had been found previously that trace alcohol on the surface of SMK prevents the reaction with alkali-silicate solutions.*" As a consequence, the preparation includes a second heat treatment at 800°C to remove all organics from the surface of the SMK-750. This additional calcination generates densification and agglomeration of the particles. In Medri et al. (2010)[14] paper, natural MK-750, brand M1200S, calcined in a flash kiln, has a d_{50}=1.7 μm. This size is similar to Kriven's SMK-750 powder. Yet, its surface area is low, only 23 m²/g, to be compared with 166 m²/g and higher, up to 600 m²/g for the synthetic metakaolins.

MAS-NMR spectroscopy

At the first glance, ^{27}Al and ^{29}Si MAS-NMR spectra of SMK-750 synthetic metakaolins are similar to those obtained for natural MK-750. They display Al resonances assigned to Al(4), Al(5) and Al(6), but the intensities are different. The relative concentrations of the aluminum species Al(6), Al(5) and Al(4) of our reference MK-750 (MetaStar 501) and SMK-750 obtain with the method by Cui according to Zheng et al. (2013)[34] are :
- SMK-750: 44.4% Al(6), 49.6% Al(5), 6.1% Al(4), with Al(5) + Al(4) = 55.7%.
- MK-750: 24.0% Al(6), 49.0% Al(5), 27% Al(4), with Al(5) + Al(4) = 76%.

The amount in Al(4) is surprisingly low in the sol-gel product and the geopolymeric reactivity of SMK-750 is less than the one of MK-750. This is due to the high amount of Al(6) species, twice the amount in synthetic MK than in natural MK-750, probably attributed to remains of precipitated nano-size Al_2O_3 particles.

It should be noted that according to Cui et al. (2010)[35], the resonances related to Al(4) and Al(5) come into sight in the spectra of the calcined powders at temperatures as low as 200°C-300°C. Their intensities increase up to 800°C, essentially for Al(5). Similar resonances usually take shape in kaolinite calcination at temperatures over 450 °C; see the paper by Sanz et al. (1988)[19] on the transition of natural kaolinite to metakaolinite.

There are also several ^{29}Si resonances in SMK mainly centred at -101.9 and -105.2 ppm, corresponding to Q_4(1Al) and Q_4(0Al), respectively. The ^{29}Si spectra for SMK-750 show a high concentration in SiO_2, which is not found in natural MK-750 when quartz SiO_2 is absent. Like for aluminum, the high concentration of SiO_2 is probably a result of precipitated nano-size colloidal SiO_2.

Reactivity

Forty years ago, we discovered the reactivity of natural metakaolin partly because of its exothermicity. We also complained about the lack of research carried out on the study of this parameter, during geopolymerization. The contact with synthetic metakaolin SMK-750 is even more dramatic. In any papers referenced in this critical review, we find a sentence similar to this one: "... *the reaction rate is very fast and the preparation process is highly exothermic. So, the reaction device should be put in ice water bath to lower the reaction rate and to prevent the rapid setting of the geopolymer*."

The high reaction rate may be due to the very high surface area of SMK-750, namely 166 m^2/g and higher, up to 600 m^2/g. This is to be compared with the values for MK-750 that are in the range of 15 to 23 m^2/g. In addition, the morphology of SMK-750 is quite different. Although SMK-750 shows sheet-like morphologies similar to natural MK-750, the layered structure of SMK is clearly not the same. Cui et al (2008)[32] observed the morphology after acidic leaching. The synthetic metakaolin has an irregular layered structure with individual layers likely composed of many particle clusters (see in Figure 1 above).

TEM morphology of the SMK-750-based geopolymer is similar to that of natural metakaolin. However the resulting nano-particulates (we called them geopolymer micelles) are smaller, in the 4 nm diameter range, to be compared with 10-20 nm for MK-750-based geopolymer. It has a denser and more homogeneous microstructure.

The mechanical properties depend on the calcination temperature. According to Cui et al. (2010)[35], the compressive strength is 4.8 MPa for SMK-750 treated at 300, 400 and 500°C, 14.2 at 600°C, reaches 28 MPa for 700-800°C and drops to nil when calcined at 900°C.

Applications

It is obvious that the SMK-750-based geopolymer technology is not targeted towards mass applications. So far, the purpose of the scientists involved in these developments was to implement a system providing high purity products. It is very difficult to employ the conventional approach with natural or waste products to prepare geopolymers which can meet the stringent requirements for medical applications. Therefore, it is of great interest to develop a synthetic protocol to prepare geopolymers with well-defined chemical composition. This was the target of Chan and his team from Taiwan as well as for Cui and his team at Nanning.

This is also the case in the development of high temperature ceramics, by following the geopolymer route already mentioned for natural MK-750-based geopolymer. Kriven' team studied the fabrication of structural leucite glass-ceramics (see the paper by Xie *et al* (2010)[38]). Jia's team at Harbin Institute of Technology, China, employs the steric entrapment method for the manufacture of pollucite ceramic. In the paper by He and Jia, (2013)[36], the results show that the whole thermal shrinkage ($\Delta L/L_0$) of the Cs-based geopolymer from 25 to 1400°C was 0.17. Thermal shrinkage before 800°C was caused by capillary shrinkage, which contributed to 32.1% of the whole shrinkage. From 800 to 1200°C, thermal shrinkage was caused by viscous sintering, corresponding to 67.9% of the whole shrinkage. The onset and ending sintering temperatures of the Cs-based geopolymer using synthetic metakaolin were much lower than those of the one using natural metakaolin. The average coefficient of thermal expansion was 2.8×10^{-6}/°C for the pollucite ceramic.

REFERENCES
[1] Davidovits J., (1979), Polymère Minéral, *French Patent Applications* FR 79.22041 (FR 2,464,227) and FR 80.18970 (FR 2,489,290); *US Patent 4,349,386*, (1982) Mineral polymers and methods of making them.
[2] Breck D.W., (1974), *Zeolite Molecular Sieves, Structure, Chemistry and Use*, John Wiley & Sons, New York, 771 pp.
[3] Davidovits J., (1991), Geopolymers: Inorganic Polymeric New Materials, *J. Thermal Analysis*, 37, 1633–1656. PDF file at: http://www.geopolymer.org/library/technical-papers/12-geopolymers-inorganic-polymeric-new-materials/
[4] Davidovits J., (1994), Geopolymers: Man-Made Rock Geosynthesis and the Resulting Development of Very Early High Strength Cement, *J. Materials Education*, Vol.16 (2&3), 91–139. PDF file at: http://www.geopolymer.org/library/technical-papers/3-geopolymers-inorganic-polymeric-new-materials/
[5] Reference book: Joseph Davidovits, *Geopolymer Chemistry and Applications*, 2nd ed. 2008, 3rd ed. 2011, 4th ed. 2015, Geopolymer Institute, ISBN 4th ed. 9782951482098.
[6] Yao X., Zhang Z., Zhu H., Chen Y., (2009), Geopolymerization process of alkali–metakaolinite characterized by isothermal calorimetry, *Thermochimica Acta*, 493, 49–54.
[7] Zibouche F., Kerdjoudj H., d'Espinose de Lacaillerie J-B., Van DammeH., (2009), Geopolymers from Algerian metakaolin. Influence of secondary minerals, *Applied Clay Science*, 43, 453–458.
[8] Zhang Yunsheng, Jia Y., Sun W., Li Z., (2009), Study of ion cluster reorientation process of geopolymerisation reaction using semi-empirical AM1 calculations, *Cement and Concrete Research*, 39, 1174–1179.
[9] Rowles M. R. and O'Connor B. H., (2009), Chemical and Structural Microanalysis of Aluminosilicate Geopolymers Synthesized by Sodium Silicate Activation of Metakaolinite, *J. Am. Ceram. Soc.*, 92 [10], 2354–2361.
[10] Medri V., Fabbri S., Ruffini A., Dedecek J., Vaccari A., (2011), SiC-based refractory paints prepared with alkali aluminosilicate binders, *Journal of the European Ceramic Society,* 31, 2155–2165.
[11] Zhang B., MacKenzie K. J. D, Brown I. W. M., (2009), Crystalline phase formation in metakaolinite geopolymers activated with NaOH and sodium silicate, *J. Mater. Sci.*, 44, 4668–4676.

[12] Favier A, Habert G., d'Espinose de Lacaillerie J.B., Roussel N., (2013), Mechanical properties and compositional heterogeneities of fresh geopolymer pastes, *Cement and Concrete Res.*, 48, 9–16.

[13] Favier A, Habert G., Roussel N., d'Espinose de Lacaillerie J.B., (2015), A multinuclear static NMR study of geopolymerization, *Cement and Concrete Research* 75, 104–109.

[14] Medri V., Fabbri S., Dedecek J., Sobalik Z., Tvaruzkova Z., Vaccari A., (2010), Role of the morphology and the dehydroxylation of metakaolins on geopolymerization, *Applied Clay Science*, 50,538–545.

[15] Lippmaa E., Mägi M., Samoson A., Engelhardt G. and Grimmer A.-R., (1980), Structural Studies of Silicates by Solid-State High-Resolution 29Si NMR, *J. Am. Chem. Soc.,* 102, 4889–4893.

[16] Mäller D., Gressner W., Behrens H.J. and Scheler G., (1981), Determination of the Aluminum coordination in Aluminum-oxygen compounds by Solid-State High Resolution 27Al NMR, *Chem. Phys. Lett.*, 79 [1], 59–62.

[17] Meinhold R.H., MacKenzie K.J.D. and Brown I. W. M.., (1985), Thermal reactions of kaolinite studied solid state 27-AI and 29-Si NMR, *Journal of Materials Science Letters* 4, 163–166.

[18] Mackenzie K.J.D., Brown I.W.M, Meinhold R.H. and Bowden M.E., (1985), Outstanding Problems in the Kaolinite-Mullite Reaction; Sequence Investigated by 29Si and 27Al Solid-State Nuclear Magnetic Resonance: I, Metakaolinite, *J. Am. Ceram. Soc.*, 68 [6], 293–297.

[19] Sanz J., Madani A., Serratosa J.M., Moya J.S. and Aza S., (1988), Aluminum-27 and Silicon-29 Magic-Angle Spinning Nuclear Magnetic Resonance Study of Kaolinite-Mullite Transformation, *J. Am. Ceram. Soc.*, 71 [10] C-418-C-421.

[20] Davidovits J., (1976), Solid phase synthesis of a mineral blockpolymer by low temperature polycondensation of aluminosilicate polymers, *IUPAC International Symposium on Macromolecules Stockholm; Sept. 1976*; Topic III, New Polymers of high stability. PDF file at: http://www.geopolymer.org/library/technical-papers/20-milestone-paper-iupac-76/

[21] North M.R. and Swaddle T.W., (2000). Kinetics of Silicate Exchange in Alkaline Aluminosilicate Solutions, *Inorg. Chem.,* 39, 2661–2665.

[22] Duxson P. and Provis J. L., (2008), Designing Precursors for Geopolymer Cements, J. *Am. Ceram. Soc.*, 91 [12] 3864–3869.

[23] Bauer A. and Berger G., (1998), Kaolinite and smectite dissolution rate in high molar KOH solutions at 35° and 80°C, *Appl. Geochem.,* Vol. 13, No. 7, 905– 916.

[24] Zoulgami M., Lucas-Girot A., Michaud V., Briard P., Gaudé J. and Oudadesse H., (2002), Synthesis and physico-chemical characterization of a polysialate-hydroxyapatite composite for potential biomedical application, *Eur. Phys. J. AP*, 19, 173–179.

[25] Kriven W.M., Bell, J., Gordon, M., (2003), Microstructure and Microchemistry of Fully-Reacted Geopolymers and Geopolymer Matrix Composites, *Ceramic Transactions*, 153, 227–250.

[26] Kriven W.M., Bell J., Gordon M. (2006), Microstructure and nanoporosity of as-set geopolymers, *Ceramic Engineering and Science Proceedings*, 27 (2), pp. 491–503.

[27] Kriven W.M., Bell J., Gordon M. and Wen G., (2005), Geopolymers, more than just a cement, *Geopolymer 2005 Proceedings*, 179–183.

[28] He P., Jia D., Wang M. and Zhou Y., (2010), Effect of cesium substitution on the thermal evolution and ceramics formation of potassium-based geopolymer, *Ceramics International*, 36, 2395–2400.

[29] He P., Jia D., Wang M. and Zhou Y., (2011), Thermal evolution and crystallization kinetics of potassium-based geopolymer, *Ceramics International*, 37, 59–63.

[30] De Kimpe C.R., Kodame H., Rivard R., (1981), Hydrothermal Formation of a Kaolinite-like Product from Noncrystalline Aluminosilicate Gels, *Clays and Clay Minerals,* Vol. 29, No. 6, 446-450.

[31] Gordon M., Bell L., Kriven W. M., (2005), Comparison of naturally and Synthetically-Derived, Potassium-Based Geopolymers, Advances in Ceramic Matrix Composites X, *Ceramic Transactions*, 165, 95-165. See also M. A. Gülgün, W. M. Kriven and M. H. Nguyen, Processes for Preparing Mixed-Oxide Powders, US patent number 6,482,387 issued Nov 19th 2002.

[32] Cui X-M, Zheng G-J, Han Y-C, Feng S., Ji Z., (2008), A study on electrical conductivity of chemosynthetic Al_2O_3-$2SiO_2$ geopolymer materials, *Journal of Power Sources* 184, 652–656.

[33] Tsai Y-L, Hanna J. V., Lee Y-L, Smith M. E., Chan J. C. C., (2010), Solid-state NMR study of geopolymer prepared by sol–gel chemistry, *Journal of Solid State Chemistry*, 183, 3017–3022.

[34] Zheng G., Cui X., Huang D., Pang J., Mo G., Yu S., Tong Z., (2015), Alkali-activation reactivity of chemosynthetic Al_2O_3-$2SiO_2$ powders and their 27Al and 29Si magic-angle spinning nuclear magnetic resonance spectra, *Particuology*, 22, 151–156.

[35] Cui X-M., Liu L-P, Zheng G-J, Wang R-P, Lu J-P, (2010), Characterization of chemosynthetic Al_2O_3-$2SiO_2$ geopolymers, *Journal of Non-Crystalline Solids*, 356, 72-76.

[36] He P. and Jia D., (2013), Low-temperature sintered pollucite ceramic from geopolymer precursor using synthetic metakaolin, *J Mater Sci.,* 48, 1812–1818.

[37] Duxson P., Provis J.L., Lukey G.C. and van Deventer J.S.V., (2005), Structural ordering in geopolymers derived from metakaolin, *Geopolymer 2005 Proceedings*, 21–25.

[38] Xie N., Bell J.L., Kriven W.M. (2010), Fabrication of structural leucite glass-ceramics from potassium-based geopolymer precursors, J. Am. Ceram. Soc., 93 [9], 2585-2590.

ON THE DURABILITY BEHAVIOR OF NATURAL FIBER REINFORCED GEOPOLYMERS

A. C. C. Trindade and F. A. Silva
Department of Civil and Environmental Engineering, Pontifícia Universidade Católica Do Rio de Janeiro (PUC-Rio)
Rio de Janeiro, RJ, Brazil

H. A. Alcamand and P. H. R. Borges
Department of Civil Engineering, Centro Federal de Educação Tecnológica de Minas Gerais (CEFET-MG)
Belo Horizonte, MG, Brazil

ABSTRACT

Geopolymers may be described as solid and stable synthesized aluminosilicate materials which properties, presented in previous studies, are generally believed to provide a superior durability behavior for components used in the construction industry. The work in hand presents the results of an experimental investigation on the mechanical and durability behavior of jute fiber reinforced geopolymers. Silica fume (SF), metakaolin (MK) and blast furnace slag (BFS) based geopolymer composites were produced. The alkaline activator solution used in the mixture consisted of sodium silicate and sodium hydroxide. X-ray diffraction was used to determine the crystalline characteristics of the geopolymer reference materials and matrices. The jute fabric reinforcement is intended to arrest and bridge cracks hence, leading to an increase in ductility and durability. The composites were subjected to compression, tensile and flexural loading in order to evaluate their mechanical response and crack formation, before and after the accelerated aging process. The results showed that the composites containing BFS presented higher strength, and that the composites subjected to the aging process did not present significant degradation.

INTRODUCTION

Geopolymers may be described as synthesized solid and stable aluminosilicates, derived through chemical reactions from the combination of an alkaline activator solution with aluminosilicates present in a geological origin material, such as metakaolin, or reusable materials, such as fly-ash [1]. A polymerization reaction occurs in this process; so the French researcher Joseph Davidovits[1] decided in the 70s to name this complex class of materials as "geopolymers", which was firstly developed as an alternative material to withstand high temperatures.

Although much of the macroscopic characteristics of the geopolymers prepared from different sources of aluminosilicates are very similar, their microstructure, chemical and mechanical properties may vary significantly depending on the raw materials and molar ratios used[2]. This class of materials is known to present itself as a beneficial alternative to the environment, as the industrial production of its raw materials can reduce by up to six times the release of CO_2 into the atmosphere when compared to the production of Portland cement[3]. A study developed by researchers in Melbourne and Illinois[4] showed that the compressive strength does not significantly changes when varying the type of alkali (sodium or potassium), or age (between 7 and 28 days). However, samples with high Si/Al ratio showed an increase in strength. They also demonstrated that the elastic modulus is fully dependent on the type of activator. When using sodium hydroxide the modulus of elasticity increases, whereas for potassium hydroxide the effect was the opposite.

This class of materials is typically characterized by low tensile strength and poor deformation capacity[5]. Fibrous reinforcement can be a great alternative to be incorporated into the matrix to overcome these vulnerabilities. Shaikh et. al[6] observed that the incorporation of short fibers (steel and polypropylene) into the geopolymer matrix significantly improves flexural strength, regardless of the fiber type and mixture. The materials studied achieved better toughness results when compared to Portland cement based composites, presenting deflection hardening and multiple cracking formation. Sankar and Kriven[7] investigated the processing, microstructure and mechanical properties of jute weave reinforced geopolymer composites, using pressurized plates with alkali treatment. The composites presented high flexural strength, although indicating poor fiber-matrix bonding in the jute-geopolymer composites. Additionally, crack formation in deflection also confirmed this mechanism. The same researchers also investigated the use of two-dimensional and unidirectional, alkali-treated, fique reinforced potassium geopolymers composites[8]. The results showed responses of 13.9 MPa in tensile, 11.4 MPa in flexural and 9.39 J of energy absorbed in impact tests. The investigation also revealed that the fique composite presented better properties than did alkali-treated jute composites.

The studies mentioned before didn't analyze the long-term effects related to the use of reinforcement in the geopolymer matrices. However, studies related to different geopolymer formulations showed their capacity in terms of durability. Olivia and Nikraz[9] indicated that geopolymer mixtures based on fly ash, when subjected to wetting and drying cycles after exposure to high temperatures, presented loss of mass and degradation of the outer layers. On the other hand, the specimens were able to maintain the increase of compressive strength with age. Similar behavior was presented by Pong et al.[10] The investigation demonstrated that for geopolymers based on fly ash, a small loss of mass occurs in the first cycles of wetting and drying, but soon this process stabilizes, not interfering in the resistance of service.

This work presents the preliminary results of an experimental investigation on the durability and mechanical response of jute fabrics reinforced geopolymers produced from different formulations. The composites were subjected to compression, flexural and tensile loading in order to evaluate their mechanical response and cracking formation. Durability tests were carried by submitting samples to accelerated aging conditions (wetting and drying cycles). Their long-term degradation was evaluated by comparisons of the flexural tests results performed with the composites, before and after the w/d cycles.

EXPERIMENTAL PROGRAM

Materials

The used source of silica and alumina were metakaolin (MK), silica fume (SF), and blast furnace slag (BFS) supplied by Metacaulim do Brasil, Tecnosil and Central IBEC, respectively. The chemical compositions of the materials were obtained by X-ray fluorescence (XRF) and are reported in Table 1. The alkaline activator solution, mixed in pre-determined proportions, was composed of sodium silicate (Na_2SiO_3) and sodium hydroxide (NaOH). Their chemical compositions are also presented in Table 2.

River sand was used as a natural aggregate, with a density of 2.68 g/cm^3 and fineness modulus of 2.28, having a maximum diameter of 1.18 mm. The particle size distribution (PSD) of the metakaolin, silica fume, and blast furnace slag were obtained by laser granulometry and are presented in Figure 1, along with the PSD of the sand.

Table 1. Chemical Compositions of metakaolin, silica fume and blast furnace slag.

Chemical Compositions (wt%)	Metakaolin (MK)	Silica Fume (SF)	Blast Furnace Slag (BFS)
SiO_2	40.02%	93.40%	45.18%
Al_2O_3	34.00%	0.75%	10.78%
Fe_2O_3	2.00%	1.24%	2.30%
TiO_2	1.00%	0.02%	0.43%
CaO	0.10%	1.39%	32.73%
MgO	0.60%	1.02%	5.38%
K_2O	1.70%	1.25%	0.93%
Na_2O	0.10%	0.39%	0.14%
SO_3	0.10%	-	0.44%
P_2O_5	-	0.13%	0.04%
ZnO	-	0.04%	-
SrO	-	0.02%	0.12%
MnO	-	0.05%	1.38%
BaO	-	-	0.07%

Table 2. Chemical Compositions of Na_2SiO_3 and $NaOH$.

Chemical Compositions (wt%)	Sodium Silicate (Na_2SiO_3)	Sodium Hydroxide ($NaOH$)
SiO_2	32.20%	-
Na_2O	14.70%	-
H_2O	53.10%	50.00%
$NaOH$	-	48.03%
SO_4	-	00.02%
Cl	-	0.005%
Na_2CO_3	-	1.094%

Plain weave jute fabrics obtained from Castanhal Companhia Têxtil (Castanhal, Pará, Brazil) were used as reinforcement. The jute fiber is extracted from the stem of the plant *Corchorus capsularis* by a combination of processes comprised of the following steps: cutting, retting, shredding, drying, packing, and classification. The chemical composition of the jute fiber comprehends approximately 72% cellulose, 12.8% hemicellulose and 8.1% lignin. The properties of the jute yarn are presented in Table 3.

Table 3. Physical and mechanical properties of the jute yarn.

Properties	Jute yarn
Fineness (tex)	326
Diameter (mm)	0.785
Number of filaments	141
Tensile Strength (MPa)	104
Strain-to-failure (%)	2.11
Young's Modulus (GPa)	5.68

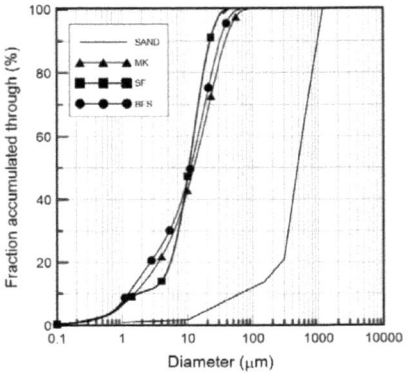

Figure 1. Particle size distribution curves.

Processing

Three geopolymer matrices were developed in this research. Table 4 shows their mix design (in mass, per 1000g of binder), as well as the activating parameters used (molar ratios). Borges et al. [11,12] studied the activation of the same matrices with different parameters. They achieved the best mechanical behaviors when using the molar ratios indicated in this study. The reference matrix, resulted from the activation of a binder material consisting 100% by MK, with a SiO_2/Al_2O_3 molar ratio equal to 3.0. The second and third matrices were activated with equal molar ratios, $SiO_2/Al_2O_3 = 3.9$, but their binder materials consisted of, respectively, 20%SF/80%MK, and 40%MK/60%BFS.

These three matrices had as fine aggregate sand with a maximum diameter of 1.18 mm, and mass ratio of 1:1 (between binders/aggregate materials). The amounts of silicate and sodium hydroxide used in each mixture varied, in order to ensure the indicated molar ratios presented in Table 4. Once again, following the results obtained by Borges et al. [11,12]: $H_2O/Na_2O = 11$ and $Na_2O/SiO_2 = 0.25$ for all mortars; Na_2O/Al_2O_3 equal to 0.75 or 0.98, respectively for $SiO_2/Al_2O_3=3.0$ and 3.9, and $SiO_2/CaO = 4.03$ and $Al_2O_3/CaO = 1.03$ for the mixture containing BFS. From the previously mentioned characteristics, the matrices were named as F3.0-100MK, F3.9-80MK20SF, and F3.9-60MK40BFS.

The preparation of the geopolymer mixtures was performed in a 5L capacity planetary mixer. The mixture process, following the instructions given by Borges et al. [12], is described as follows: manual mixing of the dry materials with a metallic spatula, to achieve maximum homogeneity in the mixture of solids; addition of the alkaline activator solution, comprising of sodium hydroxide and sodium silicate (premixed) for 1 minute; mixture during 4 minutes at 136 rpm in the planetary mixer; turning off the planetary mixer for removal of trapped solids on the walls of the container; and finally homogenizing the mixture for 3 minutes at 281 rpm.

Table 4. Solid materials (mass) and molar ratios for each geopolymer matrix (with 1000g as reference).

Matrix	Mass (g)				Molar Ratios					
	MK	SF	BFS	Sand	$SiO_2/$ Al_2O_3	$H_2O/$ Na_2O	$Na_2O/$ SiO_2	$Na_2O/$ Al_2O_3	$SiO_2/$ CaO	$Al_2O_3/$ CaO
F3.0-100MK	1000	-	-	1000	3	11	0.25	0.75	-	-
F3.9-80MK20SF	800	200	-	1000	3.9	11	0.25	0.98	-	-
F3.9-60MK40BFS	600	-	400	1000	3.9	11	0.25	0.98	4.03	1.03

Flow table tests were carried out according to the ASTM C143[13]. The results are presented in Table 5. It can be observed that as the amount of metakaolin decreases and the volume of slag or silica increases, the matrix becomes more fluid. This is due to the fact that metakaolin absorbs large amounts of water, reducing the workability of the material.

Table 5. Flow table testing spreading.

Matrix	Standard Consistency (mm)
F3.0-100MK	127.5
F3.9-80MK20SF	163.5
F3.9-60MK40BFS	201.0

Cylindrical specimens with the height of 100 mm and diameter of 50 mm were produced for compression tests. Plate specimens were manufactured for tensile and flexural tests, with dimensions equals to: 450 mm x 60 mm x 12 mm (length x width x thickness), and 270 mm x 60 mm x 12 mm, respectively. To produce the geopolymer plates reinforced with jute fibers, five layers of fibers were used (10% of volume fraction). The reinforcement arrangement was made alternating textile layers with geopolymers (2 mm each) up to a thickness of 12 mm. This process was developed in an acrylic mold designed for this research (Fig. 2.a). Due to the fluid consistency of the mortar, it was not necessary to perform any type of consolidation.

All plates and cylindrical specimens were prepared and cured at room temperature (25 ± 2°C). Then, they were demolded after 24 hours and wrapped in plastic bags to prevent moisture loss and cracking during the subsequent curing of 7 days. The cure regime was established according to the results found by Borges et al. [11]. Despite showing a small variation in mechanical results regarding the formulation containing BFS (between 7 and 28 days), they did not present any significant variation for the other formulations. Therefore, all specimens presented in this study (matrices and composites) were tested at 7 days. The composite manufacturing process is illustrated in Fig. 2.b.

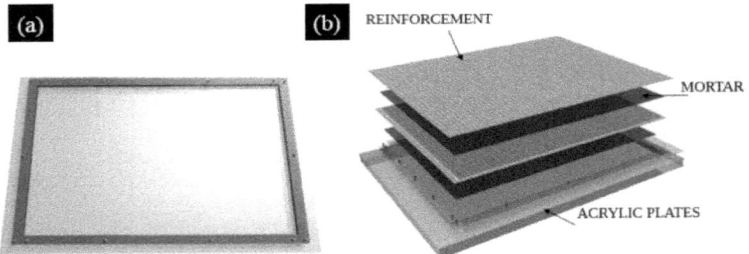

Figure 2. (a) Schematic drawing of the acrylic mold and (b) composite manufacturing process.

TESTING METHODS

X-ray diffraction was performed to characterize the crystalline phases of the used materials (metakaolin, blast furnace slag and silica fume) and geopolymer matrices (F3.0-100MK, F3.9-80MK20SF, and F3.9-60MK40BFS). To perform this procedure a Shimadzu diffractometer model 7000 XRD with copper radiation (Cu-Ka, $\lambda = 1.5418$ Å) operating at 40 kV and 30 mA was used. To determine the crystalline phases, scans were performed with an angular velocity of 0,02° per second and measuring the interval between Bragg angles (2θ) 5° and 80°.

Compression tests were performed following the methods indicated in ASTM C39[14]. The procedure occurred on a MTS testing equipment, model 810, with load capacity of 500 kN, at 7 days. Three cylindrical specimens were tested for each formulation with 100 mm height and 50 mm diameter, previously prepared in order to obtain the smoothing of the surface, avoiding the concentration of stresses in undesired points. The tests were performed at a displacement rate of 0.5 mm/min. The axial displacements were measured by two LVDTs, with length of 70 mm, coupled with acrylic rings positioned around the specimen. The displacement result considered was the average value obtained by the transducers. Fig. 3.a shows the test setup.

Tensile tests were carried on a MTS testing machine, model 311. The tests were performed under displacement control at a rate of 0.1 mm/min, at 7 days. The dimensions of the tensile samples were 450 mm x 60 mm x 12 mm (length x width x thickness). The displacements were measured by two LVDTs positioned on the sides of the specimens with 250 mm gauge length, and it was only considered the average value obtained through the readings of the two LVDTs. The specimens were fixed in steel plates with screws. Three specimens for each geopolymer composite were tested. Figure 3.b shows the experimental setup used in direct tensile tests.

Three-point bending tests were performed in a MTS testing system. The dimensions of the specimens were 270 mm x 60 mm x 12 mm (length x width x thickness). The tests were carried on under displacement control at a rate of 1.0 mm/min with a 200 mm span between end supports, once again at 7 days. The displacement was measured by two displacement transducers positioned in the middle of the span, one in each side of the specimen. Again, three specimens for each geopolymer composite were tested. Figure 4 illustrates the bending test set-up.

The durability tests were carried according to the methodology used by Toledo Filho et al.[8]. The composites that presented the best mechanical behavior in the compressive and tensile tests were submitted to flexural tests before and after accelerated aging conditions (wetting/drying cycles). The duration stablished for each cycle was 24 hours of wetting and 48 hours of drying according to the methodology proposed by Toledo Filho et al.[15]. Melo Filho et al.[16] obtained significant changes in the mechanical properties of cementitious composites reinforced with natural fibers when submitted to 10 w/d cycles. For this reason, in this research it was chosen to evaluate the degradation in geopolymeric composites by flexural tests, that were performed after submitting the specimens to 15 w/d cycles.

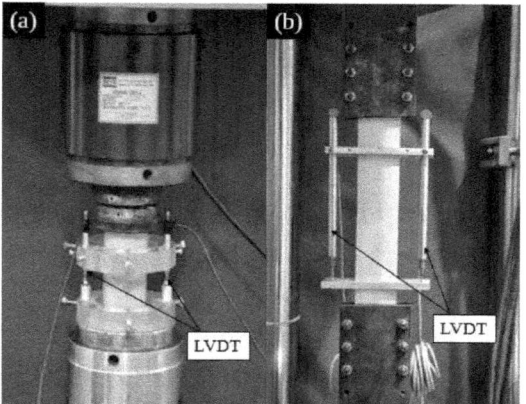

Figure 3. Setup for: (a) compression tests; and (b) tensile tests.

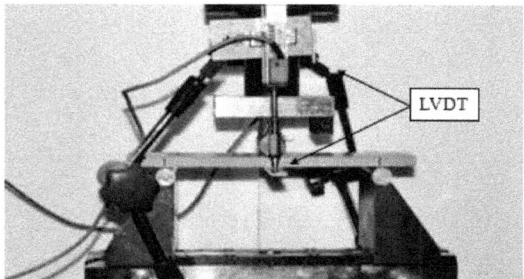

Figure 4. Setup for three-point bending tests.

RESULTS AND DISCUSSION

Figure 5 shows the X-ray diffractograms for metakaolin, blast furnace slag and silica fume. In Figure 5.a it is possible to see that silica fume is an amorphous material, so it does not show characteristic crystalline peaks. Figures 5.b and 5.c, for metakaolin and blast furnace slag, respectively, represent aluminosilicates, mostly amorphous. Metakaolin presents quartz, muscovite, kaolinite and illite peaks, resulted from impurities included in the metakaolin used in this study. BFS show crystalline peaks for mineral gehlenite, a product normally found in iron slag and akermanite.

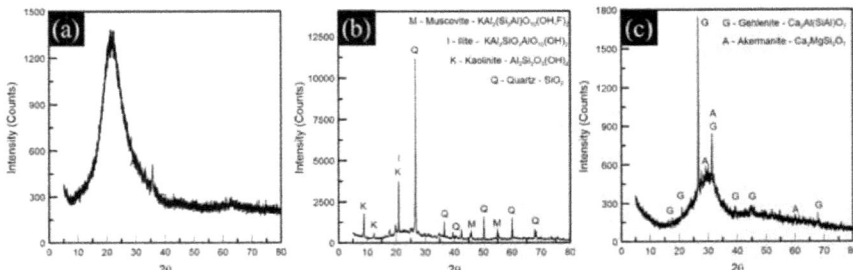

Figure 5. DRX of: a) silica fume, b) metakaolin, and c) blast furnace slag.

Figure 6 shows the X-ray diffractograms for the geopolymer matrices. It is possible to notice that all of the matrices demonstrated similar crystalline behaviors detecting that the crystalline peaks of the impurities found in metakaolin are preponderant in all of the results. It is also possible to visualize that the amorphous structure of the silica fume did not significantly change the geopolymeric crystalline behavior in Figure 6.b.

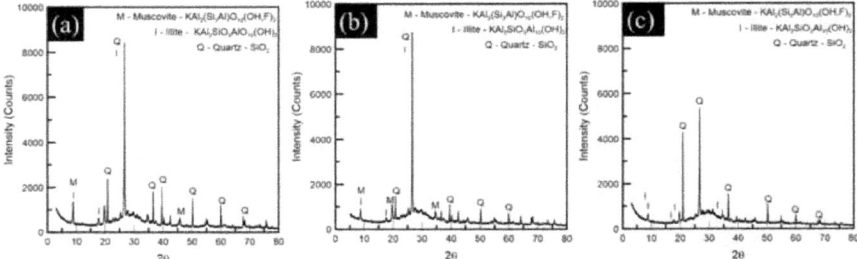

Figure 6. DRX of: (a) F3.0-100MK, (b) F3.9-80MK20SF, and (c) F3.9-60MK40BFS.

The results of the compressive tests for the three matrices are shown in Figure 7 and Table 6. There was a significant increase in the compressive strength (σ_{max}) for the matrix containing BFS, regardless of the SiO_2/Al_2O_3 molar ratio. In fact, the partial substitution of MK by BFS increases the mechanical strength due to the additional gel formation of $CaO-Al_2O_3-SiO_2-H_2O$ (C-A-S-H), related to the presence of calcium arising from the BFS. This gel fills the voids in the characteristic geopolymer matrix $Na_2O-Al_2O_3-SiO_2-(H_2O)(N-A-S-(H))$, reducing porosity and permeability[12]. It is noted, however, that the partial replacement of MK with SF reduced the compressive strength, and increased its strain capacity (ε_{max}). This fact can be attributed to the incomplete dispersion of SF during the mixing, forming small clusters of the material, which increases the porosity, decreasing the mechanical strength.

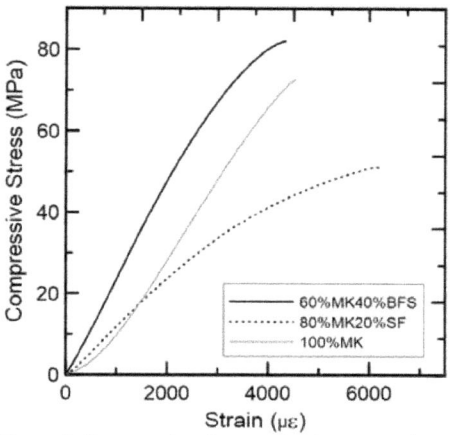

Figure 7. Compressive Stress x Strain curves of F3.0-100MK, F3.9-80MK20SF, and F3.9-60MK40BFS.

Table 6. Results of compression tests performed in three different matrices (values in parentheses refer to standard deviation).

Matrix	Compression Tests		
	σ_{max} (MPa)	ε_{max} ($\mu\varepsilon$)	E_c (GPa)
F3.0-100MK	72.7 (2.1)	4543.4 (314.3)	14.26 (1.87)
F3.9-80MK20SF	51.24 (1.4)	6223.5 (478.9)	12.05 (1.56)
F3.9-60MK40BFS	81.98 (3.2)	4342.05 (305.6)	23.94 (3.15)

The results of the tensile tests for the geopolymer composites reinforced with 5 layers of jute fabrics are shown in Figure 8 and Table 7. The composites showed a high strain capacity and formation of multiple cracks. By comparing the three formulations, it is possible to notice that there was a significant increase in first cracking (σ_{1f}), ultimate tensile strength (σ_u) and strain capacity (ε_{1f} and ε_u) for the composite based on BFS. The composite based on the 100%MK matrix presented a small decrease in crack width, but maintained a similar behavior to the previous matrix, although showing slightly lower results regarding the first cracking and ultimate strength (σ_{1f} and σ_u). The composite based on SF had the lowest results for tensile strength. H, it presented lower cracking opening and a significantly higher crack formation during the test. This mechanism may be associated with an increased bond and also to a lower matrix first crack strength. All composites exhibited strain-hardening and multiple cracking behavior. The elastic modulus was computed in the linear elastic range up to 40% of the ultimate strength, reaching values as high as 16.64 GPa (F3.9-60MK40BFS).

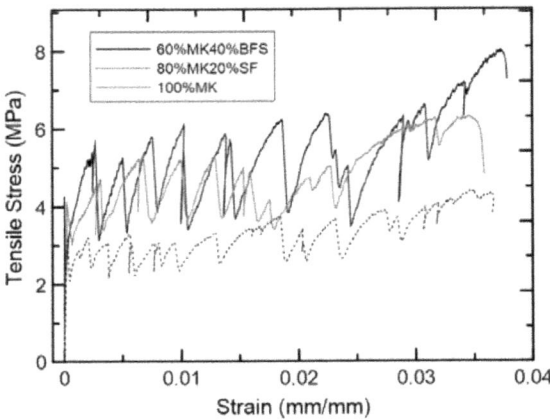

Figure 8. Tensile Stress x Strain curves of the F3.0-100MK, F3.9-80MK20SF, and F3.9-60MK40BFS.

Table 7. Results of tensile tests performed in three different matrices reinforced with five layers of jute fibers (values in parentheses refer to standard deviation).

Composite	Tensile Tests				
	σ_{1f} (MPa)	ε_{1f} (mm/mm)	σ_u (MPa)	ε_u (mm/mm)	E_t (GPa)
F3.0-100MK + 5L	4.13 (0.51)	0.0007	6.31 (0.78)	0.034	14.92 (1.37)
F3.9-80MK20SF + 5L	2.51 (0.37)	0.0004	4.34 (0.55)	0.036	13.84 (1.09)
F3.9-60MK40BFS + 5L	4.37 (0.34)	0.0012	8.02 (1.13)	0.037	16.64 (1.15)

Regarding the density of macro cracks in specimens after they were subjected to tensile testing, it is possible to observe that the composites consisting of MK+SF had the highest number of macro cracks per meter. The average values obtained were: 30 cracks/m for the F3.0-100MK+5L; 62.5 cracks/m for the F3.9-80MK20SF+5L; and 42.5 cracks/m for the F3.9-60MK40BFS + 5L. Figure 9 shows the different cracking patterns for each formulation.

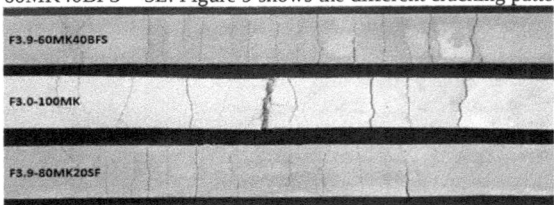

Figure 9. Composites submitted to tensile tests showing different patterns of cracking.

. The results of the flexural tests for the composites reinforced with 5 layers of jute fabric are shown in Figure 10 and Table 8. Again, the composite containing BFS showed a better performance for the first and ultimate deflection (δ_{1f} and δ_u), and consequently first and ultimate strength (σ_{1f} and σ_u), as expected. Once more, the composite containing only MK showed the lower mechanical capacity and a larger cracking opening, as expected. All composites exhibited deflection-hardening and multiple cracking behavior. The composites that presented the best mechanical behavior (100MK and 60MK40BFS) in the previous tensile and flexural tests, were then subjected to durability conditions (wetting/drying cycles).

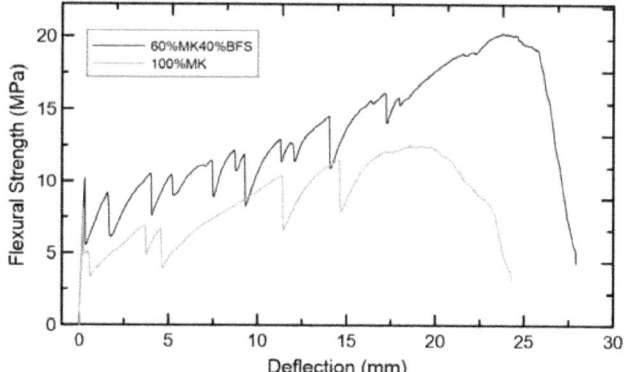

Figure 10. Flexural Strength x Deflection curves of the F3.0-100MK and F3.9-60MK40BFS composites.

Table 8. Results of flexural tests performed in two different matrices reinforced with five layers of jute fibers (values in parentheses refer to standard deviation).

Composite	Flexural Tests				
	P_{1f} (kN)	σ_{1f} (MPa)	δ_{1f} (mm)	σ_u (MPa)	δ_u (mm)
F3.0-100MK + 5L	0.215	7.49 (0.72)	0.20	12.33 (0.61)	21.57
F3.9-60MK40BFS + 5L	0.275	9.57 (0.56)	0.31	18.36 (1.80)	19.74

After 15 w/d cycles the specimens were also submitted to three-point bending tests and the results are presented in Figure 11 and Table 9. It is possible to notice that although there was a considerable decrease in the value of the first crack strength (δ_{1f}) for both composites (close to 25%), the ultimate strength did not vary significantly (close to 10%). This behavior indicates that there was no significant modification on the composites ultimate mechanical capacities after being subjected to 15 wetting/drying cycles. The most interesting difference observed occurred in the composite cracking pattern. It seems that to some extent the exposure to the accelerated aging cycles improved the fiber matrix bond resulting in smaller crack openings and formation of higher number of cracks during the test. The results after 15 cycles did not present significant degradations, suggesting a superior durability of geopolymers reinforced with jute fabrics when compared to composites based on Portland cement matrices[17]. All composites exhibited deflection hardening behavior with the formation of multiple cracks after the accelerated aging regime.

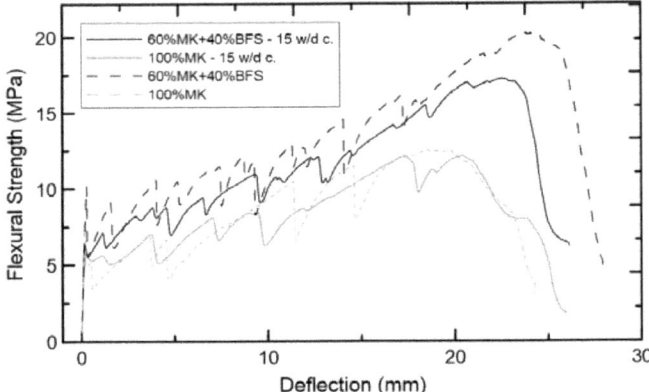

Figure 11. Flexural Strength x Deflection curves of the F3.0-100MK and F3.9-60MK40BFS composites before and after being subjected to 15 cycles of wetting and drying.

Table 9. Results of flexural tests performed in two different matrices reinforced with five layers of jute fibers after being subjected to 15 cycles of wetting and drying (values in parentheses refer to standard deviation).

Composite	Flexural Tests				
	P_{1f} (kN)	σ_{1f} (MPa)	δ_{1f} (mm)	σ_u (MPa)	δ_u (mm)
F3.0-100MK + 5L	0.16	5.61 (0.30)	0.27	11.32 (1.51)	21.72
F3.9-60MK40BFS + 5L	0.20	6.99 (0.58)	0.81	17.09 (0.68)	24.68

CONCLUSIONS

Modifications in the metakaolin matrix compositions through its partial replacement by BFS and SF, changes the physical and mechanical behavior of the matrices. The presence of BFS significantly increases the compressive strength of the material, related to the additional gel formation $CaO\text{-}Al_2O_3\text{-}SiO_2\text{-}H_2O$ (C-A-S-H). Due to the incomplete dispersion of silica fume, there was no strength increase when using SF. However, all three geopolymer matrices exhibited high compressive strength, reaching values as high as 82 MPa at 7 days.

The composites reinforced with 5 layers of jute fabrics showed strain and deflection hardening behaviors with multiple crack formation under tensile and flexural loading. The addition of jute fibers as reinforcement modifies the mechanical behavior of the matrices, making them ductile materials and changing their crack pattern. In both tests, composites based on BFS showed higher strength values when compared to the ones based on MK and SF. The results after the aging process suggested that the composites didn't suffer significant deteriorations after 15 w/d cycles. This behavior demonstrate a great durability capacity and an interesting increase in toughness, a feature that can be very beneficial in many structural applications.

REFERENCES

[1] DAVIDOVITS, J. et al. (1998). Alkaline alumino-silicate geopolymeric matrix for composite materials with fiber reinforcement and method for obtaining same. *U.S. Patent*. n. 5,798,307, 25 ago.

[2] VICKERS, L. et al. (2015). Fire-Resistant Geopolymers Role of Fibres and Fillers to Enhance Thermal Properties. *Springer*.

[3] WHITE, C. E. et al. (2010). The effects of temperature on the local structure of metakaolin-based geopolymer binder: A neutron pair distribution function investigation. *Journal of the American Ceramic Society,* v. 93, n. 10, p. 3486–3492.

[4] DUXSON, P. et al. (2007). The effect of alkali and Si/Al ratio on the development of mechanical properties of metakaolin-based geopolymers. *Colloids and Surfaces A: Physicochemical and Engineering Aspects*, v. 292, n. 1, p. 8–20.

[5] GANESAN, N. et al. (2013). Engineering properties of steel fibre reinforced geopolymer concrete. *Advances in Concrete Construction,* v. 1, n. 4, p. 305-318.

[6] SHAIKH, F. U. A. (2013). Review of mechanical properties of short fibre reinforced geopolymer composites. *Construction and Building Materials*, v. 43, p. 37-49.

[7] SANKAR, K.; KRIVEN, W. M. (2014). Sodium geopolymer reinforced with jute weave. *Ceramic Engineering and Science Proceedings*, v. 35, n. 8. p. 39-60.

[8] SANKAR, K.; KRIVEN, W. M. (2014). Potassium geopolymer reinforced alkali-treated fique. *Ceramic Engineering and Science Proceedings*, v. 35, n. 8. p. 61-78.

[9] OLIVIA, M.; NIKRAZ, H. (2012). Properties of fly ash geopolymer concrete designed by Taguchi method. *Materials & Design*, v. 36, p. 191-198.

[10] SUKSIRIPATTANAPONG, C. et al. (2015). Compressive strength development in fly ash geopolymer masonry units manufactured from water treatment sludge. *Construction and Building Materials*, v. 82, p. 20-30.

[11] BORGES, P.H.R. et al. (2013). Andreasen particle packing method on the development of geopolymer concrete for civil engineering. *Journal of Materials in Civil Engineering*, v. 26, p. 692-697.

[12] BORGES, P.H.R. et al. (2016). Performance of blended metakaolin/blastfurnace slag alkali-activated mortars. *Cement and Concrete Composites*, v. 71, p. 42-52.

[13] ASTM C143/C143M-09, Standard Test Method for Slump of Hydraulic-Cement Concrete.

2009. *ASTM International: West Conshohocken*, PA.

[14]ASTM C39/C39M-15a, Standard Test Method for Compressive Strength of Cylindrical Concrete Specimens. 2015. *ASTM International, West Conshohocken*, PA.

[15] TOLEDO FILHO, Romildo Dias et al. (2009). Durability of compression molded sisal fiber reinforced mortar laminates. *Construction and Building Materials*, v. 23, n. 6, p. 2409-2420.

[16] MELO FILHO, João et al. (2013). Degradation kinetics and aging mechanisms on sisal fiber cement composite systems. *Cement and Concrete Composites*, v. 40, p. 30-39.

[17]FIDELIS, Maria Ernestina Alves et al. (2016). The effect of accelerated aging on the interface of jute textile reinforced concrete. *Cement and Concrete Composites,* v. 74, p. 7-15.

PERFORMANCE AND DURABILITY OF Fe-RICH INORGANIC POLYMER COMPOSITES WITH BASALT FIBERS

A. Peys[a], M. Peeters[a], A. Katsiki[b], L. Kriskova[a], H. Rahier[b] and Y. Pontikes[a]

[a]KU Leuven Department of Materials Engineering, Kasteelpark Arenberg 44, 3001 Leuven, Belgium
[b]Department of Materials and Chemistry, Vrije Universiteit Brussel, Pleinlaan 2, 1050 Brussels, Belgium

ABSTRACT

The use of inorganic instead of organic polymers in composites receives increasing interest from research institutions and industry because of their fire resistance and lower energy input upon manufacturing. It remains however unclear whether the fibers can resist the alkalinity of the inorganic polymer in a larger timeframe. This is investigated in the present work for basalt fibers. Inorganic polymers were synthesized from an iron-silicate slag, originating from non-ferrous metallurgy, and potassium silicate solutions (SiO_2/K_2O = 1.6-2.0, H_2O/K_2O = 16-24), which were mixed in a solution/slag mass ratio of 0.3. To this mixture, 3 wt.% of basalt fibers (12.7 mm long, 13 µm diameter) was added. Curing occurred at room temperature. The 3-point flexural strength as a function of time increased for all batches. The highest flexural strength was recorded when an activating solution with SiO_2/K_2O = 2.0 and H_2O/K_2O = 16 was used; values were 29 ± 1 MPa and 37 ± 5 MPa, after 7 and 90 days, respectively. Without fibers, the strength was 13 ± 1 MPa after 90 days. The results show that the basalt fibers retain their structural performance in time, suggesting that basalt fiber reinforced inorganic polymers can be durable. However, a repetition of a similar mix with a different precursor (requiring more solution to be mixable), lead to the contradicting conclusion of having degraded during the curing period. The amount of solution is therefore considered to be an important factor for the durability of fiber reinforced inorganic polymers.

INTRODUCTION

Geopolymers, or inorganic polymers in general, are rising in importance as an alternative for Portland cement because of their lower CO_2 output upon manufacture[1]. Like Portland cement, inorganic polymers perform poorly under tensile stress. Fibers are used as reinforcement of the inorganic polymer matrix, by analogy with fiber reinforced organic polymers. The replacement of the organic matrix in fiber reinforced organic polymers with an inorganic polymer matrix results in interesting properties, certainly in the aspect of high temperature resistance. While organic polymers or Portland cement fail at temperatures below 300 °C, inorganic polymer materials, in certain formulations, withstand temperatures up to 1000 °C[2]. Apart from the properties, the price and ecological impact of inorganic polymers are orders of magnitude smaller than that of organic polymers. However, the strong alkaline conditions during synthesis do not allow the use of just any type of fiber. More knowledge on the inorganic polymer-fiber interaction during and after synthesis is needed.

Short fiber reinforcement is most common in inorganic polymer science or cementitious composite technology in general. A wide range of different fibers have been investigated. Tensile strength and toughness has been increased by the introduction of steel, polymer, glass, basalt, carbon, and natural fibers. In general, the increase in tensile or flexural strength is of the order of magnitude of 20 %[3, 4, 5, 6]. Carbon fibers can reach an order of magnitude higher, a maximum increase of 575 % in flexural strength is observed up to 96.6 MPa (16.8 MPa for the matrix)[7, 8]. Lin et al.[7, 8] also showed a large increase in work of fracture by the introduction of short carbon fibers in the inorganic polymer matrix (100 times increase). It should be noted that the strength of the inorganic polymer matrix of the latter cited reference was already much higher in comparison with other references. Alomayri and Low[9] observed an increase in compressive strength of more than 100% for the addition of 0.5 wt.% of cotton fibers. This increase was also observed for the impact strength. Short basalt fiber reinforced metakaolin based inorganic polymer composites obtained a 3-point flexural strength of 27.1 MPa (2.2 MPa for the matrix)[10, 11]. Apart from short fibers, inert particulate addition has been proven beneficial for refractory applications. The addition of alumina, chamotte, or granite resulted in an increase in flexural strength and a reduction of thermal shrinkage[2]. Because of the latter, the strength is maintained at higher temperatures (up to 800-1000 °C). Crack deflection[3] and multiple cracking[12, 13] has been observed, leading to large strain upon fracture or even strain hardening. Strains up to 5% have been reported[13].

The referred papers show promising mechanical and thermal properties of inorganic polymer composites. However, the discussion on durability or mechanical properties after a longer time period is often avoided. This is of high importance for engineering materials as they need to keep their properties for the life-time of the product. Therefore, the present study investigates the properties of basalt fibers in a longer time-frame. First the choice of basalt fibers is motivated, after which a short pre-optimization of the processing of basalt fiber reinforced Fe-rich inorganic polymers is executed. One set of processing conditions is selected and used to investigate the degradation of the basalt fibers in the composites, using different molarities and alkalinities of the activating solution. The latter is performed in the attempt to minimize the degradation of the fibers.

EXPERIMENTAL

As a precursor an iron-silicate slag was used, with a composition similar to previous work in the research group[14] and with major elements presented in Table I.

Table I. Major elements of chemical composition of precursor slag in wt.%, from XRF in Iacobescu et al.[14]

FeO	SiO2	ZnO	Al2O3	CaO
57	26	6	3	2

This was mixed with a potassium silicate solution for the synthesis of the inorganic polymer paste. Table II shows the studied compositions of potassium silicate solutions. In first instance, only solution 1.6 – 16 was used for roughly tuning the processing parameters.

Table II. Chemical composition of the activating solutions.

	SiO$_2$/K$_2$O molar ratio	H$_2$O/K$_2$O molar ratio
Solution 1.6 – 16	1.6	16
Solution 1.6 – 24	1.6	24
Solution 2.0 – 16	2.0	16
Solution 2.0 – 24	2.0	24

Basalt fibers from Basaltex with two different coatings were investigated, by the company branded as "dry" and "wet". Both coatings were silane based of which one had a higher moisture content (wet) to adapt the hydrophilicity of the surface. This was designed for lowering the surface energy with cementitious paste, increasing the dispersion of the fibers in the paste. The fibers had a length of ½ inch (12.7 mm).

The composite mixture had a solution/slag ratio of 0.3 and 3 wt.% of basalt fibers. Two different mixers were investigated: a Hobart mixer and a high shear mixer (Dispermat VMA-Getzmann GMBH D-51580). The slag was mixed with the solution for different times (exact times discussed later) and subsequently, the fibers were added to the paste and further mixed for various times. Samples of 2x2x8 cm^3 were cast and cured at room temperature (22 ± 2 °C).

The flexural strength of the samples was tested using 3-point bending tests on an Instron, model 5985, with a load cell of 250 kN and a crosshead speed of 0.5 mm/min. The fracture surface of the samples and the basalt fibers themselves were studied using scanning electron microscopy (SEM) on a FEI Nova NanoSEM 450. A coating of 5 nm of platinum was applied prior to the investigations.

RESULTS AND DISCUSSION

Single fiber tests

Motivation for the selection of basalt fibers was provided by dissolution tests in comparison with E-glass and carbon fibers. Figure 1 qualitatively shows the effect of (left to right) 1 hour, 1 day, and 1 week of exposure to the alkaline solution with a SiO$_2$/K$_2$O molar ratio of 1.6. E-glass, known for its weak performance in alkaline environments, dissolves extensively, exhibiting a pelletizing effect. Carbon fibers stay intact, only sticking to each other weakly after 1 week. Basalt fibers seem to behave similar to carbon fibers in this study, making them a good alternative reinforcement for cements and inorganic polymers. However, previous studies showed that damage in the form of surface roughness was induced when exposed to NaOH[15]. This was not observed for the milder silicate solution. No damage was detected using SEM (indicative image shown in Figure 2 left) and only a minor loss of mass, 7 wt.%, was measured (Figure 2 right).

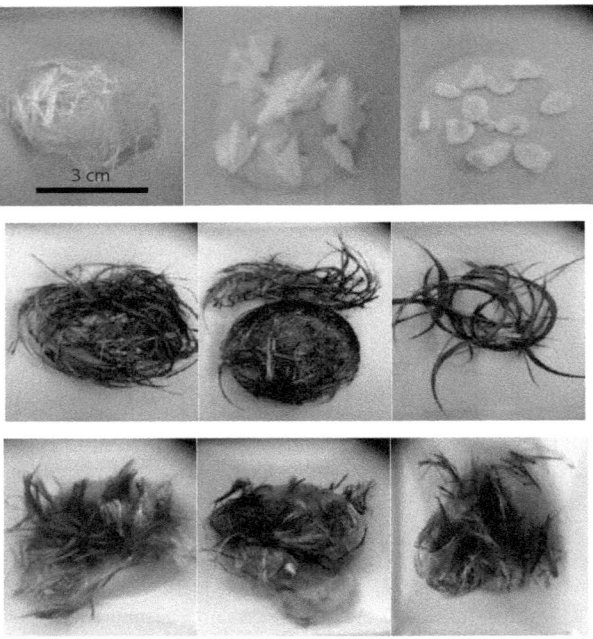

Figure 1. Fibers after dissolution test. Left to right: 1 hour, 1 day, 1 week. Top to bottom: glass, carbon, basalt fibers.

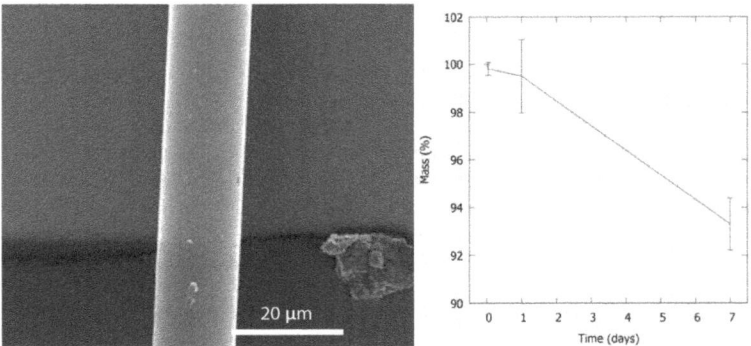

Figure 2. Electron microscopy image of basalt fibers exposed for 1 week (left) and the mass loss of basalt fibers during the dissolution test.

Figure 3 shows that the tensile strength of the fibers deteriorated to ± ¼ of the virgin strength. As in SEM analysis, no damage initiation points were observed, it is assumed that the surface roughness observed by Wei et al.[15] was present in a small extent, not observable by SEM, but influencing the tensile strength of the fibers.

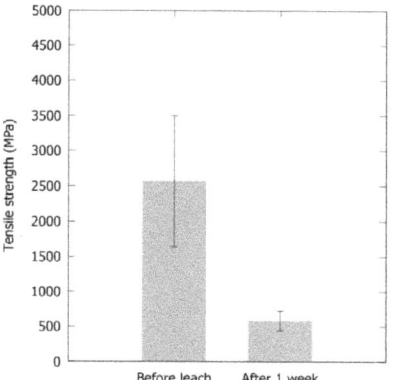

Figure 3. Tensile strength of single basalt fiber before and after 1 week of exposure to the alkaline solution.

These tests showed that the fibers deteriorated when exposed to the alkaline solution. However, it was not yet unequivocal proof for the non-durability of inorganic polymer composites, as the following differences between the dissolution test and the inorganic polymer composite also have to be considered:

- The liquid/solid ratio during the dissolution tests was much higher than in an inorganic polymer paste, enabling a larger extent and speed of dissolution.
- In the composite, the alkalinity dropped because of the dissolution of slag particles, making the environment less aggressive for the fibers.
- During the production of the composite, the slag was mixed first with the solution, resulting in an immediate drop of the alkalinity. Therefore, the fibers were not in contact with the solution before its alkalinity has dropped.
- When the composite would eventually be used in dry environments, the possibility arose that the fiber-matrix interface dried out, stopping dissolution completely.

The performance in inorganic polymer composites is studied in the next sections.

Behavior of inorganic polymer composites: Processing

The influence of the parameters that were discussed in the experimental section on the 3-point flexural strength can be observed in Figure 4. The strength of a sample without fibers was added as a reference. No significant influence was observed for the coating of the fiber or the mixing time of the paste. The interaction between the type of mixer and the mixing time of the fibers showed a significant influence (p-value 0.0535). Short mixing times were beneficial when mixing with a Hobart mixer, whereas longer times were needed when using a high shear mixer. The Hobart mixer seemed to be more aggressive, causing the fibers to disperse faster, but they also broke. The high shear mixer had more difficulty dispersing the fibers, but did not show an extensive breaking of the fibers. As the optimization of the processing was not the goal of the present paper, the parameters were not fine-tuned. Since the sample "Hobart wet 1-1" provided the highest flexural strength result, these conditions (Hobart mixer, wet fiber, short mixing time) were used in the continuation of this paper.

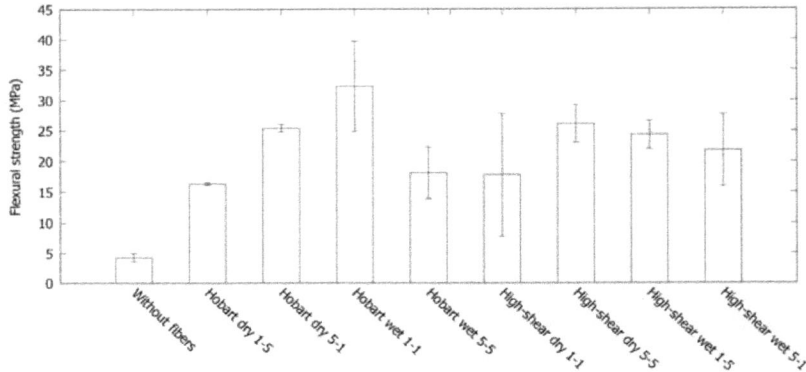

Figure 4. 3-point flexural strength of fiber reinforced inorganic polymers with different processing conditions. The virgin strength without fibers is added.

Behavior of inorganic polymer composites: Activating solution and long-term behavior

The influence of the lowering of concentration (higher H_2O/K_2O) and alkalinity (higher SiO_2/K_2O) on the short and long-term flexural strength is presented in Figure 5. The strength increased for all studied compositions from 7 to 90 days. The maximal strength, 37 MPa, was reached using the activating solution with molar ratios SiO_2/K_2O of 2.0 and H_2O/K_2O of 16. The high strength of this sample is associated with the high strength of its matrix (see Table 2). The flexural strength showed a large decline when subduing the samples to an autoclave treatment. The rise in temperature and humidity enabled the remaining alkalis in the matrix to dissolve the basalt fibers. The attack of the fibers could be visually observed on the fracture surface of the tested samples, as no fibers were observed. For these reasons, the autoclave cannot be used for accelerated durability testing.

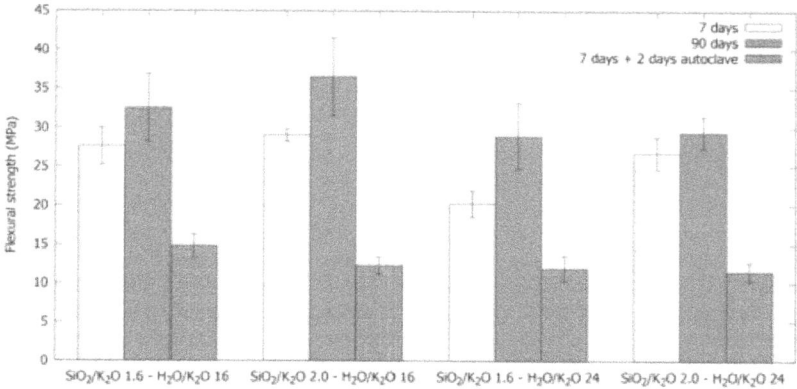

Figure 5. 3-point flexural strength of fiber reinforced inorganic polymers with different activating solutions after 7 and 28 days at room temperature and 2 days autoclave at 150 °C after 7 days at room temperature.

To assess the quality of the presented results, the theoretical strength is calculated using the rule of mixtures on the strength of the matrix and the strength of the fibers. By comparing these theoretical strengths when starting from the tensile strength of the virgin fibers and damaged fibers (those exposed to the activating solution for 1 week), knowledge can be gained on the extent of deterioration of the fibers in the composites. The calculations are performed on the results after 90 days of the samples using an activating solution with a molar ratio H_2O/K_2O of 16. Table III shows the comparison.

Table III. Comparison of theoretical and measured flexural strength of the composites after 90 days. The "damaged" fibers represent the fibers after being exposed to the alkaline solution for 1 week.

Material	Tensile/3pt flexural strength (MPa)
Single virgin basalt fiber	2568
Single damaged basalt fiber	581
Inorganic polymer matrix (1.6 – 16)	10
Inorganic polymer matrix (2.0 – 16)	13
Theoretical strength composite (virgin fiber – 1.6)	87
Theoretical strength composite (virgin fiber – 2.0)	90
Theoretical strength composite (damaged fiber – 1.6)	27
Theoretical strength composite (damaged fiber – 2.0)	30
Actual strength composite (1.6 – 16)	33
Actual strength composite (2.0 – 16)	37

The theoretical strength when using the tensile strength of the damaged fibers (after the dissolution tests) is below the actual strength that was measured for the composite. Normally, the theoretical strength should always be higher than the actual. Therefore, it can be concluded that the fibers in the inorganic polymer composite are less damaged than after exposure to the pure activating solution.

The experiment with the solution 2.0 – 16 was repeated using a different slag. Higher solution/slag ratio was required in order to assure the paste workability. The results of this trial are presented in Figure 6 (the last point is estimated). Already after 28 days, a degradation in strength could be observed. The excess of alkalis caused the fibers to be attacked in a significant extent. The relative difference between the theoretical strength using undamaged fibers and the actual strength of slag 1 and between the theoretical strength of the damaged fibers and the actual strength using slag 2 shows to be similar. Probably, there is a threshold for the amount of alkali solution, beyond which the flexural strength at later stages sensitively drops and the durability of the composite is endangered. This can also be influenced by the chemistry and reactivity of the slag, resulting in a larger/smaller consumption of the alkalis.

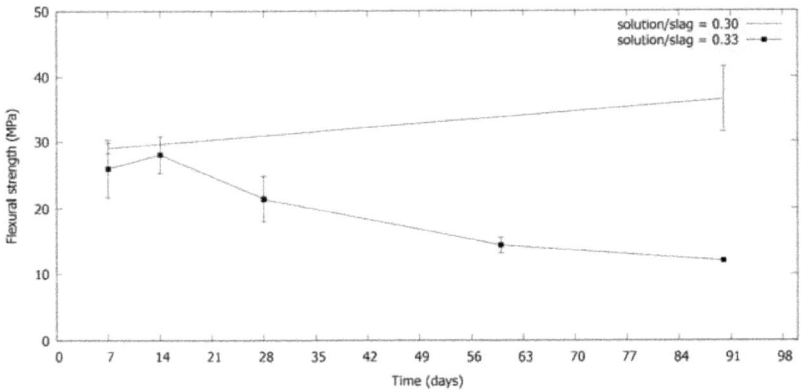

Figure 6. Comparison of flexural strengths as a function of time of composites using 2 different solution/slag ratios.

CONCLUSIONS

Basalt fibers were shown to be an interesting reinforcement for Fe-rich inorganic polymers. 3 wt.% of 12.7 mm long, 13 μm diameter basalt fibers were used throughout the paper. An inorganic polymer paste with a solution/slag mass ratio of 0.3 and a solution with molar ratios $SiO_2/K_2O = 2.0$ and $H_2O/K_2O = 16$ delivered the maximum strength for the composites. A 3-point flexural strength was reached of 29 ± 1 MPa after 7 days of curing and 37 ± 5 MPa after 90 days. It was shown that the fibers were still intact after 90 days. When using the same solution, fibers and processing conditions, but a different solution/slag ratio (changed from 0.3 to 0.33), the flexural strength declined after 28 days. Therefore, this is considered to be an important factor in the synthesis of inorganic polymer composites as well as the importance of testing the samples at a later age.

ACKNOWLEDGEMENTS

A. P. is grateful to the Research Foundation - Flanders (FWO) for the PhD grant. The authors would like to thank Basaltex and J. Debruyne in particular for the supply of the fibers and colleague K. Hendrickx for the assistance with the single fiber tests.

REFERENCES

[1] McLellan, B. C., Williams, R. P., Lay, J., van Riessen, A., and Corder, G. D. (2011). Cost and carbon emissions for geopolymer pastes in comparison to ordinary portland cement, *Journal of Cleaner Production 19*, 1080–1090.
[2] Kriven, W. M. (2015). Potassium geopolymer reinforced with granite and chamotte powder, *1st Geopolymer Conference, Hernstein, Austria.*
[3] Nematollahi, B., Sanjayan, J., and Shaikh, F. U. A. (2014). Comparative deflection hardening behavior of short fiber reinforced geopolymer composites, *Construction and Building Materials 70*, 54–64.
[4] Nazari, A., Maghsoudpour, A., and Sanjayan, J. G. (2015). Flexural strength of plain and fibre-reinforced boroaluminosilicate geopolymer, *Construction and Building Materials 76*, 207–213.
[5] Dias, D. P., and Thaumaturgo, C. (2005). Fracture toughness of geopolymeric concretes reinforced with basalt fibers, *Cement and Concrete Composites 27*, 49–54.
[6] Chen, R., Ahmari, S., and Zhang, L. (2014). Utilization of sweet sorghum fiber to reinforce fly ash-based geopolymer, *Journal of Materials Science 49*, 2548–2558.
[7] Lin, T., Jia, D., Wang, M., He, P., and Liang, D. (2009). Effect of fibre content on mechanical properties and fracture behaviour of short carbon fibre reinforced geopolymer matrix composites, *Bulletin of Material Science 32*, 77–81.
[8] Lin, T., Jia, D., He, P., Wang, M., and Liang, D. (2008). Effect of fibre length on mechanical properties and fracture behavior of short carbon fibre reinforced geopolymer matrix composites, *Materials Science and Engineering A 497*, 181–185.
[9] Alomayri, T., and Low, L. M. (2013). Synthesis and characterization of mechanical properties in cotton fiber-reinforced geopolymer composites, *Journal of Asian Ceramic Societies 1*, 30–34.
[10] Rill, E., Lowry, D. R., and Kriven, W. M. (2010). Properties of basalt fiber reinforced geopolymer composites, *Proceedings of the 34th International Conference and Expo on Advanced Ceramics and Composites, Daytona Beach (FL), USA.* doi: 10.13140/2.1.1763.3287.
[11] Musil, S. S., Kutyla, G., and Kriven, W. M. (2012). The Effect of Basalt Chopped Fiber Reinforcement on the Mechanical Properties of Potassium Based Geopolymer, chapter in *Developments in Strategic Materials and Computational Design III* (eds W. M. Kriven, A. L. Gyekenyesi, G. Westin, J. Wang, M. Halbig and S. Mathur), John Wiley & Sons, Inc., Hoboken, NJ, USA. doi: 10.1002/9781118217542.ch4.
[12] Shaikh, F. U. A. (2013). Review of mechanical properties of short fibre reinforced geopolymer composites, *Construction and Building Materials 43*, 37–49.

[13] Ohno, M., and Li, V. C. (2014). A feasibility study of strain hardening fiber reinforced fly ash-based geopolymer composites, *Construction and Building Materials 57*, 163–168.

[14] Iacobescu, R. I., Cappuyns, V., Geens, T., Kriskova, L., Onisei, S., Jones, P. T. and Pontikes, Y. (2017). The influence of curing conditions on the mechanical properties and leaching of inorganic polymers made of fayalitic slag, *Frontiers of Chemical Science and Engineering*, doi: 10.1007/s11705-017-1622-6.

[15] Wei, B., Cao, H., and Song, S. (2010). Tensile behavior contrast of basalt and glass fibers after chemical treatment, *Materials and Design 31*, 4244-4250.

WETTING ANGLE: NEW PARAMETER INDICATING THE REACTIVITY OF ALKALINE SOLUTIONS AND GEOPOLYMER BINDERS

Ameni Gharzouni, Laeticia Vidal, Robin Stocky, Julie Peyne and Sylvie Rossignol

Science des Procédés Céramiques et de Traitements de Surface (SPCTS), Ecole Nationale Supérieure de Céramique Industrielle, 12 rue Atlantis, 87068 Limoges Cedex, France. corresponding author: sylvie.rossignol@unilim.fr, tel.: 33 5 87 50 25 64

ABSTRACT

Geopolymers are inorganic aluminosilicate materials synthesized at room temperature by the activation of metakaolin with an alkaline solution. The comprehension of the structure and the parameters controlling the reactivity of alkaline solutions as well as their interaction with aluminosilicate species is primordial. In this study, based on Raman and FTIR data spectroscopy, wetting angle measurement was investigated as a new technique to evaluate the depolymerization degree of the alkaline solution. Different solutions with different alkali cations (K, Na, Li) and different Si/M molar ratios were studied. The results showed that the cation influences the formation of various reactive species (rings, chains, oligomers). Moreover, the increase of the depolymerization degree of the solution was evidenced by the increase of the wetting angle. The wetting angle of geopolymer mixtures based on different solutions in presence of the same metakaolin was also investigated. Consequently, for a given metakaolin, lower wetting angle value corresponds to the formation of a single geopolymer network. However, higher wetting angles are attributable to the existence of several networks. Thus, the wetting angle can be considered as a parameter to determine the reactivity of alkaline solutions and their interaction with metakaolin.

INTRODUCTION

Alkaline silicate solutions, or water glass, are of great interest for several industrial applications[1,2]. Recently, one of the main applications is their use as alkaline reactants for the preparation of geopolymer materials[3,4,5]. There are different method used for the preparation of these solutions i.e. commercial or laboratory solutions[2,6,7]. Svensson et al. also evidenced this fact by studying various solutions prepared using four different methods[8]. Previous works showed that the presence of various silicate species in silicate solutions depends on several parameters, such as the Si/M molar ratio, the alkaline cation and the dilution level[3,5]. According to NMR data, the Si/M molar ratio controls the polymerization degree of the silicate species in the solutions[9]. An increase in this ratio leads to the polymerization of the silicate solution. Several authors demonstrated that for solutions with Si/M molar ratios higher than 1, the major species are the Q^2 and Q^3 units. Nevertheless, the Q^4 species are also observed. For Si/M molar ratios lower than 1, the Q^2 and Q^3 species disappeared to the profit of the Q^0 and Q^1 entities[10]. However, this phenomenon is associated with the appearance of Q^2 and Q^3 cyclic species (Q^{2c} and Q^{3c})[11,12]. Another parameter influencing the silicate species is the dilution level[13] which induces a slight polymerization of the silicate anions to higher polymeric species[2,14]. Various analyses are used to determine the presence of the various silicate species in the solutions and thus, their reactivity. Fourier transform infrared (FTIR) spectroscopy allowed determining the silicate species in the solutions by following the broad peak attributed to the asymmetric stretching of the Si-O-Si bonds[15]. However, it is not a quantitative technique. Raman spectroscopy was used in several studies on silicate glasses[16, 17,18] and on silicate solutions[19,20,21] to investigate the effects of different parameters on the various species such as the monomers and oligomers[22,23] and the entities such as the 3-, 4- or 5-membered rings vibrations[18, 23, 24]. Moreover, Malfait et al.[25] evidenced the variation of Q^n species depending on the amount of M_2O (M = Na or K) by Raman spectroscopy. Indeed, the authors observed a decrease of Q^4 and Q^3 species and an increase of Q^2 and Q^1 species when the Na_2O or the K_2O content increases in glasses. Certainly, spectroscopic characterizations are useful to understand the structure of alkaline solutions and geopolymer materials. Nevertheless, additional informations can be obtained by wetting angle measurements. Indeed, the wetting angle allows determining the ability of a liquid to wet a smooth and homogeneous surface[26]. High contact angle values denote low wettability and inversely[27]. At thermodynamic equilibrium, the wetting angle θ depends on the superficial energies of solid-liquid (γSL), solid-vapor (γSV) and liquid- vapor (γLV) according to young equation[28]:

$$\gamma SV = \gamma SL + \gamma LV \cos \theta \qquad (1)$$

For example, wetting angles were measured in order to study the interaction between a carboxymethylcellulose binder droplet and different powders of limestone and/ or teawaste[29]. It was found that limestone has higher wettability than the teawaste. Consequently, changing teawaste fraction may change the wettability of the binary mixture. Contact angle measurements were also useful in the processing of aluminum powder aqueous slurries[30]. In fact, it was demonstrated that the wetting behavior of binders is one of the most important parameters affecting the slurries stability and film surface quality. The wetting behavior of metallic powders was also studied using the sessile drop technique and Washburn's test. It was evidenced that for mineral and metallic powder, considered as non ideal surfaces, kinetics are more important than contact angle[31]. Recently, the wettability of tungsten carbide by liquid binders with various carbon contents was investigated[32]. It was shown that, with low carbon content, tungsten carbide wettability by liquid Co-based binder is complete and becomes incomplete when increasing the carbon content. However, there is no focus on the wetting angle of the alkaline solutions as well as the geopolymer binders. Duan et al.[33] tried to modify the geopolymer surface to make it hydrophobic. The modification of the geopolymer surface led to the increase of the wetting angle from 36° to 132°. Recently, Vidal et al.[34] studied the interactions between geopolymer binders and different substrates by wetting angles measurement in order to apply a coating on metal or agglomerated sand based on a geopolymer binder for different applications.

So, the aim of the present study is to determine the reactivity of various alkaline silicate solutions and geopolymer reactive mixture by wetting angle meassurments and spectroscopic analysis.

EXPERIMENTAL
Sample preparation
The alkaline solutions were prepared according to two methods: (i) sodium and / or potassium commercial solutions having a Si/M molar ratio (M = Na and / or K) varying between 0.7 and 1.7. Then, sodium, potassium or lithium hydroxide pellets were dissolved in the solution. The final molar ratios were 0.5, 0.6 and 0.7 (ii) Laboratory solutions were obtained by dissolving amorphous silica in a solution of distilled water and sodium and / or potassium hydroxide. Only a Si/M of 0.7 was investigated in this case. The different solutions were denoted as $_wM^x_Ny$ with w is the preparation method (w = C or L for commercial or laboratory respectively), M is the initial cation of the solution (M = Na and / or K), N is the cation added to the solution (N = Na, K or Li), X is the initial Si/M ratio of the solution (x = 1.7, 1.6 and 0.7) and Y is the final silica to cation ratio of the solution (y = 0.5, 0.6 and 0.7). The metakaolin (M1, Si/Al =1.17[35]) was then added to eight selected solutions with different Si/M ratios to obtain geopolymer reactive mixtures at room temperature. The mixtures were magnetically stirred for 5 min at 500 rpm.

Characterization

The alkaline solutions were characterized immediately after the preparation. The pH values of the different alkaline solutions were measured using a Schott Instrument Lab860 pH-meter at 25°C equipped with a specific electrode for alkaline media, U402 S7 A120, fisherbrand, supplied by Thermo Fisher Scientific.

FTIR spectroscopy in ATR (attenuated total reflectance) mode was used to characterize the solutions. FTIR spectra were obtained using a Thermo Fisher Scientific Nicolet 380 infrared spectrometer. The IR spectra were collected over a wavenumber range of 500–4000 cm^{-1} with a resolution of 4 cm^{-1}. The atmospheric CO_2 contribution was removed using a straight line between 2400 and 2280 cm^{-1}. To allow comparisons of the spectra, they were corrected using a baseline and were normalized.

Raman spectroscopy was performed on liquid samples using a T64000 Horiba-Jobin-Yvon spectrophotometer with a 514 nm laser excitation operating at a power of 30 mW at the sample in the triple substractive mode using 1800 grooves/mm grating. Scattered light was collected in backscattering mode using a specific "pseudo-macro" stage for liquid samples. The spectral range was 300 to 1300 cm^{-1}, and 60 scans of 15 s each were carried out for the acquisition. The acquired spectra were corrected by subtracting the baseline, which was modeled by a 5° polynomial curve. Then, the spectra were decomposed using wire software.

The wetting angles of the solutions and the reactive mixtures were also determined using a Digidrop MCAT from GBX composed of a camera and a lamp to record the deposit of a drop. The various solutions and reactive mixture (immediately after mixing) were deposited on HDPE substrate. Then, the wetting angles were measured at t = 0 on the first image where the drop was placed on the substrate (Figure 1), using GBX's Visiodrop software. Wetting angle were measured 5 times with an estimated error of 4%.

Figure 1. (a) Diagram of the wetting angle measurement θ and (b) an example of obtained image.

RESULTS AND DISCUSSION
Alkaline solutions
Physical and chemical characterizations

At first, the physical and chemical characteristics of the studied solutions were analyzed. Table I shows the silicon and alkaline cations concentrations as well as the densities and pH values of the different solutions.

The silicon concentrations range from 3.08 to 6.16 mol.L^{-1}. The sodium, potassium and lithium cations concentrations vary from 1.42 to 6.58, from 1.54 to 7.73 and from 2.73 to 5.04 mol.L^{-1}, respectively. The density value ranges from 1.26 g/cm^3 to 1.56 g/cm^3 and the pH value is between 11.3 and 14.0. The increase of silicon or alkaline cation concentration leads to the increase of the density value. Indeed, for silicon concentrations equal to 3.26 and 6.03 mol.L^{-1}, the densities of the solutions are 1.26 and 1.55 g/cm^3 respectively. Similarly, an increase in the total concentration of alkali cations gives rise to the increase of the density. The pH values depend on the Si/M molar ratio as well as the added alkali cation. In fact, solutions having high Si/M ratios exhibit lower pH values. For example, $_cNaK^{1.6}$ and $_cNaK^{1.6}{}_{Na}{}^{0.7}$ solutions have pH values equal to 11.8 and 13.8 respectively. The solutions, in which lithium hydroxide has been added, have slightly lower pH values. On the other hand, the addition of sodium or potassium does not lead to any notable differences. For example, the $_cK^{1.7}{}_{Li}{}^{0.7}$, $_LK^{1.7}{}_{Na}{}^{0.7}$ and $_LNa^{1.7}{}_{K}{}^{0.7}$ solutions have pH values of 13.3, 13.9 and 14.0 respectively.

Table I. Physical and chemical characteristics of the studied solutions

solution	[Si] (M)	[Na] (M)	[K] (M)	[Li] (M)	ρ (g/cm³)	pH value
$_cK^{1.7}$	3.54	-	2.11	-	1.31	11.4
$_cK^{1.7}{}_{K}{}^{0.7}$	3.08	-	4.42	-	1.31	13.8
$_cK^{1.7}{}_{K}{}^{0.6}$	3.15	-	5.46	-	1.40	14.0
$_cK^{1.7}{}_{Na}{}^{0.7}$	3.27	2.74	1.93	-	1.31	13.6
$_cK^{1.7}{}_{Na}{}^{0.6}$	3.21	3.69	1.92	-	1.34	14.0
$_cK^{1.7}{}_{Li}{}^{0.7}$	3.26	-	1.93	2.73	1.26	13.3
$_LK^{1.7}{}_{Na}{}^{0.7}$	4.69	3.98	2.79	-	1.45	13.9
$_cK^{0.7}$	6.03	-	6.20	-	1.51	14.0
$_cK^{0.7}{}_{K}{}^{0.5}$	5.75	-	7.73	-	1.56	14.0
$_cK^{0.7}{}_{Na}{}^{0.5}$	5.93	1.83	6.09	-	1.56	14.0
$_cNa^{1.7}$	6.16	3.64	-	-	1.36	11.3
$_cNa^{1.7}{}_{Li}{}^{0.7}$	5.98	3.49	-	5.04	1.43	13.0
$_cNa^{1.7}{}_{K}{}^{0.7}$	5.55	3.25	4.68	-	1.52	14.0
$_LNa^{1.7}{}_{K}{}^{0.7}$	4.57	2.71	3.87	-	1.44	14.0
$_cNaK^{1.6}$	6.09	1.83	2.08	-	1.38	11.8
$_cNaK^{1.6}{}_{Na}{}^{0.7}$	6.03	6.58	2.04	-	1.55	13.8
$_cNaK^{1.6}{}_{K}{}^{0.7}$	5.57	1.65	6.30	-	1.54	14.0
$_LNaK^{1.6}$	4.86	1.54	1.54	-	1.31	11.9
$_LNaK^{1.6}{}_{K}{}^{0.7}$	4.69	3.98	2.79	-	1.45	13.8

FTIR and Raman investigation

This part of the study focuses on the influence of mixing alkaline cations and the preparation method (laboratory vs. commercial) on the structure of alkaline solutions. So, to facilitate the comparison of the solutions, only data of the solutions having a final Si/M ratio of 0.7 were presented. For more accurate structural informations, the solutions were analyzed by FTIR and Raman spectroscopies. Figure. 2 shows the FTIR and Raman spectra of $_LNaK^{1.6}$ and $_LNaK^{1.6}{}_{Na}{}^{0.7}$ solutions.

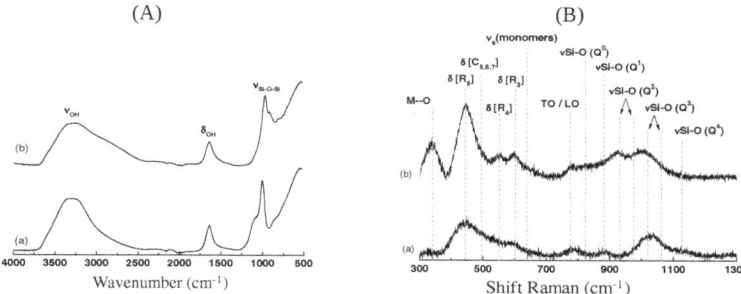

(A)

v_{OH}

δ_{OH}

$v_{Si-O-Si}$

(b)

(a)

| 4000 | 3500 | 3000 | 2500 | 2000 | 1500 | 1000 | 500 |

Wavenumber (cm^{-1})

(B)

v_s(monomers)

δ [C$_{6,6,7}$]

δ [R$_5$]

δ [R$_4$] δ [R$_3$]

δ [R$_4$]

M--O

TO / LO

vSi-O (Q^0)

vSi-O (Q^1)

vSi-O (Q^2)

vSi-O (Q^3)

vSi-O (Q^4)

(b)

(a)

| 300 | 500 | 700 | 900 | 1100 | 1300 |

Shift Raman (cm^{-1})

Figure 2. (A) FTIR and (B) Raman spectra of (a) $_L$NaK$^{1.6}$ and (b) $_L$NaK$^{1.6}$$_{Na}$$^{0.7}$ solutions

The FTIR spectra (Figure 2. A) of $_L$NaK$^{1.6}$ and $_L$NaK$^{1.6}$$_{Na}$$^{0.7}$ laboratory solutions exhibit a broad band around 3300 cm^{-1} corresponding to v_{OH} bonds. The contribution towards 1640 cm^{-1} and the broad band around 1000 cm^{-1} can be attributed to the deformation of water molecules (δ_{OH}) and to the elongation of the Si-O-Si bonds ($v_{Si-O-Si}$) of the different silicate species, respectively[15,36]. The $_L$NaK$^{1.6}$ solution shows a peak located around 1000 cm^{-1} with a shoulder at 1100 cm^{-1} which are characteristic of the Q^2 (SiO$_2$O^{2-}) et Q^3 (SiO$_{3/2}$O$^-$) species respectively[32,37]. The $_L$NaK$^{1.6}$$_{Na}$$^{0.7}$ solution has a peak at about 980 cm^{-1} with a shoulder towards 920 cm^{-1} corresponding to SiO$_{1/2}$O^{3-} (Q^1) and Si(O$^-$)$_4$ (Q^0) species, respectively. The shift of the Si-O-Si peak at lower wavenumbers as well as the appearance of the contribution of Si(O$^-$)$_4$ species, are due to the depolymerization of the solution[38].

The Raman spectra (Figure 2. B) relative to the two solutions present different contributions with various intensities. For $_L$NaK$^{1.6}$ solution, the observed bands, positioned at 450, 490, 540, 600, 787, 880, 975, 1030, 1070 and 1150 cm^{-1} are attributable to δ [R$_5$] (rings with 5 tetrahedra [SiO$_4$] and more), δ [C$_{5,6,7}$] (chains with 5, 6 and 7 tetrahedra), δ [R$_4$] (rings with 4 tetrahedra), δ [R$_3$] (rings with 3 tetrahedra), polar TO/LO of the silica and vSi-O- of Q^1, Q^2, Q^3 et Q^4 species, respectively. The band at 325 cm^{-1} is associated to the M--O vibration (vibration due to the network modifying cation). The major contributions are attributed to the vibrations of Q^3 and Q^4 species as well as δ [R$_5$] demonstrating the significant polymerization degree of this solution. The decrease of the Si/M ratio (in the case of $_L$NaK$^{1.6}$$_{Na}$$^{0.7}$) leads to the appearance of a contribution at 930 cm^{-1} and the increase of those at 595, 485 and 325 cm^{-1} attributed to the Q^2 species, δ [R$_{3, 4 and 5}$] and M--O vibration, respectively. In addition to that, the contributions attributed to δ [C$_{5, 6 and 7}$], TO/LO modes and Q^3 species are less intense. The contribution associated to Q^4 species is no longer observed. These facts evidence a depolymerization phenomenon. Indeed, the increase in the quantity of alkaline ions induces the formation of Si-O- entities within the tetrahedra [SiO$_4$] leading to the depolymerization of the species. The variation of the various contributions intensities and in particular the increase in the M--O vibration with the decrease of the Si/M ratio reveal an important depolymerization of the solutions which is in agreement with the increase of the pH value.

The Raman spectra of the various solutions were then classified according to the molar percentages of alkaline cations (Figure 3). Regardless of the solution, the alkaline cation molar percentage controls the species contained in the solution. However, variations in intensities are noticed for the same contribution. For a percentage varying between 3 and 5% (Figure. 3A), $_L$NaK$^{1.6}$ and $_C$NaK$^{1.6}$ solutions are polymerized and contain predominantly polymerized silicate species (Q^4 and Q^3) characterized by high TO/LO mode contribution. On the other hand, the intensity of Q^4 species appears to be lower for the laboratory solution than in the case of the commercial solution because of a lower polymerization for laboratory solutions. The increase of the cations molar percentage (8%) leads to the depolymerization of the solutions $_C$K$^{1.7}$$_{Na}$$^{0.7}$ and $_C$K$^{1.7}$$_{Li}$$^{0.7}$ (Figure. 3B). Indeed, the disappearance of Q^4 species is observed with the decrease of the TO/LO mode as well as an increase of the M--O vibration. The addition of lithium in the potassium silicate solution, in the case of $_C$K$^{1.7}$$_{Li}$$^{0.7}$ solution, appears to decrease slightly the M--O contribution. The observed data are directly related to the size of cation and therefore the ionic forces and polarizing effect (Z/r^2 with Z is the charge of cation and r is the radius). In fact, Li, smaller cation [39], in presence of potassium silicate solution will exchange with potassium. The replacement of a cation with a smaller one leads to the formation of less non bridging oxygens [40] and therefore the decrease of M--O vibration.

For a rate of 12% (Figure. 3C), $_C$NaK$^{1.6}$$_{Na}$$^{0.7}$, $_C$NaK$^{1.6}$$_K$$^{0.7}$, $_C$Na$^{1.7}$$_K$$^{0.7}$ and $_C$Na$^{1.7}$$_{Li}$$^{0.7}$ solutions are depolymerized and the contribution attributed to the rings with three tetrahedra (R$_3$) seems more important. The band associated with M--O vibration is lower when lithium is added to the solution which is in accordance with previously explained data (Figure 3.B). On the other hand, the addition of sodium in a mixed solution, contrary to potassium cation, leads to an increase in the M--O contribution. This phenomenon can be explained by a double exchange of potassium and sodium in the mixed solution [41]. According to literature [42], sodium is preferentially located in the bridging/non-bridging oxygen while potassium is located in various oxygen tri-clusters

Finally, the same observations are noted for $_L$K$^{1.7}$$_{Na}$$^{0.7}$, $_L$Na$^{1.7}$$_K$$^{0.7}$, $_L$NaK$^{1.6}$$_{Na}$$^{0.7}$ and $_L$NaK$^{1.6}$$_K$$^{0.7}$ solutions (Figure. 3D) having alkaline cations content greater than 14%. The addition of potassium cation in a sodium-based solution has little influence on silicate species. On the other hand, when sodium is added to potassium silicate solution, an increase in the band attributed to the 3-tetrahedral rings is noticed. This variation is explained by the lower sodium cation size compared to potassium. Consequently, the increase in the amount of alkali cations induces the disappearance of Q^4 species, a decrease in Q^3 species contributions and an increase in M--O vibration, R$_3$ and R$_4$ rings revealing a depolymerization phenomenon.

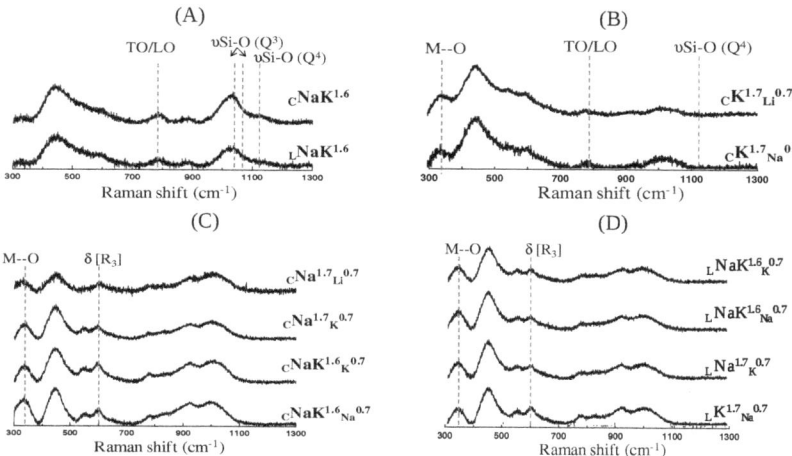

Figure 3. Raman spectra of the solutions with the alkali cation molar percentages (A) varying between 3 and 5 % for $_L$NaK$^{1.6}$ and $_C$NaK$^{1.6}$, (B) 8 % for $_C$K$^{1.7}$Na$^{0.7}$ and $_C$K$^{1.7}$Li$^{0.7}$, (C) 12 % for $_C$NaK$^{1.6}$Na$^{0.7}$, $_C$NaK$^{1.6}$K$^{0.7}$, $_C$Na$^{1.7}$K$^{0.7}$ and $_C$Na$^{1.7}$Li$^{0.7}$ and (D) 14 % for $_L$K$^{1.7}$Na$^{0.7}$, $_L$Na$^{1.7}$K$^{0.7}$, $_L$NaK$^{1.6}$Na$^{0.7}$ and $_L$NaK$^{1.6}$K$^{0.7}$ solutions.

Wetting angle

In this part, it is investigated the alkaline solutions wetting angle as a determining factor of alkaline solution depolymerization state. Only commercial solutions with different Si/M molar ratios are presented.

Figure 4 shows the evolution of alkaline solutions wetting angle on polyethylene as a function of the alkali cation concentration. Whatever the solution, the wetting angle seems to increase with the increase of the alkali cation concentration. For example, $_C$K$^{1.7}$ solution, having low concentration (2.1 Mol.L^{-1}), exhibits lower wetting angle value (83.3°) compared to $_C$K$^{1.7}$K$^{0.6}$ showing a wetting angle of 98,1° for a concentration equal to 5.46 Mol.L^{-1}. Indeed, the increase of cation concentration, playing the role of network modifier, induces easier and faster breaking of Si-O-Si bonds leading to higher depolymerization state. The high amount of interactions within the solution is responsible of an increase of hydrophobic repulsive forces[26]. Thus, the wetting angle of alkaline solutions can be considered as a determining factor of their depolymerization state and therefore their reactivity. Indeed, the increase of the depolymerization degree of the solution is evidenced by the increase of the wetting angle.

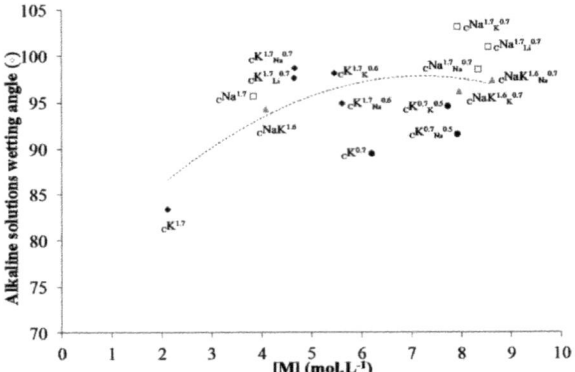

Figure 4. The wetting angle on polyethylene as a function of alkali cation concentration [M] (M= Na, K, Li) for solutions based on (♦) $K^{1.7}$ (•) $K^{0.7}$, (□) $Na^{1.7}$ and (▲) $NaK^{1.6}$ starting silicate solutions.

Interactions with metakaolin

In this section, eight solutions with different Si/M (0.5, 0.6 and 0.7, Table II), were selected in order to study their interaction with the same metakaolin (M1) by wetting angle measurement and FTIR spectroscopy. The structural evolution of the different synthesized reactive mixtures over time was studied using FTIR spectroscopy. This technique allows following the shift of the Si-O-M band position initially situated at approximately 980 cm^{-1} to lower wavenumbers. The shift value was demonstrated to be a proof of the substitution of silicon with aluminum. This value also reveals the number of formed networks. For example, for a shift value between 15 cm^{-1} and 30 cm^{-1}, a single geopolymer network was evidenced. Higher shift values were attributed to the formation of different network (Si-rich and Al- rich phases)[9]. Very low shift value corresponds to the formation of gel network. On the other hand, the wetting angle of the reactive mixture traduces the interaction between the aluminosilicate species and the alkaline one. Figure 5 presents the evolution of shift values determined by FTIR spectroscopy versus the reactive mixture wetting angle on polyethylene.

Two linear trends, depending on the starting silicate solution, reveal the change of the shift value with the variation of the wetting angle: i) In the case of reactive mixtures based on $_cK^{1.7}$ starting solution, the shift value decreases from 47 cm^{-1} to 36 cm^{-1} with the decrease of the wetting angle from 91.9° to 86.6° for $_cK^{1.7}{}_K^{0.6}$ and $_cK^{1.7}{}_{Na}^{0.7}$ solutions, respectively. The shift values are relatively high and denote the formation of different networks. This fact is due to the high amount of water of the solution hindering the exchange between the species[30,36].

ii) For mixtures based on $_cK^{0.7 \text{ or } 1.7}$ and $_cNa^{1.7}$ starting solutions, the shift value decreases from 32 cm^{-1} to 15 cm^{-1} with the decrease of the wetting angle from 98.9° to 87.8° for $_cK^{0.7}{}_K^{0.5}$ and $_cNa^{1.7}{}_{Na}^{0.7}$ solutions, respectively. The shift values are characteristic of a single geopolymer network. The lower amount of water in $_cK^{0.7}$ and $_cNa^{1.7}$ compared to $_cK^{1.7}$ facilitates and enhances the exchange between the species[30,37].

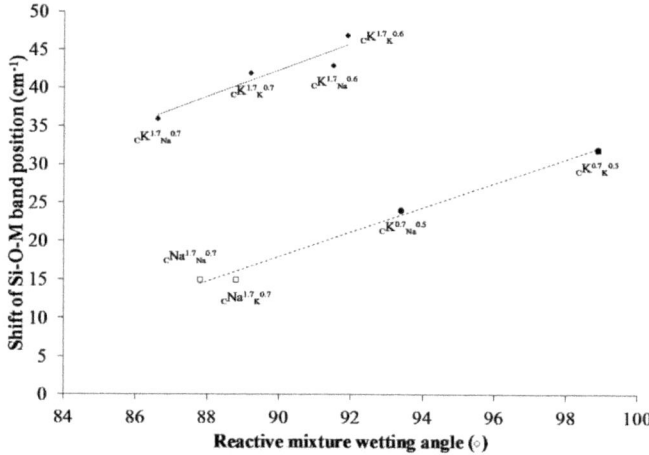

Figure 5. The evolution of shift values determined by FTIR spectroscopy versus the wetting angle on polyethylene for reactive mixtures based on ($\blacklozenge$) $K^{1.7}$, ($\bullet$) $K^{0.7}$ and ($\square$) $Na^{1.7}$ starting silicate solutions.

Table II. Rings and chains intensities ratio and the amount of negative charges for the selected alkaline solutions

Solution	I (R_5)	I (R_4)	I (R_3)	I ($C_{5,6,7}$)	I ($R_{3,4,5}$)/ I($C_{5,6,7}$)	Σ(-) charges
$_cK^{1.7}{}_K{}^{0.7}$	30.5	8.3	5.8	8.7	5.1	7.1
$_cK^{1.7}{}_K{}^{0.6}$	66.1	16.3	14.1	15.0	6.4	6.4
$_cK^{1.7}{}_{Na}{}^{0.7}$	74.4	29.0	24.5	26.9	4.8	10.2
$_cK^{1.7}{}_{Na}{}^{0.6}$	87.4	25.4	20.8	21.9	6.1	7.3
$_cK^{0.7}{}_K{}^{0.5}$	112.9	26.3	25.7	22.9	7.2	6.1
$_cK^{0.7}{}_{Na}{}^{0.5}$	139.7	39.8	75,2	31.4	6,1	8.7
$_cNa^{1.7}{}_{Na}{}^{0.7}$	75.7	24.6	32.1	23.8	5.6	9.6
$_cNa^{1.7}{}_K{}^{0.7}$	85.8	25.3	31.5	24.3	5.9	8.5

Based on Raman data (Table II), whatever the used solution, the decrease of the wetting angle is associated to the decrease of rings to chains intensities ratio ($I_{R3,4,5}/I_{C5,6,7}$) and the increase of negative charges. The negative charges of the solutions are calculated considering that the negative charges are 14, 8 and 6 for chains and R_4 and R_3 rings respectively as previously established by vidal et al,[22,30]. The high amount of rings (high $I_{R3,4,5}/I_{C5,6,7}$ in the case of $_cK^{1.7}Na^{0.6}$ and $_cK^{0.7}K^{0.5}$ solutions for example), in relation with the increase of the amount of negative charges of the solutions, leads to steric hindrance within the mixture and therefore higher repulsive forces and higher wetting angle. However, the formation of chains in detriment of rings (low $I_{R3,4,5}/I_{C5,6,7}$ in the case of $_cK^{1.7}Na^{0.7}$ and $_cK^{1.7}K^{0.7}$ solutions for example) related to the decrease of the amount of negative charges of the solutions reduces this phenomenon inducing lower wetting angle.

CONCLUSION

In recent years, alkaline solutions have regained interest since they have been used as precursor for geopolymer synthesis. So, the understanding of their structure and the parameters controlling their reactivity is primordial. Many spectroscopic techniques (FTIR, NMR and Raman) were useful. In this study, based on Raman and FTIR data, wetting angle measurement was investigated as a new technique to evaluate the depolymerization degree of the alkaline solution and therefore their reactivity. Different solutions with different alkali cations (K, Na, Li) and different initial and final Si/M molar ratios were studied. The results show that the high amount of interactions within the solution due to depolymerization is responsible of higher wetting angle value. Thus, the wetting angle of alkaline solutions can be considered as a determining factor of their depolymerization state and therefore their reactivity.

Similarly, wetting angle measurements were undertaken for geopolymer mixtures based on different alkaline solutions in presence of the same metakaolin. A relation was evidenced between the wetting angle and the capacity of the solutions to react with aluminosilicate species to create Si-O-Al bonds. For a given metakaolin, the variation of reactive mixture wetting angle is dependent on the number of formed network (one or several networks). Indeed, lower wetting angle value corresponds to the formation of a single geopolymer network. However, larger wetting angles are attributable to the existence of several networks. This fact is in relation with the amount and type of siliceous species coming from the alkaline solution especially rings and chains controlling the resulting negative charges. Once again, the wetting angle can be a new factor indicating the reactivity of the alkaline solution in presence of aluminosilicate source.

REFERENCES

[1] A.C.J.H. Johnson, P.Greenwood, M. Hagström, Z. Abbas, S. Wall (2008). Aggregation of nanosized colloidal silica in the presence of various alkali cations investigated by the electrospray technique, *Langmuir*, *24*, 12798-12806.

[2] M.T. Tognonvi, D. Massiot, A. Lecomte, S.Rossignol, J.-P. Bonnet (2010). Identification of solvated species present in concentrated and dilute sodium silicate solutions by combined ^{29}Si NMR and SAXS studies, *J. Colloid Interface Sci. 352*, 309-315.

[3] A. Bourlon, 2011, Physico-chimie et rhéologie des géopolymères frais pour la cimentation des puits pétroliers. Ph.D. Thesis, University of Pierre et Marie Curie.

[4] A. Autef, E. Joussein, G. Gasgnier, S. Rossignol (2012). Role of the silica source on the geopolymerization rate, *J. Non-Cryst. Solids. 358*, 2886-2893.

[5] P. Steins, 2014, Influence des paramètres de formulation sur la texturation et la structuration des géopolymères. Ph.D. Thesis, University of Limoges.

[6] R.K. Iler, 1979, The chemistry of silica: solubility, polymerization, colloid and surface properties, and biochemistry, Wiley-Interscience, p. 63, New York.

[7] R.K. Harris and C.T.G. Knight (1983). Silicon-29 nuclear magnetic resonance studies of aqueous silicate solutions. Part 5. First-order patterns in potassium silicate solutions enriched with silicon-29, *J. Chem. Soc. 79*, 1525-1538.

[8] I.L. Svensson, S. Sjöberg, L.-O. Öhman (1986). Polysilicate equilibria in concentrated sodium silicate solutions, *J. Chem. Soc. 82*, 3635-3646.

[9] R. Couty and L. Fernandez (1998). Etude du passage de l'état colloïdal à l'état ionique de solutions de silicates sodiques par spectroscopie RMN ^{29}Si et infrarouge, *J. Chim. Phys. Phys.-Chim. Biol. 95*, 384-387.

[10] J.L. Bass and G.L. Turner (1997). Anion distributions in sodium silicate solutions. Characterization by ^{29}Si NMR and infrared spectroscopies and vapor phase osmometry, *J. Phys. Chem. B. 101*, 10638-10644.

[11] C.F Weber and R.D Hunt (2003). Modeling alkaline silicate solutions at 25 °C, *Ind. Eng. Chem. Res. 42*, 6970-6976.

[12] J.L. Provis, P. Duxson, G.C Lukey, F. Separovic, W.M. Kriven, J.S.J. van Deventer (2005). Modeling speciation in Highly concentrated alkaline silicate solutions, *Ind. Eng. Chem. Res. 44*, 8899-8908.

[13] D. Böschel, M. Janich, H. Roggendorf (2003). Size distribution of colloidal silica in sodium silicate solutions investigated by dynamic light scattering and viscosity measurements, *J. Colloid Interface Sci. 267*, 360-368.

[14] M.A. McGarry and J.F Hazel (1965). Electron microscopy of labile alkali silicate solutions, *J. Colloid Sci. 20*, 72-80.

[15] M.T. Tognonvi, J. Soro, S. Rossignol (2012). Physical-chemistry of silica/alkaline silicate interactions during consolidation. Part 1: Effect of cation size, *J. Non-Cryst. Solids*. 358, 81-87.

[16] J. Tan, S. Zhao, W. Wang, G. Davies, X. Mo (2004). The effect of cooling rate on the structure of sodium silicate glass, *Mater. Sci. Eng. B. 106*, 295-299.

[17] B.O Mysen and G.D.Cody (2005). Solution mechanisms of H_2O in depolymerized peralkaline melts, *Geochim. Cosmochim. Ac. 69*, 5557-5566.

[18] C. Le Losq and D.R. Neuville (2013). Effect of the Na/K mixing on the structure and the rheology of tectosilicate silica-rich melts. Chem. Geol. 346, 57-71.

[19] I. Halasz, M. Agarwal, R. Li, N.Miller (2007). Vibrational spectra of aqueous Na_2SiO_3 solutions, *Catal. Lett. 117*, 34-42.

[20] L. Vidal, E. Joussein, M. Colas, J. Cornette, J. Sanz, I. Sobrados, J.-L. Gelet, J. Absi, S. Rossignol (2016). Controlling the reactivity of silicate solutions: a FTIR, Raman and NMR study, *Colloid Surf. A, 503*, 101-109.

[21] J.D.Hunt, A. Kavner, E.A. Schauble, D. Snyder, C.E. Manning (2011). Polymerization of aqueous silica in H_2O-K_2O solutions at 25-200 °C and 1 bar to 20 kbar, *Chem. Geol. 283*, 161-170.

[22] L.Vidal, E. Joussein, I.Sobrados, J. Absi, S. Rossignol (2016). How to counter act the low reactivity of an alkaline solution, *J. Non-Cryst.Solids, 45*, 2220–230.

[23] H. Aguiar, J. Serra, P. González, B. León (2009). Structural study of sol-gel silicate glasses by IR and Raman spectroscopies. *J. Non-Cryst. Solids*, 355, 475-480.

[24] A.E. Geissberger, F.L. Galeener (1983). Raman studies of vitreous SiO_2 versus fictive temperature. *Phys. Rev. B*. 28, 3266-3271.

[25] W.J. Malfait, V.P. Zakaznova-Herzog, W.E. Halter (2008). Quantitative Raman spectroscopy: speciation of Na-silicate glasses and melts, *Am. Mineral*, 93, 1505-1518.

[26] O. Dezellus, Contribution à l'étude des mécanismes de mouillage réactif, Thèse de doctorat, Institut National Polytechnique de Grenoble, Grenoble, 2005.

[27] K.P. Hapgood, B. Khanmohammadi (2009). Granulation of hydrophobic powders, *Powder Technol. 189*, 253–262.

[28] A. W. Adamson. J. Wiley, 1991, physical chemistry of surfaces, fifth edition. New York, 110, Issue 4, 137.

[29] C. Mangwandi, L. JiangTao, A B. Albadarin, R.M. Dhenge, G. M. Walker (2015). High shear granulation of binary mixtures: Effect of powder composition on granule properties, *Powder Technol. 270*, 424–434.

[30] M. Amirjan, H. Khorsand (2014). Processing and properties of Al-based powder suspension/slurry: A comparison study of aqueous binder systems, stability and film uniformity, *Powder Technol. 254*, 12–21.

[31] L. Susana , F. Campaci, A.C. Santomaso (2012). Wettability of mineral and metallic powders: Applicability and limitations of sessile drop method and Washburn's technique, *Powder Technol. 226*, 68–77.

[32] I. Konyashin, A.A. Zaitsev, D. Sidorenko, E.A. Levashov, B. Ries, S.N. Konischev, M. Sorokin , A.A. Mazilkin, M. Herrmann , A. Kaiser (2017). Wettability of tungsten carbide by liquid binders in WC–Co cemented carbides: Is it complete for all carbon contents, *Int. Journal of Refractory Metals and Hard Materials. 62*, 134–148.

[33] P. Duan, C. Yan ,W. Luo , W. Zhou (2016). A novel surface waterproof geopolymer derived from metakaolin by hydrophobic modification, *Mater. Lett. 164*, 172–175.

[34] L. Vidal, E. Joussein, J. Absi, S. Rossignol (2017). Coating of unreactive and reactive surfaces by aluminosilicate binder, *Ceram. Int. 43*, 1819–1829.

[35] A. Gharzouni, I. Sobrados, E. Joussein, S. Baklouti, S. Rossignol (2016). Predictive tools to control the structure and the properties of metakaolin based geopolymer materials. *Colloid Surf. A, 511*, 212–221.

[36] M. Muroya (1999). Correlation between the formation of silica skeleton and Fourier transform reflection infrared absorption spectroscopy spectra, *Colloid Surf. A*. 157, 147-155.

[37] A. Gharzouni, E. Joussein, B. Samet, S. Baklouti, S. Pronier, I. Sobrados, J. Sanz, S. Rossignol (2014). The effect of an activation solution with siliceous species on the chemical reactivity and mechanical properties of geopolymers. *J. Sol-Gel Sci. Technol. 73*, 250-259.

[38] S. Lucas, M.T. Tognonvi, J.-L. Gelet, J. Soro, S.Rossignol (2011). Interactions between silica sand and sodium silicate solution during consolidation process, *J. Non-Cryst. Solids. 357*, 1310-1318.

[39] J. Mähler, I. Persson (2012). A study of the hydration of the alkali metal ions in aqueous solutions, *Inorg. Chem.* 51, 425-438.

[40] C-C Lin, S-F Chen, L-g Liu, C-Ching Li Size effects of modifying cations on the structure and elastic properties of Na2O–MO–SiO2 glasses (M= Mg, Ca, Sr, Ba), *Mater. Chem. Phys.*123 (2010) 569–580.

[41] TBD

[42] K. Li, R. Khanna, M. Bouhadja, J. Zhang, Z. Liu, B. Su, T.Yang, V. Sahajwalla, C.V. Singh, M. Barati, A molecular dynamic simulation on the factors influencing the fluidity of molten coke ash during alkalization with K_2O and Na_2O, Chem. Eng. J. 313 (2017) 1184–1193.

EFFECT OF TiO_2 AND ZnO NANOPOWDERS ON METAKAOLIN-SODIUM HYDROXIDE GEOPOLYMERS

D. Sarbapalli and P. Mondal
Department of Civil and Environmental Engineering, University of Illinois Urbana Champaign
Urbana, IL, USA

ABSTRACT

Nanoparticles were used as potential nucleation seeds to gain insights into the reaction mechanism of alkali-activated metakaolin binders. Commercially available titanium dioxide and zinc oxide nano-powders were added in small dosages and their effect on the reaction mechanism was studied by monitoring heat evolution though isothermal calorimetry. A selective chemical dissolution (Hydrochloric acid attack) method was used to quantify the amount of unreacted metakaolin at different ages. Calorimetry results show that TiO_2 accelerates the reaction at early ages and the same was verified through selective dissolution data. ZnO were observed to retard the reaction and adding TiO_2 in this system was seen to reverse the retardation process. This indirectly gave proof of TiO_2 acting as a nucleation seed. Microstructural improvement was seen at 24h age through compressive strength testing, although no significant differences were observed at later ages. Effects of alumina and ZnO when compared with that from literature did not show expected trends, which indicated differences in reaction mechanism of geopolymers prepared from different precursors.

INTRODUCTION

An aluminosilicate precursor such as metakaolin yields a three-dimensional aluminosilicate structure with short-ranged, ordered framework upon reaction (often referred as alkali activation) with hydroxide or silicate solution[1]. As established by Rahier et al. experimentally, the proposed reaction pathways involve a dissolution step[2] where silicate and aluminate species are released into solution, followed by a polymerization process where polymerized silicate species along with aluminosilicate oligomers are formed[3]. These oligomers further polymerize and ultimately form a three-dimensional amorphous framework[3]. Zeolite phases can also be detected in such systems given certain environmental conditions and age of specimens[4,5]. A complete review on this topic has been recently summed up by Provis[6].

Thus, it is established that a phase transformation occurs from the dissolved silicate and aluminate species in the solution to the final hardened binder. In the context of alkali-activated materials, this transformation is believed to be a nucleation and growth controlled process[4,7]. Nucleation and growth is a typical example of a discontinuous, liquid-solid phase transformation, where the new phase formed has drastically different properties from its parent but the change is highly local. The new phase consequently grows spatially out of these small nuclei[8]. Thus, adding nucleation seeds offers promise in controlling reaction kinetics and early age properties, as has been summarized by Kawashima et al. in cement based systems[9]. Addition of tobermorite and xonotlite, which are crystalline analogues of calcium silicate hydrate (C-S-H), the primary reaction product of cement hydration, have shown to increase the rate of hydration of portland cement pastes[10]. The structure of the reaction product at the nanoscale can also be potentially modified due to the presence of seeds. For example, addition of different types of synthesized C-S-H to hydrating tri-calcium silicate (C_3S) systems has been reported to influence the nano-structure and composition of the reaction product[11]. Furthermore, presence of C-S-H is shown to accelerate the reaction kinetics and increase early age strength of alkali activated slag[12].

Addition of nucleation seeds can also prove whether a reaction is nucleation and growth controlled or dissolution controlled as in the case of the latter, the addition will have little or no effect on the reaction kinetics. By adding nucleation seeds (nano Al_2O_3, ZnO and ZrO_2) to two different geopolymeric binders, a Class F fly ash geopolymer activated through NaOH[13] and a binder made by mixing geothermal silica with sodium aluminate[14], it has been proved experimentally that the formation of geopolymeric binder could involve nucleation. The kinetics of the reaction was seen to be accelerated in both cases and the former study reported that the addition of nano-aluminum in very small dosages (<<1% by wt. of precursor) can influence the nature of the product being formed. Adding seeds to the geothermal silica-sodium aluminate system resulted in a refined microstructure, which consequently led to improvements in compressive strength. Furthermore, elemental analysis revealed that the Si/Al and Na/Al ratios of the product was statistically different between the control and seeded sample, demonstrating the potential for seeds to control the nature of the reaction product.

In summary, addition of nucleation seeds provides i) an opportunity to verify whether a reaction is nucleation controlled, ii) a potential to refine the microstructure of the binder, and iii) a chance to influence the nano-structure of the final reaction product. However, to the best of the author's knowledge, any information on the nucleation seeding in metakaolin based geopolymers is lacking in the published literature. In this study, metakaolin precursor was chosen because the material is a phase pure amorphous aluminosilicate and any changes to the nano-structure and chemical properties due to seeding can be characterized easily. Alkali activation was done by adding sodium hydroxide. Nano-TiO_2 (anatase polymorph) and nano ZnO were added as potential seeding agents in low dosages. Particle size distribution of nanopowders was characterized through scanning electron microscopy (SEM) in the powder state and by using dynamic light scattering (DLS) in the dispersed state. Influence on the reaction kinetics was observed through isothermal calorimetry. Compressive strength tests were carried out at 1, 3 and 7 days' age. Selective dissolution using hydrochloric acid (HCl) extractions were performed at 4h, 24h and 7 days to quantify the amount of unreacted precursor and results were correlated with the reaction kinetics data obtained from isothermal calorimetry.

EXPERIMENTAL METHODS

Materials and Sample Preparation

Metakaolin (MK) was procured from BASF chemicals, with XRF composition being listed in Table I. Median particle size and surface area was reported to be 1.21μm and 14.2m^2/g respectively. Sodium hydroxide pellets were supplied by Fisher chemicals with a purity ≥97% and subsequently, made into solutions at the required concentration. All materials were stored at a temperature of 22°C. The metakaolin geopolymers were prepared with a 1:1:1.9:13.6 (Na_2O:Al_2O_3:SiO_2:H_2O) molar stoichiometry, using a mechanical shear mixer. It should be noted that no soluble silicates in the form of sodium silicate was added to the mix and metakaolin was the only source of silica and alumina. Metakaolin powder was added to sodium hydroxide in a plastic mixing cup and mixed for four minutes. The cup was sealed off and kept at an environmentally controlled room at 22°C prior to testing. Nanopowders were dispersed through a probe sonicator (Fisher Scientific, Model 505) in the sodium hydroxide solution prior to mixing. All the tests did not require the same quantity of binder to be prepared and the weights of metakaolin and sodium hydroxide were scaled up for compressive strength tests. To ensure that this did not affect the reaction kinetics and the mixing efficiency, calorimetry was carried out on samples prepared with the initial and scaled up proportions, and no change was observed in the rate of heat released. Henceforth, the metakaolin geopolymers prepared are denoted as 'MK-

NaOH'. Nano-pure water obtained from a Milli-Q water purification system (EMD Millipore, USA) was used throughout the study for preparing the geopolymer pastes.

Table I. XRF Composition of metakaolin

SiO$_2$	Al$_2$O$_3$	Na$_2$O	K$_2$O	TiO$_2$	Fe$_2$O$_3$	CaO	MgO	P$_2$O$_5$	LOI (%)
52.3	45.2	0.22	0.15	1.74	0.42	0.04	0.04	0.08	0.79

TiO$_2$ and ZnO nanopowders were procured from US Research Nanomaterials. The nanoparticle characteristics as provided by the manufacturer (particle size measured through transmission electron microscopy and surface area calculated by using the size and density of the materials) is given in Table II. The nanopowders were added by 1% weight of metakaolin in the mix, unless otherwise specified. Although seeds have been observed to be added at levels beyond 5%[9,10,15], the authors believe a small quantity of nano-sized particles with high surface area should provide enough sites for heterogeneous nucleation[12,13,14]. Prior to mixing, the seed was dispersed in the sodium hydroxide solution using a Probe sonicator. Hydroxide solution was taken in the mixing cup, with appropriate amount of seed and sonicated for 10 minutes. To allow for heat dissipation, the sonicator was operated in a pulsed mode while the sample container was kept in an ice-water bath. Amplitude of sonication was kept at 30% of the maximum rated amplitude for the tip used (Manufacturer recommends the level be kept below 40%). Post probe sonication, the solution was equilibrated to 22°C.

Table II. Properties of nanopowders, as specified by manufacturer

Nanopowder	Average Particle Size (nm)	Specific surface area (m^2/g)
TiO$_2$ (anatase)	5	289
ZnO	18	40-70

Scanning Electron Microscopy (SEM)
A cold field emission SEM (Hitachi S4700) was used in the secondary electron mode to characterize the particle size and morphology of both TiO$_2$ and ZnO nanopowders. Prior to imaging, the powder was pressed down on carbon tape and any loosely attached powder was blown away using compressed air. Accelerating voltage of 5 kV and 3 kV was used for TiO$_2$ and ZnO powder respectively. Both samples were sputter coated with Au-Pd. SEM images were obtained from two locations and the morphology and particle size distribution was observed to be similar across both.

Dynamic Light Scattering (DLS)
This technique is typically used to measure the size of sub-micron particles dispersed in a liquid medium. A Malvern Zetasizer Nano ZS was used in this study to determine the particle size distribution of nanopowders after dispersion in NaOH solution. Particle size detection range for the instrument is within 0.3nm and 6μm. The light source was a laser operating at 632.8 nm and scattered light was collected in the backscatter mode, at an angle of 173° to the incident light. Laser position was automatically detected based on trials carried out by the instrument before measurement. Intensity of light was also adjusted automatically by an attenuator during these trials, ensuring that the detector is neither under-saturated or over-saturated. DLS works on the basis that light scattering changes with time due to Brownian motion. Particles with higher size exhibit

slower motion and hence the changes in light scattering are slower, whereas for a smaller particle size yields rapid changes in scattered signal intensity. The scattered signal generated at the very beginning was correlated to signal obtained in the subsequent time steps through a digital auto-correlator and a correlogram was generated. The same was then processed to give two measurements of size distribution: i) Z-avg, which gives an estimate of the mean size and polydispersity index (PDI, refers to the distribution of particle sizes – a lower index indicates the width of the particle size range is low and vice versa) through cumulants analysis (described in ISO 22412:2008), ii) Particle size distribution by fitting multiple exponentials to the correlation function using a non-negative least squares method. All particle sizes from the DLS measurements refer to the hydrodynamic diameter which is essentially the diameter of a spherical particle with the same diffusion coefficient of the sample. The particle size distribution is reported against the intensity of light scattered by the particles.

The nanopowders were dispersed in solution in the same proportions using the same probe-sonication procedure as required for an actual mix. Prior to the DLS measurement using the Zetasizer, the solution was ensured to be at equilibrium with room temperature ($\sim22°C$) and then agitated on a benchtop vortex stirrer for 1 minute. Subsequently, around 2-3 ml of solution was transferred to a disposable plastic cuvette for DLS measurements. The efficacy of the Zetasizer measurements were first verified on a polystyrene latex standard of known particle size distribution. DLS measurements were then taken. Viscosity of the sodium hydroxide solution was taken to be 21.95 mPa-s, based on technical data available on the same[16].

Isothermal Calorimetry

After MK-NaOH mixtures were prepared, they were transferred into 20 mL glass ampoules and kept in a TAM Air 8 channel micro-calorimeter. The instrument is reported to have a precision of ±20 μW and can detect heat released beyond 4 μW. The signal was stabilized prior to starting the experiment, under the following conditions i) maximum value of the slope of every twenty minutes of data being less than 2 $\mu W/h$ and standard deviation less than 4 μW. Baseline data was collected for thirty minutes after the signal was stabilized. The samples were then inserted in the calorimeter within 5 minutes after mixing, and the rate of heat released from the ampoules was collected for 2-3 days. First 45 minutes of data was discarded since the mixing was done outside of the calorimeter, and the initial rate of heat release can be attributed to frictional heat generated while inserting the ampoules. Four replicates for each sample was run, and the results came out to be identical. Thus, only one curve from each sample was presented in this paper. The rate of heat and total heat released was normalized to the total amount of geopolymer paste in the ampoule, which varied between 8-13g. Total heat curves were automatically calculated by the instrument software by integrating the rate of heat evolution with time. The time t=0 in the calorimetry curves presented here denotes the time at which samples were inserted in the calorimeter and data was recorded every second.

Compressive Strength testing

Compressive strength was measured on 1" cubes using an Instron 4500 load frame. To remove trapped air voids, all the paste samples were compacted to the same degree. After filling molds with the paste, they were kept on a drop table (used for slump flow test[17]) surface, and compaction was achieved through 50 drops of the table surface. Displacement rate of 1mm/min was used for the strength testing and a LabView program was used to collect the load-displacement data. Compressive strength of cubes at 1, 3 and 7 days were measured. Six samples were crushed till failure and ultimate strength was reported as mean of the six values.

HCl Extraction

The degree of reaction or the amount of product formed in alkali activated binders has been measured quantitatively using selective dissolution techniques previously[18]. In particular, HCl solutions have been known to remove products generated through alkali-activated binders[19]. To carry out the dissolution process, the reaction needs to be stopped and all free water has to be removed. For this purpose, a solvent extraction method was employed. Geopolymers were crushed with an agate pestle and mortar at the required age (4h, 24h and 7 days) with a 1:1 mixture of acetone and methanol (by volume). While solvent extractions have been known to stop the geopolymer reaction, it has also been reported that the presence of soluble silicates can precipitate silica along with the geopolymers[20]. To ensure that there is no such precipitation, MK-NaOH powders were washed with water prior to the solvent extraction in a manner described by Chen et al.[20]. However, Fourier Transform Infrared Spectroscopy (FTIR) spectra of powders without and with the water wash was seen to be the same owing to which the solvent extraction without water wash was used throughout this study. Finally, 0.5g of a powder thus obtained was stirred in 100 ml of a 1:20 HCl solution for 3 hours to remove all reaction products[18]. Subsequently, the solution was vacuum filtered through a Whatman Grade 1 filter paper in a Buchner funnel and flushed twice with nano-pure water. The funnel was then dried at 105°C and the mass loss was recorded. The residue was verified to be metakaolin using FTIR and the mass loss data was used to compute the percent of unreacted metakaolin in a sample. Unreacted metakaolin was calculated by averaging values obtained from four separate runs.

RESULTS AND DISCUSSION

Seed Characterization

SEM images of TiO_2 and ZnO nano-powders are shown in Figures 1 and 2 respectively. TiO_2 powders are shown to have particles below 100nm and ZnO powders have particle sizes around 50nm from SEM images.

Figure 1. SEM images of TiO_2 nanopowder showing distinct particles in the range of ~60-100nm.

Figure 2. SEM images of ZnO nanopowder showing distinct particles in the range of ~50-100 nm.

The dispersion of 1%TiO_2 in NaOH results in an opaque white solution. DLS experiments on this sample failed because the opaque solution led to issues with absorbance of light and the detector was always undersaturated. Subsequently, the concentration was lowered to 0.1% TiO_2. Since the particle concentration usually does not affect the hydrodynamic diameter, this measurement is assumed to yield a similar PSD to 1.0% TiO_2 in NaOH. On the other hand, 1% ZnO was also tested on DLS; the dispersion of ZnO in NaOH yielded a more or less clear liquid and no dilution was required for the measurements. Z-avg and PDI measurements for the two systems are shown in Table III. Particle size distribution analysis from multiple exponential fits are plotted in Figure 3.

Table III. DLS results for Z-avg and polydispersity index

Sample	Z-avg (nm)	Polydispersity Index (PDI)
0.1% TiO_2-NaOH	1233	0.080
1.0% ZnO-NaOH	189	0.378

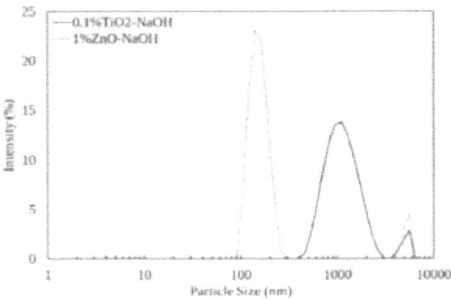

Figure 3. Particle size distribution for TiO_2 (solid) and ZnO (dotted) nanopowders in NaOH

Test results indicate that the TiO_2 is not stable in NaOH since the reported Z-avg comes out to be beyond one micron, whereas the manufacturer reported a particle size of 5nm. The software also detected the presence of flocculating particles. Z-avg reported matches closely with the peak size reported through particle size distribution which demonstrates the accuracy of the experiment. On the contrary, ZnO appears to be in a well dispersed state, with Z-avg being measured as 189 nm, against a reported 18nm. The primary particles as measured by the manufacturer through TEM appears to aggregate to form these small flocs, based on the DLS results. The polydispersity index was high, but it is probably attributed to the formation of small micron sized flocs which is seen from the particle size distribution of the sample. Once again, the agreement of Z-avg with the peak position in particle size distribution indicates a good measurement. Additionally, particle size of ZnO observed in the SEM images corroborates well with what is observed from DLS measurements. This indicates that the dispersed solution is stable. TiO_2 is clearly forming a flocculated system in NaOH since the particle size of the material through SEM can be seen to be much lesser than what is reported through DLS measurements. There is no evidence presented which shows that the aggregates break down during the geopolymerization process.

Effect of TiO_2 addition on the reaction kinetics of MK-NaOH

Figure 4(a) shows the effect of addition of 1% TiO_2 and ZnO to the control MK-NaOH sample. The control specimen has two distinct peaks in the time-frame of data presented. As part of the experimental model presented by Zhang et al.,[21] the first peak (within the first hour) is attributed to the initial heat released from the dissolution of metakaolin. Subsequently the reaction goes through an apparent induction period, followed by a second peak (seen within 5-6 hours). This peak is attributed to the polymerization process that is known to occur within alkali activated binders[21]. Addition of TiO_2 shows two changes compared to control, with the first being rate of heat released increases up to the first 24 hours. The second feature is harder to distinguish, but there is a slight decrease in the time at which the second peak takes place. It should be also noted that further increase in the dosage of TiO_2 did not cause any additional change in the reaction kinetics.

While the changes observed through calorimetry is not as dramatic compared to what was observed by Rees et al.[13], the effect of TiO_2 addition on the reaction kinetics is undeniable. Possible explanation of the effect of TiO_2 addition on the reaction kinetics of MK-NaOH is carefully considered here. It is unlikely that TiO_2 directly participated in any chemical reaction since it is known to be inert[15] and the amount added is very small. On the other hand, if the rate of reaction at the early age is controlled by the availability of nucleation sites, addition of TiO_2 nano-particles was expected to reduce or even eliminate the induction period. For example, addition of a small quantity (0.5% - 4% by weight of C_3S) of synthesized C-S-H gel accelerates hydration of tricalcium silicate and virtually eliminates the induction period[22]. This result strongly indicated that the formation of C-S-H is autocatalytic. The lack of pronounced effect of TiO_2 addition on the length of the induction period could indicate that the availability of nucleation sites is not a rate limiting factor, when geopolymer is synthesized from metakaolin owing to its extremely high surface area (14.2 m^2/g, which is around forty times higher than typical ASTM Type 1 cement). Another possibility is that the formation of geopolymeric sodium aluminosilicate hydrate (N-A-S-H) is autocatalytic[4]. Therefore, TiO_2 nano-particles offer limited opportunity for the nucleation of N-A-S-H on its surface which in turn provide more suitable nucleation sites for subsequent N-A-S-H formation. Similar phenomenon is reported when silica fume is added to C_3S based systems[22]. It is believed that pozzolanic C-S-H forms slowly on the surface of the silica fume and subsequently auto-catalyzes the reaction. Thus, a slow acceleration is observed since additional nucleation sites form gradually throughout the early period of hydration instead of being immediately available at the time of mixing. Similar behavior is observed in Figure 4 (a), where an increase in the peak heat of hydration is observed without any change in the induction period. Lastly, there is always a possibility that the lack of pronounced effect of TiO_2 addition could be attributed to the poor dispersion of TiO_2 as mentioned earlier. However, this may not be entirely true, based on effects of TiO_2 observed in the presence of ZnO as described below.

To verify that TiO_2 acts as a nucleation seed in MK-NaOH geopolymer, it was used in combination with ZnO nanoparticles. In cement-based systems, ZnO is believed to poison nucleation sites which ultimately causes retardation[24]. Ataie et al.[24] observed that the retarding effect of ZnO on cement hydration can be reduced by addition of a silicon rich, finely divided, highly amorphous rice and wheat husk ash which are verified to act as nucleation sites for C-S-H. Following similar analogy, ZnO was added at 1% and 3% by wt. of MK-NaOH. Figure 4(a) clearly shows that the addition of 1%ZnO has a retarding effect on the MK-NaOH system. It decreases the rate of heat release and delays the appearance of the second peak. The figure also illustrates the retardation increases with the increase in dosage of ZnO. In a parallel study carried out by the authors, addition of ZnO to an alkali activated fly-ash slag binder also caused similar retardation. The evidence of ZnO nanoparticles being a retarder does contradict the accelerating effect observed by Hajimohammadi et al.[14] however these observations could stem from differences in the reaction mechanism when different precursors are used. For example, addition of Al_2O_3 nanoparticles to Class F fly ash activated through sodium hydroxide has been reported to accelerate the reaction significantly[13]. However, the same did not show any effect on MK-NaOH. When 1%TiO_2 was added to the system containing 1%ZnO, it was observed that the extent of retardation is arrested to some extent (illustrated in Figure 4(b)). Increasing the dosage of TiO_2 to 3% further reduces the retardation, compared to 1%TiO_2. Since ZnO is known to poison nucleation sites, the fact that TiO_2 can reverse the action of ZnO provides indirectproof that it is indeed acting as a nucleation seed.

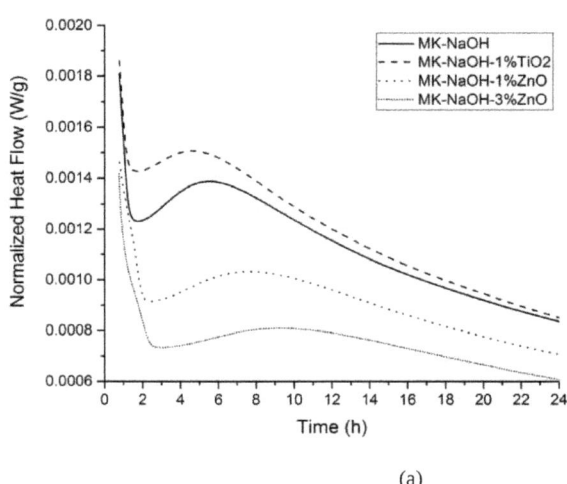

(a)

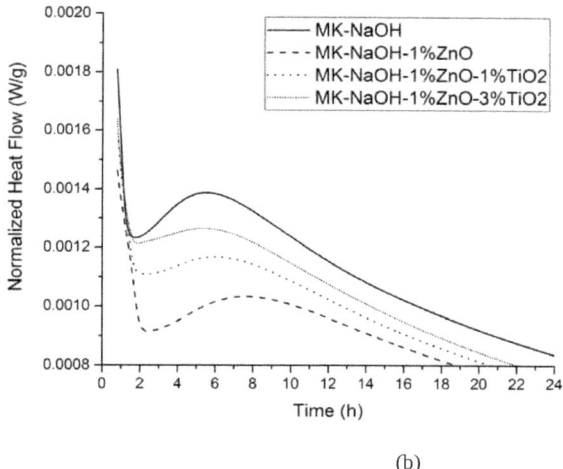

(b)

Figure 4. Rate of heat evolution from MK-NaOH systems with (a) 1%TiO$_2$, 1%ZnO and 3%ZnO and (b) a combination of the TiO$_2$ and ZnO nanopowders

Effect of TiO$_2$ on degree of reaction and compressive strength

Amount of unreacted metakaolin at various ages can be estimated through HCl extraction as explained earlier. Geopolymer was extracted at 4 hours, which is right before the onset of the second peak observed in calorimetry, at 24 hours and at 7 days' age. Figure 5 shows that at 4h, the amount of unreacted metakaolin in the TiO$_2$ seeded binder is lesser (with statistical significance) compared to the unseeded control binder, however, the amount of unreacted metakaolin is practically the same at 24h age. This corroborates well with the chemical kinetics information provided by isothermal calorimetry (Figure 4) where the reaction rates across seeded and unseeded system after 24h is more or less the same. The results also prove that TiO$_2$ is promoting the dissolution of metakaolin at early ages (4h).

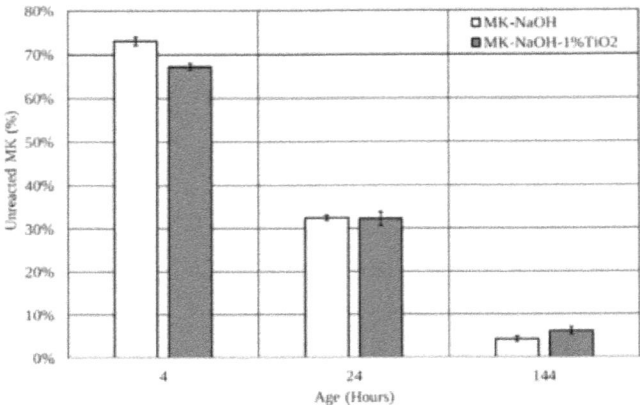

Figure 5. Amount of unreacted metakaolin, as measured by HCl dissolution. Error bars represents one standard deviation from the mean.

It was discussed earlier, nucleation seeding offers some chances of refinement in microstructure, which could manifest as an increase in compressive strength. Addition of TiO_2 was seen to improve the compressive strength of MK-NaOH systems marginally at the age of 24h, as shown in Figure 6. However, at subsequent ages of 3d and 7d, no statistically significant data was obtained to conclude whether microstructure refinement was taking place.

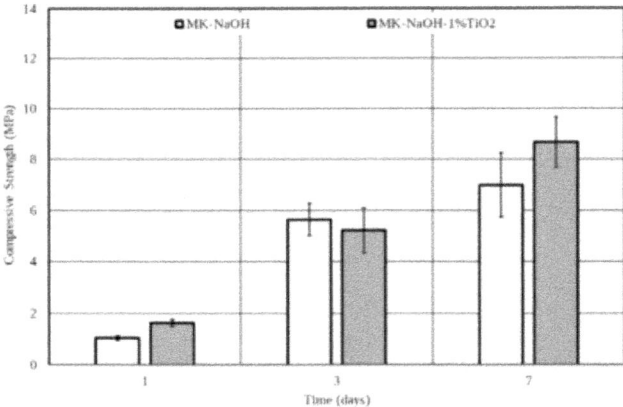

Figure 6. Compressive strength of 1" cubes with error bars representing one standard deviation from mean.

SUMMARY AND CONCLUSIONS

Adding high surface area TiO_2 has been proven to accelerate the rate of reaction of metakaolin activated with sodium hydroxide. On the other hand, ZnO nano-powders were observed to retard the reaction rate. Using the two nanopowders at the same system provided indirect experimental evidence that the phase transformation of species in solution to the solid binder is a nucleation and growth controlled process. The calorimetry results correlated well with the unreacted metakaolin present in the system at early ages which furthermore proved that TiO_2 accelerates the reaction rate. However, in terms of compressive strength development, no changes were seen except at the age of 24 hours. The differences in the reaction mechanism between alkali activation of different precursor materials were highlighted from the observation that ZnO retarded the MK-NaOH reaction whereas an acceleration was seen in another case.

ACKNOWLEDGEMENTS

The study was funded through an NSF Grant (No. #1538432). Scanning electron microscopy and dynamic light scattering experiments were carried out in part in the Frederick Seitz Materials Research Laboratory Central Facilities, University of Illinois. The authors would like to thank Robbie Damiani for generating some of the SEM data.

REFERENCES

[1]P. Duxson, J.L. Provis, G.C. Lukey, F. Separovic, and J.S.J. Van Deventer, "29Si NMR study of structural ordering in aluminosilicate geopolymer gels," *Langmuir*, **21** [7] 3028–3036 (2005).

[2]H. Rahier, B. Van Mele, M. Biesemans, J. Wastiels, and X. Wu, "Low-temperature synthesized aluminosilicate glasses Part I Low-temperature reaction stoichiometry and structure of a model compound," *J. Mater. Sci.*, **31** [1] 71–79 (1996).

[3]J.L. Provis and J.S.J. van Deventer, "Geopolymerisation kinetics. 2. Reaction kinetic modelling," *Chem. Eng. Sci.*, **62** [9] 2318–2329 (2007).

[4]J.L. Provis, G.C. Lukey, and J.S.J. Van Deventer, "Do geopolymers actually contain nanocrystalline zeolites? a reexamination of existing results", *Chem. Mater.*, **17** [12] 3075–3085 (2005).

[5]M. Król, J. Minkiewicz, and W. Mozgawa, "IR spectroscopy studies of zeolites in geopolymeric materials derived from kaolinite," *J. Mol. Struct.*, **1126** 200–206 (2016).

[6]J.L. Provis, "Geopolymers and other alkali activated materials: why, how, and what?," *Mater. Struct.*, **47** 11–25 (2014).

[7]P. Duxson, G.C. Lukey, J.S.J. Van Deventer, S.W. Mallicoat, and W.M. Kriven, "Microstructural characterisation of metakaolin-based geopolymers;" pp. 71–85 in *Ceram. Trans.* 2005.

[8]R. O'Hayre, *Materials Kinetics Fundamentals*; pp. 190-192. John Wiley & Sons, Hoboken, NJ, 2015.

[9]S. Kawashima, P. Hou, D.J. Corr, and S.P. Shah, "Modification of cement-based materials with nanoparticles;" *Cem. Concr. Compos.*, **36** [1] 8-15 (2013).

[10]G. Land and D. Stephan, "Controlling cement hydration with nanoparticles," *Cem. Concr. Compos.*, **57** 64–67 (2015).

[11]R. Alizadeh, L. Raki, J.M. Makar, J.J. Beaudoin, and I. Moudrakovski, "Hydration of tricalcium silicate in the presence of synthetic calcium–silicate–hydrate," *J. Mater. Chem.*, **19** [42] 7937 (2009).

[12]M.H. Hubler, J.J. Thomas, and H.M. Jennings, "Influence of nucleation seeding on the hydration kinetics and compressive strength of alkali activated slag paste," *Cem. Concr. Res.*, **41** [8] 842–846 (2011).

[13]C.A. Rees, J.L. Provis, G.C. Lukey, and J.S.J. van Deventer, "The mechanism of geopolymer gel formation investigated through seeded nucleation," *Colloids Surfaces A: Physicochem. Eng. Asp.*, **318** [1-3] 97–105 (2008).

[14]A. Hajimohammadi, J.L. Provis, and J.S.J. van Deventer, "Time-resolved and spatially-resolved infrared spectroscopic observation of seeded nucleation controlling geopolymer gel formation," *J. Colloid Interface Sci.*, **357** [2] 384–392 (2011).

[15]A.R. Jayapalan, B.Y. Lee, S.M. Fredrich, and K.E. Kurtis, "Influence of Additions of Anatase TiO_2 Nanoparticles on Early-Age Properties of Cement-Based Materials," *Transp. Res. Rec. J. Transp. Res. Board*, **2141** 41–46 (2010).

[16]"Caustic Soda Solution (NaOH 50%)," Product Information, Vinnolit GmBH & Co. KG (2011).

[17]ASTM International, "Standard Specification for Flow Table for Use in Tests of Hydraulic Cement (C 230)." West Conshohocken, 2014.

[18]S. Puligilla and P. Mondal, "Co-existence of aluminosilicate and calcium silicate gel characterized through selective dissolution and FTIR spectral subtraction," *Cem. Concr. Res.*, **70** 39–49 (2015).

[19]M. Granizo, S. Alonso, M. Blanco-Varela and A. Palomo, "Alkaline Activation of Metakaolin: Effect of Calcium Hydroxide in the Products of Reaction", *J. Am. Ceram. Soc.*, **85** [1] 225-231 (2002).

[20]X. Chen, A. Meawad, and L.J. Struble, "Method to Stop Geopolymer Reaction," *J. Am. Ceram. Soc.*, **97** [10] 3270-3275 (2014).

[21]Z. Zhang, H. Wang, J.L. Provis, F. Bullen, A. Reid, and Y. Zhu, "Quantitative kinetic and structural analysis of geopolymers. Part 1. the activation of metakaolin with sodium hydroxide," *Thermochim. Acta*, **539** 23–33 (2012).

[22]J. Thomas, H. Jennings and J. Chen, "Influence of Nucleation Seeding on the Hydration Mechanisms of Tricalcium Silicate and Cement", *J. Phys. Chem. C*, **113** [11] 4327-4334 (2009).

[23]H. Rahier, J.F. Denayer, and B. Van Mele, "Low-temperature synthesized aluminosilicate glasses: Part IV. Modulated DSC study on the effect of particle size of metakaolinite on the production of inorganic polymer glasses," *J. Mater. Sci.*, **38** [14] 3131–3136 (2003).

[24]F.F. Ataie, M.C.G. Juenger, S.C. Taylor-Lange, and K.A. Riding, "Comparison of the retarding mechanisms of zinc oxide and sucrose on cement hydration and interactions with supplementary cementitious materials," *Cem. Concr. Res.*, **72** 128–136 (2015).

ECO-FRIENDLY GEOPOLYMER COMPOSITE FOR WINTER SEASON PAVEMENT POTHOLE PATCHING

M. Sarkkinen[a], K. Kujala[b] and S. Gehör[b]
[a]Department of Mechanical and Mining Engineering, Kajaani University of Applied Sciences, Finland
[b]Solid Liner Ltd, Finland

ABSTRACT

The purpose of the study was to examine the usability of geopolymer materials for pavement pothole patching. The special focus of the study was on the use of materials in cold (5…-10°C) weather conditions, such as at the end of winter when the problems faced while conducting repairs are most serious. The influence of the chosen factors on the performance of the material was studied using the Taguchi design of experiments method. The experiment was conducted using 13 control factors, with special attention given to strength development and weather resistance characteristics. The results indicated that the magnesium phosphate material studied had good weather resistance, as a result of the low-grade reactive waste magnesia used. According to the results using bitumen as an additive and a high activator to magnesia ratio had a positive impact on the performance of the patching material. In addition, relations between the responses were analysed using structural equation modelling. According to the model, higher water demand indicated an increase in water absorption and further lowered water and freeze-thaw resistance.

INTRODUCTION

The potholes denote small, typically sharp edged holes in the pavement hampering traffic. They have become more common due to climate change and consequent increasing amount of freeze-thaw cycles, for example in Finland. There are several different identified pavement damage types, which have been reported in the connection of earlier studies (e.g. Aspahlt Institut, 2016; Orr, 2006; Belt, 2002). The reasons are diverse and interrelated, the most common being traffic and weather stress (temperature, freezing and thawing, precipitation), and the own structural weight of the road structures. In addition to the traffic load, water ingress through the pavement is a prerequisite for the pothole forming. The holes are typically initiated after the damage in the pavement allowing water ingress through the pavement. It is commonly believed that repeating freezing and thawing cycles accelerate the pothole forming due to increased pressure during the freezing period and oversaturated state under the pavement during the warmer period.

The aim of the work was to study the usability of geopolymer materials for the repairs of paved roads, especially during the cold winter and spring seasons when the need for repair is most serious and the use of hot asphalt is not possible. The objective was to a find material which is both more cost-efficient and durable than cold asphalt. In addition, properties like rapid strength development, good bonding with old pavement material, durability and low material costs are required.

Typically, dead burnt magnesia (DBM) has been used in patch applications. In this study, the usability of low-grade reactive waste magnesia as the main binder component due to its lower costs compared to DBM was investigated. It is known that geopolymer based materials have good durability in harsh climate conditions as well as good mechanical characteristics. They can also be ecological due to their ability to utilize high amounts of the constituents of industrial by-products[1].

Magnesium phosphate binders are based on the concept known as phosphate ceramics, CBPC[2]. Phosphate binders are typically comprised of a phosphate based activator and magnesia source. In addition, binder components can include additives such as coal fly ash. They share the characteristics of ceramic materials and hydraulically bound Portland cements, that is, they harden without heating in normal temperatures. The characteristics of magnesium phosphate cements depend on various factors. According to some earlier studies[3], activator and magnesia ratio, amount of retarder, fineness of magnesia, type and amount of additives and temperature influence the binding time, strength development and bonding strength. CBPC materials can also bound with many industrial by-products to form composite matrices with various characteristics[4].

EXPERIMENTAL

Secondary Raw Materials Used in the Study

The principal secondary raw material used in this study is the magnesite-talc tailings derived from Mondo Minerals talc mine in Sotkamo, middle Finland. Minor additives are coal fly ash, a by-product of coal power production and steel slag.

Mondo´s talc ore contains pentlandite in a measure that the froth flotation is viable. Thus the tailings is mineralogically simple for magnesia synthesis. The two major phases being magnesite and talc, an additional phase is clinochlore which counts only few percent of the tailings mineralogy. The XRF-analysis (Table 1) indicate the chemical compositions of Mondo tailings (LGMgO), coal fly ash (FA) and steel slag (SS).

Table 1. Chemical composition (%) by XRF.

	MgO	SiO_2	Al_2O_3	SO_3	K_2O	CaO	Cr	Mn	Fe_2O_3	P_2O_5
LGM	55.12	23.80	0.00	1.40	0.05	1.12	0.25	0.15	13.03	0.00
SS	1.70	11.80	1.50	0.20	0.07	51.7	0.25	2.00	24.88	0.96
FA	n.d.	58.85	22.47	0.47	2.62	5.37	0.01	0.05	7.91	0.74

Loss on ignition: steel slag 3.6 % and fly ash 3.7 %.

Noteworthy, Mondo tailings contains 13 % Fe_2O_3 in whole rock composition. Iron is mainly bounded to magnesite lattice, where it substitutes magnesium. MgO in whole rock composition is considerable, 55.12 %, anyhow, into some extent that amount of iron reduces the quality of magnesite in binder purpose, but it does not preclude it from using for that purpose. Iron is oxidized in thermal treatment and forms ferrihydroxide compounds, which give the yellow tint for the calcined product (Fig. 1). Related to the reaction affinity of the additives, the coal ash stands out with considerable high SiO_2 and Al_2O_3. XRD-analysis of coal ash reveal two major phases in coal ash; mullite and quartz. Amorphous SiO_2 is an important phase to participate the binder reactions. Steel slag, in turn, stands out by elevated CaO that carried by the reactive portlandite, brownmillerite and lamite.

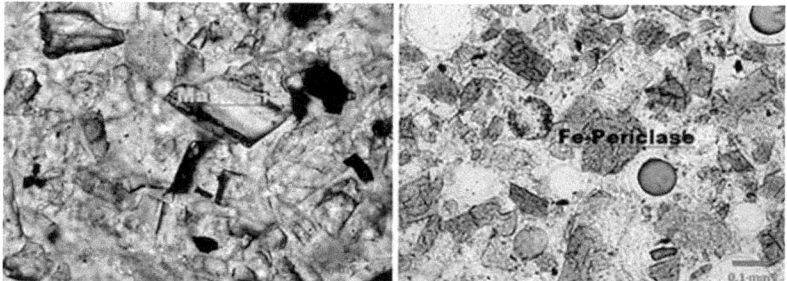

Figure 1. Microphotographs of Mondo tailing, transmitted light. Figure on the left is the sample dried at 105°C and the right one of the sample heated to 700°C. Decarbonization of magnesite to periclase is associated to the oxidation of the magnesite bounded Fe^{2+} to a Fe^{3+} compound, thus giving the yellow tint for the synthetized periclase .

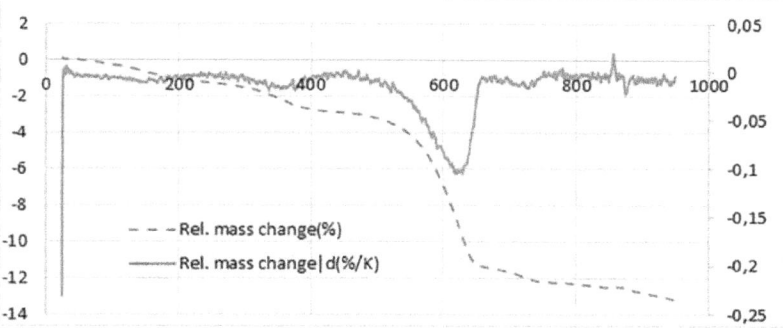

Figure 2. TGA/DT analysis on LGMgO.

The TGA/DT analysis (Fig. 2) indicates that magnesite decarbonization takes place at temperature interval from 400 to 700 °C. To obtain chemically active magnesium oxide usable as magnesia binder in solid phase reactions the burning temperature is adjusted at 700 °C. In thermal treatment magnesite decomposes to periclase (MgO) while talc and clinochlore remain inert in the binder reactions. The holding time in laboratory conditions is in laboratory tests adjusted to 2 hours in order to ensure that decarbonatization reaction proceeds to completion.

Design of Experiment

The experimental study applied the Taguchi method which separates system, parameter and tolerance design[5]. The system design in this study refers to the optimisation of the material concept, in particular the optimisation of the chosen parameters in relation to the determined functional factors. The aim of the parameter design was to identify the optimal material composition. In Taguchi design, robustness is achieved by identifying the control factors that reduce variability in the material and minimize the effects of noise. Noise factors are manipulated to force variability during the Taguchi experiment in order to identify the optimal setting to gain robustness and resistance[6]. Higher SN ratios indicate optimal control factor settings and reduce the effects of noise. Extreme response values can be achieved by optimal grouping of the noise factors. Compounding can reduce the number of tests when the primary objective is not to estimate the effects of individual noise factors. Taguchi design uses orthogonal arrays which help to reduce the number of experiments required. The test series used in this study is based on orthogonal matrix L^{27}, with 3 levels.

Patching material can be considered as a system where different constituents interact with each other. Thus the impact of the geopolymer binder depends on the other material components. In addition to the magnesia source and activator, additive, admixture and aggregate variables were studied in order to find the optimal combination considering the desirable performance characteristics of the patching material. The first test series comprised of low grade reactive magnesia (LGMgO) as the main binder component, coal fly ash and steel slag as additives, bitumen emulsion, plastic fibers, 3 plasticizers, 2 retarders, air entraining agent (AEA) and viscosity modifiying admixtures. Potassium dihydrogen phosphate, MKP (KH_2PO_4) was used as the activator. The original crushed granite aggregate and crushed reclaimed asphalt (RAP) were used as aggregate alternatives (Table 2). The first test series described in Table 3 is comprised of the 9 investigated responses.

Table 2: Factors and their levels.

	Factor	Level 1	Level 2	Level 3
A	MKP/LGMgO	1:2	1:3	1:4
B	Additive (%/A)	0	10	20
C	Additive type (ss=steel slag, fa=fly ash)	ss	fa	ss+fa
D	Superplasticizer (%/A)	0	1	2
E	Fiber (%)	0	0,5	1
F	Retarder (%/LGMgO)	3	4	5
G	Binder/aggregate	1:0	2:1	4:1
H	Crushed granite aggregate/RAP	1:0	0:1	1:1
I	Bitumen/binder (A)	0:1	1:5	1:3
J	Retarder (r=iron sulfate, b=boric acid)	b	rs	b+rs
K	Plasticizer: a= polycarboxylic polymer, b= melamin polymer, c= lignosulfonate	a	b	c
L	Viscosity modifier (%/A)	0	1	2
M	Air entraining agent (%/A)	0	0,15	0,3

Table 3: Studied responses.

Investigated Response	Method
Water demand	w%
Setting time	DIN EN 196-3[7]
Temperature	5 min after water addition
Compressive strength	1d, 7d, 28d (SFS-EN 12390-3[9])
Flexural str.: flexural/compressive str.	40x40x160mm (SFS-EN 12390-5[12])
Freeze-thaw resist: change in mass and comp. str.	50 cycles (SFS-5447[20])
Water resist.: change in mass and comp. str.	20°C/7d + 60°C/7d
Frost-salt resist.: change in mass and comp. str.	25 cycles (SFS-5449[21])
Water absorption	Change in w% during 0-168 h

Batching and the Testing Specimen Preparation

The trial mixes were mixed using a Hobart blade mixer. The Mix amounted to ca. 4-5 kg, sufficient enough to fill two (100 x 100 x 100 mm) cube moulds. The dry mix components were mixed before adding water and admixture solutions. The test mixes were compacted for two minutes using a vibrating table. The test specimens were stored in room temperature ca. 20°C without coverage.

Setting Time, Compressive Strength and Water Demand

Initial and final setting time is determined by vicat apparatus[7]. The initial setting time indicates workability period of the patching material, this is especially important when using manual patching. The final setting indicates the time when the patch can be driven over. The setting time typically depends upon the temperature of the substrate and the environment. In addition, it can be influenced by admixtures and thermal treatment. The temperature of the fresh patching material indicates its setting rate as well. In this study the temperature was measured 5 minutes after the addition of water.

Compressive strength development is an important indicator of materials durability against the mechanical stresses caused by traffic. Rapid hardening and strength development is also required from pavement patching materials in low temperatures conditions of -5°C or below. Requirements for the early compressive strength values vary. Compressive strength of rapid hardening pavement patching materials should be evaluated no later than 24 hours according to some guidelines[8]. Generally, the setting time should be less than 1 hour and the compressive strength should be at least 3.5 MPa after 3 hours. Compressive strength was evaluated applying the standard SFS-EN 12390-3[9] by using cubes 100x100x100 mm stored at 20°C in air without any coverage. Water demand is typically related to the compressive strength of the cement material. In the study, it was measured to obtain information on the effect of different constituents.

Flexural Strength

Elastic modulus indicates the transformation ability of the material under loading. Materials with higher elastic modulus are stiffer and not able to adapt to movements of the substrate. Flexural strength was evaluated using specimens 160x40x40mm by applying the standard SFS-EN 12390-5[12]. Flexural strength can be affected, for example, by adding different types of fibers or additives with a needle like structure (e.g. wollastonite)[4,10]. A brittle structure is typical for rapid hardening cements. Some studies have also reported combined use of bitumen emulsion and magnesium phosphate cement[11]. Use of wollastonite and basalt fibers in metakaolin based geopolymer materials has been studied by Dias and Thaumaturgo[13] and Silva and Thaumaturgo[14] with positive results. According to the studies by Pera and Ambroise[15], the elasticity of magnesium phosphate cement can also be improved by using polypropylene and metal fibers. The addition of crushed rubber to geopolymer materials has also been studied. Increased elasticity and reduced compressive strength were reported by Skoba et al.[16]. The use of bitumen emulsion with a magnesium phosphate based cement mix has been investigated by Li et al.[11]. According to the study, bitumen reduced compressive and flexural strength, in addition to abrasion resistance.

Water Permeability and Weather Resistance

Geopolymer materials are porous and thus vulnerable to frost deterioration. The pore structure, volume, shape and distribution influence the behavior of water in the material. The pores are comprised of gel pores (0.5-2.5 nm), which are too small for water molecules and larger capillary pores, which affect the permeability of water and other solutions[17]. It has been stated that the permeability of geopolymer materials in general is lower than that of normal cement. There is still little research data on the pore structure of geopolymer materials. However, it is known that several factors such as activator type, the ratio between solution and solid binder, curing conditions (temperature, relative humidity, time), and water content affect the pore structure. Water permeability was investigated by measuring the water absorption of cube specimens. The cube sides were sealed with epoxy and the top of the cube was covered with plastic. After that the specimens were placed in a water pool and changes in mass were measured at 0-168 hour intervals.

Results on the frost resistance of geopolymer materials vary[18]. A reason may be changes in the microstructure as a result of the differences in the constituents used and curing conditions. The frost resistance of geopolymer materials can be enhanced by optimizing pore structure and increasing compressive strength[17]. Optimal pore structure can prevent deterioration due to frost stress. A sufficient number of protective pores and limited distribution are required to guarantee good frost resistance. The most common method to optimize pore structure is to use an air entraining agent (AEA). However, opinions on the efficiency of AEA with geopolymer materials are diverse and the research results vary according to the type of AEA used. According to some studies the need for AEA may be higher, but it does not guarantee better frost resistance[19].

Freeze thaw cycles typically lead to internal cracking and strength loss in patching materials. The freeze-thaw test indicates the durability of patching materials in conditions where the temperature changes quickly and under freeze-thaw stress. The test helps to predict the sensitivity of the materials to disintegrate as a result of the climate. Damage is typically assessed based on changes in elastic modulus and strength. This study applied freeze-thaw resistance tests designed for concrete structures where temperature varied between 30 and -20°C, and the RH between 0 and 100%. The test was executed in a climate chamber. The deterioration caused by chlorides typically occurs as surface scaling, which can be measured as mass loss. Both damage types are influenced by water content and pore structure. Deterioration typically occurs in saturated material where the pore structure is insufficient to resist tensions generated by ice formation and expansion. The chloride-frost test predicts resistance against the effects of deicing chemicals used on roads. Scaling of the patching material in the test indicates weak resistance to the chemicals. The weather resistance of the geopolymer materials were tested by applying standard SFS-5447[20].

Chloride-frost resistance was studied by applying the test standard SFS-5449[21] for concrete structures. In addition, water stress resistance was tested by storing geopolymer material specimens in water; first for 7 days at a temperature of 20°C and after that for 7 days at 60°C. The deterioration grade was estimated by comparing change in compressive strength and mass before and after the tests.

RESULTS AND DISCUSSION
Factorial Analysis
      The effect of the factors was evaluated using Minitab software, in relation to means and signal to noise ratios. The signal to noise (SN) ratio describes the relation between the varieties of the factor and the average response based on the delta value. The mean value in Taguchi design indicates the average response for each combination of control factor levels in the design. The delta value indicates the change of the value between the highest and lowest average-response level in relation to the factor in question. The ranking order is based on the delta values, a rank of 1 indicates the highest delta value. Ranking orders and significant p coefficients are depicted in Table 4. The statistical significance (p value) of each factor related to the response data was determined by the analysis of variance (ANOVA).

      The ranking order of the SN ratios and means indicate the relative importance of each factor to the response. In the table, all the S/N ratios with a level of 0.10 or higher are marked (*). In most cases, the ranking order and p-values were the same for SN ratios and means. According to the results of the experiment the bitumen to binder ratio had the most significant influence regarding early (1d) compressive strength development. When considering compressive strength values after 7 and 28 days, the most significant factor was also the bitumen to binder ratio. The bitumen content also had a statistically significant impact on flexural strength. Furthermore, the bitumen to binder ratio clearly reduced water demand. The setting time was mostly influenced by the activator to magnesia ratio, with a higher activator indicating an increased setting time. On the other hand, retarder type had the strongest impact on the temperature of the fresh mix. According to the experiment, the effect of iron sulfate as a retarder was weak. In addition, the use of RAP as an aggregate significantly reduced the initial temperature.

      Freeze-thaw resistance was mostly affected by retarder type. In particular the change of compressive strength and activator to magnesia ratio on mass after freeze-thaw cycles. However, their influence was not statistically significant. None of the factors had a clear influence on freeze-thaw resistance. All the tested cubes were in a good condition after 50 freeze-thaw cycles. The test cubes had no visible damages after 25 chloride-freeze-thaw cycles. There was no statistically significant impact regarding changes in compressive strength. However, a change was influenced by the bitumen to binder and activator to magnesia ratios. The result indicated a positive impact of higher bitumen and magnesia content on chloride-frost resistance. Regarding water resistance, the most significant impact to compressive strength was a higher bitumen to binder ratio and lack of AEA. However, the impact was not statistically significant concerning SN ratios. In addition, the bitumen to binder ratio had the most significant impact on water resistance and mass change. Water absorption was significantly increased by a lower activator to magnesia ratio and higher fiber content.

Table 4. Control factor ranking orders in relation to responses.

	MKP/ LGMgO	Additive (%)	Additive Type	SP (%)	Fiber (%)	Retarder (%)	Binder/ aggreg.	Aggreg./ RAP	Bitumen/ binder	Retarder Type	SP Type	Viscosite mod. (%)	AEA (%)
Control Factors													
Response: 1d Compressive Strength													
SN	4	11	7	8	3	13	2	5	1*	10	6	12	5
Response: 7d Compressive Strength													
SN	2*	7	8	12	5*	10	6*	3*	1*	4*	13	9	11
Response: 28d Compressive Strength													
SN	2	12*	9*	3*	5*	4*	11	13	1*	8	10	6	7
Response: Water Demand													
SN	3	9	11	6	2	13	8	5	1*	4	12	7	10
Response: Setting time													
SN	1*	11	4	6	7	5	13	9	3	10	2	12	8
Response: Temperature 5 minutes after mixing													
SN	13	3	12	11	8	7	4	2*	10	1*	6	9	5
Response: Freeze-thaw resistance; change in compressive strength (RA/RB)													
SN	13	10	12	3	9	1	4	7	11	8	6	5	2
Response: Freeze-thaw resistance; mass change (w%)													
SN	1	4	2	3	5	11	7	10	13*	6	9	8	12
Response: Chloride-frost resistance; change in compressive strength (RA/RB)													
SN	11	1	10	7	12	9	6	8	3	13	4	2	5
Response: Chloride-frost resistance; mass change (RA/RB)													
SN	2*	13	12	10	4	7	8	5	1*	3	11	9	6
Response: Resistance to water stress; change in compressive strength (RA/RB)													
SN	13	11	9	3	10	2	5	6	8	4	12	7	1
Response: Resistance to water stress; mass change (w%)													
SN	2	12	11	9	7	10	3	13	1*	5	4	6	8
Response: Water absorption													
SN	1*	9	10	12	2*	6	3	7	4	5	13	11	8
Response: Flexural / Compressive strength													
SN	11	3	10	4	2*	8	7	9	1	5	6	12	13

Relations between responses

Partial least squares structural equation modelling (PLS-SEM) was used as a methodological tool to explain the relations between the responses examined above. Six groups comprising of eight responses based on laboratory test results were used to develop a model using SmartPLS 3 software. A model was implemented using reflective indicators of compressive strength at day 1 and 7 (Comp.strength), freeze-thaw test results (FT RA/RB), magnesite and periclase amounts in the binder based on XRF-analysis (Binder), water demand, water absorption, and the results of the water resistance test (Water stress) (Fig. 3).

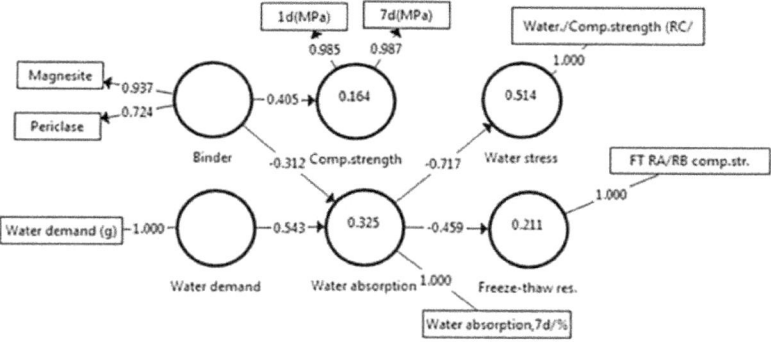

Figure 3. Structural equation model with path coefficients, indicator loadings and R^2 values.

According to the inner model, Binder had a positive effect on compressive strength (0.405) and a negative effect on water absorption (-0.312). Furthermore, Water demand had a positive relationship to water absorption (0.543), and Water absorption had negative effect on freeze-thaw resistance (-0.549) and water resistance (-0.717). The statistical significance and quality of the relationships are depicted in Table 5.

Table 5: Results summary for outer model.

Response	Indicators	Loadings	CR	AVE	R^2	Q^2
Binder	Magnesite	0.937	0.822	0.701		
	Periclase	0.724				
Compressive strength	1d	0.985	0.986	0.972	0.164	0.133
	7d	0.987				
Water demand	(%)	1.000	1.000	1.000		
Water absorpt.	(%)	1.000	1.000	1.000	0.325	0.168
Freeze-thaw res.	RA/RB	1.000	1.000	1.000	0.211	0.080
Water stress	RA/RB	1.000	1.000	1.000	0.514	0.483

Loadings describe the correspondence of the indicators to the related latent variables. They should be at least 0.40 and preferably over 0.70[22]. Table 4 shows that the indicator reliability was sufficient (> 0.4). Composite reliability (CR) shows the level of internal consistency among all reflective latent variables. The values were above 0.6, indicating good reliability[23,24]. Average Variance Extracted (AVE) describes convergent validity. The values were greater than 0.5, which is sufficient[24].

The R^2 values indicate the relationship of the variance of a latent variable to total variance. An R^2 level higher than 0.67 is substantial, higher than 0.33 is moderate, and a value above 0.19 represents a low level. Thus, water absorption and water stress R^2 values was moderate, freeze-thaw resistance was low, and compressive strength below that. This denotes that water absorption explained 51.4 percent of the variance in water stress resistance, and the binder and water demand explained 32.5 percent of the variance in water absorption. On the other hand, the binder provided a weak explanation for compressive strength.

The Stone-Geisser Q^2 values (cross-validated redundancy) indicate the predictive relevance of the latent variables. SmartPLS uses a blindfolding procedure to obtain Q^2 values. Values should be above zero, with a higher value indicating a stronger predictive relevance[25]. The predictive relevance was positive in all the factors, with the strongest being for water stress (0.483).

Table 6: Discriminant validity based on Fornell-Larcker criterion.

	Binder	Comp. str.	FT res.	Water abs.	Water dem.
Binder	0.837				
Comp.str.	0.405	0.986			
FT res.	-0.090	-0.209	1.000		
Water abs.	-0.205	-0.413	-0.459	1.000	
Water dem.	0.198	0.068	-0.430	0.481	1.000
Water stress	-0.096	0.093	0.716	-0.717	-0.608

Discriminant validity describes the difference between the latent variables[26]. It can be estimated through the square root of AVE, which should be larger than other correlation values among the latent variables[27]. According to the values presented in Table 6 the model indicates discriminant validity.

Path coefficients, corresponding T-values and effect sizes are depicted in Table 7. The magnitude of the path coefficient indicates the strength between the two responses. Path coefficients should be close to 0.2 and preferably above 0.30[28]. SmartPLS uses the bootstrapping procedure for model significance testing.

Table 7: Path coefficients, T-statistics and effect sizes of the relations.

	Path coef.	T	Effect Size f^2
Binder -> Comp.strength	0.405	1.922*	0.196
Binder -> Water absorption	-0.312	2.290**	0.139
Water absorption -> Freeze-thaw res.	-0.459	2.210**	0.267
Water absorption -> Water stress	-0.717	6.896***	1.059
Water demand -> Water absorption	0.543	2.369**	0.419

*>10%, **>5%, ***>1% significance

T-statistics denote the significance level of the path coefficients. The highest significance level (< 1 %) was measured from the water absorption to water stress (6.896) relationship. A 5% probability was identified for binder to water absorption (2.290), from water absorption to freeze-thaw resistance (2.210), and from water demand to water absorption (2.369). The effect of the binder on compressive strength was a 10% significance level (1.922).

The model's Effect Size (f^2) was above the minimum recommended value (0.02) in all cases, and above the medium level (0.15) except from binder to water absorption (0.139) and from water absorption to freeze-thaw resistance (0.267), and high (0.35) in the water absorption to water stress (1.059) and in the water demand to water absorption (0.419) relationships.

CONCLUSIONS

It can be concluded that use of geopolymer material based on industrial by-product provides ecologically sustainable and economic alternative for pavement patching. The results indicate that the studied geopolymer material can achieve good performance and durability in harsh weather conditions, unlike conventionally commonly applied cold placed bitumen based materials. It could be observed that the use of bitumen together with a geopolymer compound had a high impact on several performance properties related to durability and strength. Secondly, a higher activator had an important role in achieving sufficient strength development and durability. Thirdly, according to the experiment, a lower amount of retarder and higher amount of superplasticizer impacted positively on water and freeze-thaw resistance. On the other hand, it could be concluded that the viscosity modifier, SP type and additive type had a very low impact with respect to the investigated properties. According to the experiment, the recommended geopolymer patching material should include sufficiently high activator to magnesia ratio (0.5), moderate amount of fibers (0.5 %), boric acid (3 %), bitumen to MgO ratio of 0.3, and superplasticizer (2 %).

In the study, causal connections between the responses were analysed using structural modeling. According to the model, higher water demand indicated an increase in water absorption and further lowered water and freeze-thaw resistance. When considering periclase and magnesite in binder composition there was a positive impact on compressive strength development and a negative (reducing) impact on water absorption. The results emphasize the important role of water concerning the performance of repair material.

Further experiments included for example research on factors influencing abrasion resistance, slip resistance, bonding to old pavement and transformation of materials. In addition, strength development was closer studied in cold conditions (-10°C).

ACKNOWLEDGEMENTS

The authors would like to thank Tekes – the Finnish Funding Agency for Innovation for their support.

REFERENCES

[1] Tayabji, S.;Smith, K. D.;& Van Dam, T. (2010). Advanced High-Performance Materials for Highway Applications: A Report on the State of Technology. US Department of Transportation, Federal Highway Administration.
[2] Kingery, W. D. (1950). Fundamental study of phosphate bonding in refractories: cold-setting properties. *Journal of American Ceramics Society*, 242.
[3] Yang, Q.;& Wu, X. (1999). Factors influencing properties of phosphate cement-based binder for rapid repair of concrete. *Cement and Concrete Research*, 389-396.
[4] Wagh, A. S. (2004). Chemically Bonded Phosphate Ceramics, Twenty-first century materials with diverse applications. Elsevier.
[5] Wysk, R.A.; Niebel, B.W.; Cohen, P.H., & Simpson T.W. (2000) Manufacturing Processes: Integrated Product and Process Design, McGraw Hill, New York.
[6] Irad, B.-G. (2005) On the use of data compression measures to analyze robust designs, *IEEE Transactions on Reliability*, Vol. 54, No.3, 381-388.

[7] DIN EN 196-3 Methods of testing cement- Part 3: Determination of setting times and soundness.

[8] Ipavec, A. (2012). Deliverable No. 3 – Summary of existing standards, techniques, pothole repair materials and experience with them, draft 02, Project No. 832700 Durable Pothole Repairs, ZAG Slovenian National Building and Civil Engineering Institute, Slovenia, 1-55.

[9] SFS-EN 12390-3 Testing hardened concrete. Part 3: Compressive strength of test specimens.

[10] Frantzis, P.;& Baggot, R. (2003). Transition points in steel fibre pull-out tests from magnesium phophate and accelerated calcium aluminate binders. *Cement and Concrete Composites,* 11-17.

[11] Li, J.;Xu, G.;Chen, Y.;& Liu, G. (2014). Multiple scaling investigation of magnesium phosphate cement modified by emulsified asphalt for rapid repair of asphalt mixture pavement. *Construction and Building Materials,* 346-350.

[12] SFS-EN 12390-5 Testing hardened concrete. Part 5: Flexural strength of test specimens.

[13] Dias, D. P.;& Thaumaturgo, C. (2005). Fracture toughness of geopolymeric concretes reinforced with basalt fibers. *Cement and Concrete Composites,* 49-54.

[14] Silva, F. J.;& Thaumaturgo, C. (2002). Fiber reinforcement and fracture response in geopolymeric mortars. *Fatigue & Fracture of Engineering Materials & Structures,* 167-172.

[15] Pera, J.;& Ambroise, J. (1997). Fiber-reinforced magnesia-phosphate cement composites for rapid repair. *Cement and Concrete Composites,* 31-39.

[16] Skoba, O.;Bednarik, V.;Vondruska, M.;Slavik, R.;& Hanzlieek, T. (2005). Solidification of waste tire-shreds by geopolymerization. Green Chemistry and sustainable Development Solutions, Proceedings of the World Congress, Geopolymer 2005. Saint-Quentin: Institut Geopolymere.

[17] Pacheco-Torgal, F.;Labrincha, J.;Leonelli, A.;Palomo, A.;& Chindaprasit, P. (2015). Handbook of Alkali- Activated Cement Mortars and Concretes. Cambridge: Woodhead Publishing.

[18] Provis, J.;& van Daventer, J. S. (2009). Geopolymers: structure, processing, properties. Cambridge: Woodhead Publishing Limited.

[19] Byfors, K.;Klingsted, T.;Lehtonen, V.;Pyy, H.;& Romben, L. (1989). Durability of concrete made with alkali-activated slag. Third International Conference on the Use of Natural Pozzolans, Fly Ash, Blast Furnace Slag and Silica Fume in Concrete (ss. 1429-1466). Trondheim: ACI.

[20] SFS 5447 Concrete. Durability. Freeze-thaw resistance.

[21] SFS 5449 Concrete. Durability. Frost-salt resistance.

[22] Hulland, J. (1999). Use of partial least squares (PLS) in strategic management research: a review of four recent studies. *Strategic Management Journal,* 195-204.

[23] Urbach, N. ;& Ahlemann, F. (2010). Structural equation modeling in information systems research using partial least squares. *Journal of Information Technology Theory and Application,* 5-40.

[24] Bagozzi, R.P. ;& Yi Y. (1988). On the evaluation of structural equation models. *Journal of the Academy of Marketing Science,* 74-94.

[25] Hair, J.F.; Ringle, C.M.; & Sarstedt, M. (2011). PLS-SEM: Indeed a silver bullet. *Journal of Marketing Theory and Practice,* 139-151.

[26] Lowry, P.B. ; & Gaskin, J. (2014). Partial least squares (PLS) structural equation modeling (SEM) for building and testing behavioral causal theory: when to choose it and how to use it. *IEEE Transactions on Professional Communication,* 123-143.

[27] Fornell, C. ; & Larcker, D.F. (1981). Evaluating structural equation models with unobservable variables and measurement error. *Journal of Marketing Research*, 39-50.
[28] Chin, W.W. (1998). Issues and opinion on structural equation modeling. *MIS Quart*, 7-16.